BUSINESS ISSUES, COMPETITION AND ENTREPRENEURSHIP

PROGRESS IN MANAGEMENT ENGINEERING

Business Issues, Competition and Entrepreneurship

Improving Internet Access to Help Small Business Compete in a Global Economy
Hermann E. Walker (Editor)
2009. ISBN: 978-1-60692-515-7

Multinational Companies: Outsourcing, Conduct, and Taxes
Loran K. Cornejo (Editor)
2009. ISBN 978-1-60741-260-1

Private Equity and its Impact
Spencer J. Fritz (Editor)
2009. ISBN 978-1-60692-682-6

Progress in Management Engineering
Lucas P. Gragg and Jan M. Cassell (Editor)
2009. ISBN 978-1-60741-310-3

BUSINESS ISSUES, COMPETITION AND ENTREPRENEURSHIP

PROGRESS IN MANAGEMENT ENGINEERING

LUCAS P. GRAGG
AND
JAN M. CASSELL
EDITORS

Nova Science Publishers, Inc.
New York

For permission to use material from this book please contact us:
Telephone 631-231-7269; Fax 631-231-8175
Web Site: http://www.novapublishers.com

LIBRARY OF CONGRESS CATALOGING-IN-PUBLICATION DATA

Progress in management engineering / [edited by] Lucas P. Gragg and Jan M. Cassell.
p. cm.
Includes index.
ISBN 978-1-60741-310-3 (hardcover)
1. Industrial engineering--Research. I. Gragg, Lucas P. II. Cassell, Jan M.
T56.42P76 2009
658--dc22 2009016557

Published by Nova Science Publishers, Inc. ✦ New York

CONTENTS

PREFACE

Management engineering is a new field, which is quickly becoming a specific branch of engineering that takes a comprehensive approach to management. The underlying assumption is that the organization can be modeled as an interacting system, with cause-and-effect chains, feedback loops and other structures that behave like those in other systems. Management engineering tools are used to make the system visible so that managers can understand it and guide it better. This book presents current research in this new field.

Chapter 1 discusses cultural differences between international project partners which are held responsible for cost overrun, time delays, and the failure of many complex megaprojects. If partners are unable to cope with diverse management styles and cultures within these projects, decision-making processes can slow down and tensions are likely to emerge. In the academic debate on cross-cultural differences, national cultural differences have attracted most public and academic attention. A majority of the publications on managing cultural differences in projects is based upon Hofstede's (1980) multiple values model. This model has received criticism for its singular focus on nation-state cultures and for the absence of power issues, ambiguity and situational behavior. Megaprojects are based on informal, boundary spanning networks of (international) organizations. To perceive organizations and nation-states as homogeneous entities is out of touch with daily practices in a globalizing world. Therefore, Söderberg and Holden (2002) propose a social constructionist approach on studying management of multiple cultures; national cultures, regional cultures, industrial cultures, organizational cultures, professional cultures and departmental cultures. Such an interpretative perspective focuses at processes of meaning, sense making and social construction of culture by actors and comes to a 'verstehen' of the constructed social reality (Weick, 1995). To explore the multiple culture approach the case of the Environ Megaproject is being studied. This multi billion euro project is one of the largest and most ambitious infrastructural projects in The Netherlands. The project is an international Public Private Partnership in which a complex network of public and private organizations cooperates under the supervision of the Environ Megaproject organization. Data has been collected between September 2003 and September 2004 by a team of four internal and two external researchers under the researcher's supervision. The exploration of multiple cultures in the Environ case show which new direction is needed for studying and understanding cross-cultural cooperation in project management.

In a perfect world, changes will be confined to the planning stages. However, late changes often occur during project processes, and frequently cause serious disruption to the project. The need to make changes in a project is a matter of practical reality. Even the most thoughtfully planned project may necessitate changes due to various factors. The fundamental idea of any change management system is to anticipate, recognize, evaluate, resolve, control, document, and learn from past changes in ways that support the overall viability of the project. Learning from past changes is imperative because the professionals can then improve and apply their experience in the future. Primarily, the chapter proposes six principles of project change management. Based on these principles, a theoretical model for project change management system (PCMS) is developed. The theoretical model consists of six fundamental stages linked to two main components, i.e., a knowledgebase and a controls selection shell for making more informed decisions for effective project change management. Further, the framework for developing an information technology based project change management system is also discussed. Chapter 2 argues that the information technology can be effectively used for providing an excellent opportunity for the professionals to learn from similar past projects and to better control project changes. Finally, the chapter briefly presents an information technology based project change management system (PCMS) for the management of changes in building projects. The PCMS consists of two main components, i.e., a knowledgebase and a controls selection shell for selecting appropriate controls. The PCMS is able to assist project managers by providing accurate and timely information for decision making, and a user-friendly system for analyzing and selecting the controls for change orders for projects. The PCMS will enable the project team to take advantage of beneficial changes when the opportunity arises without an inordinate fear of the negative impacts. By having a systematic way to manage changes, the efficiency of project work and the likelihood of project success should increase. The chapter would assist professionals in developing an effective change management system. The system would be helpful for them to take proactive measures for reducing changes in projects. Furthermore, with further generic enhancement and modification, the PCMS will also be useful for the management of changes in other types of projects, thus helping to raise the overall level of productivity in the industry. Hence, the system developed and the findings from this study would also be valuable for all project management professionals in general.

Traditionally projects are considered means for getting things done, simultaneously striving for efficient and accurate methods – that is, doing more in less time. A consequence, not often discussed, is that doing more things in less time with a closer focus on cost, will inevitably lead to a more complex and tightly connected project execution system which is more sensitive to deviations. Following a "project-as-practice" perspective this paper explores and analyses how deviations are managed. The findings suggest that even though the company under consideration manages about 120 projects per year deviations cannot be avoided. The deviations were found initially to decouple (a process of creating loosely coupled activities) from the overall project process and later on to recouple (a process of tightly coupling activities) when the deviation was resolved. Chapter 3 suggests that the management of deviations is dynamic and changing and that the concept of coupling is a fruitful way of exploring the process.

A major concern among managers and administrators has been the lack of cost assessment/financial implications associated with process improvement and process capability. The impact of process control frequently gets treated more as good will than actual cost savings. In Chapter 4 the authors provide methods for quantifying cost savings through use of the metrics used to assess and improve process performance and capability. Initially the authors develop the general relationship between process capability indices and financial costs using the process capability index *Cpw* and various loss functions. The relationship between the unified approach for some common process capability indices (PCIs) through the use of a non-stochastic weight function and the expected weighted squared error loss provides an intuitive interpretation of *Cpw*. Using different values of the non-stochastic weights, w, the distributions of the estimated loss associated with the measures of process capability indices can be determined. Upper confidence limits for the expected loss associated with *Cpw* as well as its generalization *Cpw**, and special cases such as *Cp*, *Cp**, *Cpm*, *Cpm**, *Cpk* and *Cpk** are discussed. Quality practitioners and manufacturers need only specify the target, maximum loss, the estimated process mean and standard deviation, in order to determine an estimate of the expected loss associated with the process. Examples are demonstrated.

Nowadays, construction projects grow in complexity and size. So, finding feasible schedules which efficiently use scarce resources is a challenging task within project management.

Project scheduling consists of determining the starting and finishing times of the activities in a project. These activities are linked by precedence relations and their processing requires one or more resources. The resources are renewable, that is, the availability of each resource is renewed at each period of the planning horizon.

The objective of the well-known resource constrained project scheduling problem is minimizing the makespan. While the exact methods are available for providing optimal solution for small problems, its computation time is not feasible for large-scale problems [20].

Chapter 5 presents two approaches for the project scheduling problem. The first approach combines a new implementation of a genetic algorithm with a discrete system simulation. This approach generates non-delay schedules. This study also proposes applying a local search procedure trying to yield a better solution (GA-RKV-ND).

The second approach combines a new implementation of a genetic algorithm with a discrete system simulation. This approach generates active schedules. This study also proposes applying a local search procedure trying to yield a better solution (GA-RKV-AS).

The chromosome representation of the problem is based on random keys. The dynamic behaviour of the system simulation is studied by tracing various system states as a function of time and then collecting and analysing the system statistics. The events that change the system state are generated at different points in time, and the passage of time is represented by an internal clock which is incremented and maintained by the simulation program. The simulation strategy is the event oriented simulation [27].

The good computational results on benchmark instances enlighten the interest of the best approach (GA-RKV-AS).

Blood donation and transfusion service is an indispensable part of contemporary medicine and healthcare. It involves collecting, processing, storing and providing human blood intended for transfusion, performing pre-transfusion testing, cross-matching, and finally infusing into the patients. In view of the life-threatening nature of blood and blood components, it entails the rigorous controlling, monitoring and the complete documentation of the whole procedure from blood collection to blood infusion. The introduction of information and computer technology facilitates the overall procedure of blood donation and transfusion service, and improves its efficiency as well.

In general, a computerized blood bank information system refers to acquiring, validating, storing, and circulating various data and information electronically in blood donation and transfusion service. With regard to its unique service objects, the blood bank information system should pay enough attention on the following characteristics of blood bank data and information: information credibility, information integrity, information coordination, and information security. Chapter 6 firstly surveys the development of computerized blood bank information systems, elucidates their rationale and infrastructures, and then exemplifies a real-world blood bank information system. The relevant engineering implementation will be discussed too.

Other than consistency and security, another challenge in computerized blood bank information management is, in face of explosive data and information, how to make good use of them for decision making support. In this chapter, the authors will further address the underlying mechanisms of decision making support in blood bank information systems. The unique properties of blood bank data and decisions are firstly examined. Then, with special concerns on blood donation and transfusion service, the authors shift to the development of computerized decision making support. Finally, a case study will be presented to evidence our understanding on computerized decision making support in blood bank information systems.

Vietnam's economic growth has led to a demand for infrastructure facilities, residential and commercial buildings, and hi-tech parks. This has resulted in a high volume of construction activities. With Vietnam's membership in the World Trade Organization, foreign architectural, engineering and construction (AEC) firms now have the opportunity to operate in Vietnam. However, undertaking overseas construction projects is usually considered a high risk business due to a lack of information and overseas experience. Risk management is thus an important aspect of international construction.

In Chapter 7, to investigate risks associated with managing construction projects in Vietnam and to examine how *foreign firms* manage those risks, a case study was conducted. The main objectives of this case study were to find out the different type of risks encountered by *foreign* players and various risk response strategies adopted by them. These include political and legal risks, financial and economic risks, design risk, construction related risk and cultural risk. The case study relates to the development of a yeast factory in the southern part of Vietnam. Data for the case study were obtained by interviewing experts from different firms that undertook important parts in this project.

The research revealed that Vietnam has a complex government administrative system. Foreigners overcome the political risk by transferring it to a local joint venture partner who is in a better position to deal with local government officials and to obtain the necessary approvals. Negotiation is found to be the best way to settle disputes instead of suing each other in the court of law because Vietnam's legal framework is not robust. Prequalification of bidders is found to be the most effective and practical way to ensure that the contractor engaged to carry out the work is financially sound and competent, thereby reducing financial risk. Design risk was severe in this project and it caused many disputes among project team members. Design risk was mitigated through negotiation and by having many coordination meetings. The project faced some construction related risks, such as low quality of workmanship, low safety consciousness, unavailability of sophisticated materials, plant and equipment. These were solved by engaging a safety supervisor, training workmen, and changing specifications to locally available products. The project faced many cultural risks due to different mindsets between foreigners and Vietnamese, and different working styles. This risk can be overcome if foreigners strive to adapt to the local environment, and be mindful and watchful of how locals behave.

Cooling, Heating and Power (CHP) systems have been recognized as a key alternative for thermal energy and electricity generation at or near end-user sites. CHP systems are a form of distributed generation that can provide electricity while recovering waste heat to be used for space and water heating, and for space cooling by means of an absorption chiller. Although CHP technology seems to be economically feasible, due to the constant fluctuations in energy prices, CHP systems cannot always guarantee economic savings. However, a well-designed CHP system can guarantee energy reduction. This energy reduction could be increased depending on the CHP system operational strategy employed. CHP systems operational strategy defines the goal of the system's response to the energy demand, which is one of the factors that characterize the energy performance of the system. CHP systems are usually operated using a cost-oriented operational strategy. However, an operational strategy based on primary energy would yield better energy performance. In Chapter 8 the CHP system energy performance is evaluated based on primary energy consumption and a primary energy operational strategy is implemented to optimize energy consumption. To determine the energy performance, a model has been developed and implemented to simulate CHP systems in order to estimate the building-CHP system energy consumption. The novel characteristic of the developed model is the introduction of the Building Primary Energy Ratio (BPER) as a parameter to implement a primary energy operational strategy, which allows obtaining the best energy performance from the building-CHP system. Results show that the BPER operational strategy always guarantees energy savings. In addition, the BPER operational strategy is compared with a cost oriented operational strategy based on energy cost. Results from a cost-oriented operational strategy show that for some operation conditions, high economic savings can be obtained with unacceptable increment of the energy consumption. This chapter also considers how the application of the BPER operational strategy can improve the Energy Star Rating and the Leadership in Energy and Environmental Design (LEED) Rating, as well as reduce the emission of pollutants.

Rheology is regarded as the science of flow behavior, where, based on isothermic equations, the deformation of fluids and plastic bodies subjected to external stresses may be described. Hooke's law of elasticity, Newton's law for ideal fluids (viscosity), Mohr-Coulomb's equation, and finally, Bingham's yielding are well known relationships and parameters in the field of rheology.

Rheometry is a well established measurement technique to determine the specific rheological properties of fluid and plastic bodies. In order to explain point contact processes and strength an extrapolation of such findings to data of triaxial, direct shear or oedometer tests is still missing.

A parallel-plate-rheometer MCR 300 (Modular Compact Rheometer, Paar Physica, Ostfildern, Germany) has been used to conduct oscillatory tests. From the stress-strain relationship parameters and specific characteristics as storage modulus G' and loss modulus G", loss factor tan δ (= G''/G'), viscosity η, yield stress τ_y and the linear viscoelastic deformation (LVE) range including a limiting value γ_L were determined and calculated, respectively.

Thus, Chapter 9 aims to introduce rheometry as a suitable method to determine the mechanical behavior of soils, as viscoelastic material, and mineral suspensions when subjected to external stresses. To do this a Na-bentonite, Ibeco Seal-80, has been used for preliminary tests; the suspensions were equilibrated with NaCl solutions in different concentrations in order to determine the ionic strength effects on interparticle strength and changes in mechanical properties. Furthermore, a Dystric Planosol, a Calcaric Gleysol from North Germany and loess material from Israel, saturated with NaCl and/or $CaCl_2$ in several concentrations were analyzed.

In order to demonstrate clay mineralogical and/or textural effects as well as of leaching of organic matter and iron oxides, the degree of stiffness and structural stability of clay rich substrates from Brazil, - a smectitic Vertisol and a kaolinitic Ferralsols - were quantified. In addition, scanning electron microscopy was applied for visualizing structural characteristics. Due to the modification of microstructural analysis by such visual investigations, structural changes and consequences for upscaling considerations become evident as well as the need of research in soil mechanical processes on the particle-particle scale. It is shown that rheometry is an applicable method to detect microstructural changes by using a rotational rheometer.

Project Management (PM) has emerged from different fields of application and it entails planning, organizing, and managing resources to bring about the successful completion of specific project goals and objectives, while controlling the resources (time and money) and the quality. The operational research contribution for PM has mainly been done through providing tools (model, methods, and algorithms) to solve project scheduling problems. The project scheduling problem involves the scheduling of project activities subject to precedence constraints and resource constraints. Although this problem has been the subject of extensive research since the late fifties, there have been some publications reporting extreme budget over runs and/or extreme time delays, thus proving that there is still the need for further research. Chapter 10 intends to be a guided tour through the most important recent developments in algorithmic methods to solve the project scheduling problem. Since these problems are NP-hard, our main focus is on heuristic methods, particularly on meta-heuristics. The paper concludes with an examination of areas that in the opinion of the author would particularly benefit from further research.

In: Progress in Management Engineering
Editors: L.P. Gragg and J.M. Cassell, pp. 1-41

ISBN: 978-1-60741-310-3

Chapter 1

TOWARDS A NEW UNDERSTANDING OF CROSS-CULTURAL MANAGEMENT IN INTERNATIONAL PROJECTS: EXPLORING MULTIPLE CULTURES IN ENVIRON MEGAPROJECT

Alfons van Marrewijk[*]
VU University Amsterdam, Faculty of Social Sciences, Department Culture, Organization and Management, Amsterdam, The Netherlands

Abstract

This chapter discusses cultural differences between international project partners which are held responsible for cost overrun, time delays, and the failure of many complex megaprojects. If partners are unable to cope with diverse management styles and cultures within these projects, decision-making processes can slow down and tensions are likely to emerge. In the academic debate on cross-cultural differences, national cultural differences have attracted most public and academic attention. A majority of the publications on managing cultural differences in projects is based upon Hofstede's (1980) multiple values model. This model has received criticism for its singular focus on nation-state cultures and for the absence of power issues, ambiguity and situational behavior. Megaprojects are based on informal, boundary spanning networks of (international) organizations. To perceive organizations and nation-states as homogeneous entities is out of touch with daily practices in a globalizing world. Therefore, Söderberg and Holden (2002) propose a social constructionist approach on studying management of multiple cultures; national cultures, regional cultures, industrial cultures, organizational cultures, professional cultures and departmental cultures. Such an interpretative perspective focuses at processes of meaning, sense making and social construction of culture by actors and comes to a 'verstehen' of the constructed social reality (Weick, 1995). To explore the multiple culture approach the case of the Environ Megaproject is being studied. This multi billion euro project is one of the largest and most ambitious

[*] E-mail address: a.van.marrewijk@fsw.vu.nl. +31 (0)20 598 6740, fax +31 (0)20 598 6765. De Boelelaan 1081, 1081 HV Amsterdam, The Netherlands.

infrastructural projects in The Netherlands. The project is an international Public Private Partnership in which a complex network of public and private organizations cooperates under the supervision of the Environ Megaproject organization. Data has been collected between September 2003 and September 2004 by a team of four internal and two external researchers under the researcher's supervision. The exploration of multiple cultures in the Environ case show which new direction is needed for studying and understanding cross-cultural cooperation in project management.

Keywords: project management, cross cultural, multiple cultures, megaprojects

Introduction

The management of cross-cultural differences has become a major issue in the academic debate on project management (f.e. Dafoulas and Macaulay, 2001; Chevrier, 2003; Kendra and Taplin, 2004; Mäkilouka, 2004). Henrie and Sousa Poza (2005) looked at the state of research within leading project management academic level journals and project management books and concluded that attention for a culture perspective on project management has increased significant last decade. It is now widely recognized that national cultures influences the success of (global) projects (Kendra and Taplin, 2004; Mäkilouka, 2004; Staples and Zhao, 2006).

Cultural differences are held responsible for the collapse of many project-based alliances and projects (Spekman *et al.*, 1996; Söderberg and Vaara, 2003; Van Marrewijk and Veenswijk, 2006). If partners are unable to cope with diverse management styles and cultures within the project, decision-making processes can slow down and tensions are likely to emerge. According to Van Oudenhoven and Van der Zee (2002), similarities in national and corporate cultures are associated with successful cooperation but dissimilarities are more critical to success. Makilouko (2004), for example, discussed the difficulty Finnish project leaders had with managing multicultural projects. Zwikael, Shimizu and Globerson (2005) studied differences in project management styles between Japanese and Israeli cultures. Israeli project managers were more focused on performing scope and temporal processes, while communication and cost management were frequently used by Japanese project management. The researchers found that Japanese project managers used clear and measurable success measures for each project, while the Israeli project objectives were quite vague (Zwikael *et al.*, 2005).

Without a doubt, multiple values models on national cultural differences have attracted the most public and academic attention (f.e. Hall, 1976; Hofstede, 1980; Adler, 1986; Trompenaars, 1993). The debate on national cultures is dominated by the work of both Hofstede (1980) and Trompenaars (1993). These models, based upon bipolar dimensions, indicated the cultural 'distance' between nations (Morden, 1999). However, Jackson and Aycan (2006) make an appeal to cross-cultural researchers and managers to move away from cultural values research. The cultural values perspective has received criticism for its singular focus on nation-state cultures and for the absence of power issues, sub-cultures, regional differences, ambiguity and situational behavior (Low, 2002; Söderberg and Holden, 2002; Sackmann and Phillips, 2004; Jacob, 2005; Jackson and Aycan, 2006; Sackmann and Friesl, 2007). Cultural heterogeneity, local management concepts and cultural imperialism make cross-cultural management too complex to be explained by a cultural value model (Jacob, 2005).

When moving away from the multiple values model what new directions can be found in cross-cultural management? And how are these new theoretical insights applied in an empirical case study of a complex megaproject? To answer these questions this chapter explores new developments in cross cultural studies in complex megaprojects. Megaprojects have become more and more popular with national governments. It is the scale, complexity, number of international partners, high degree of uncertainty and duration that distinguish megaprojects from traditional projects (Van Marrewijk *et al.*, 2008). Megaprojects are perceived as aggregations of employees temporarily enacting on a common cause (Hodgson and Cicmil, 2006). Although the societal impact of these large-scale projects is enormous, academic interest in this subject has been modest and has mainly focused on themes related to the rational organization and (political) control in terms of policy programs, contracting, perceived outcomes, and especially risk and economic failure (Flyvbjerg *et al.*, 2002; Flyvbjerg *et al.*, 2003).

The empirical findings presented in this chapter are based upon an in-depth qualitative ethnographic study of Environ Megaproject, which was one of the largest infrastructural projects in The Netherlands. Data has been collected as an integral part of a larger evaluation research on the organization and management model used in the Environ Megaproject. In-depth case studies provide a good understanding of daily work floor practices (Yin, 2003). The evaluation research was executed between September 2003 and September 2004 by a team of four internal and two external researchers under the author's supervision. An interpretative method was needed to understand how particular cross-cultural practices work in a certain context (Yanow and Schwartz-Shea, 2006). Furthermore, two studies have been done on the project control and international collaboration in 2005. Peterson (2007: 374) acknowledges that qualitative ethnographic studies can be helpful in providing greater depths in cross cultural analysis.

The chapter has been structured as follows. Firstly, the academic debate on multiple value models for management of cultural differences is discussed. Secondly, a critical perspective on projects is explored with three new directions for studying cross-cultural management; local management concepts, hybridization and multiple cultures. Each of the concepts is described. Thirdly, the concept of power is introduced in relation to cross-cultural management because cross-cultural cooperation does not take place in a power free context. Fourthly, three groups of distinct strategies for handling multiple cultures are presented. Fifthly, an organization anthropological model of studying cross cultural in projects is being developed from earlier discussed concepts. This model is based upon descriptive theory and grounded in empirical ethnographic studies on human interaction in large projects. The model is executed in the study of Environ Megaproject. After discussing methodological reflections on the study, an introduction on the megaproject is given. Then research results are discussed at industrial level, national level, organizational level, project level, professional level and departmental level. Finally, conclusions are given.

Managing Cultural Differences in Projects

Single and multiple dimensional value models have dominated the debate on cross-cultural management (Morden, 1999). Studies on cross-cultural collaboration in projects are dominated by cultural value models such as Hofstede (1980), Adler (1986), Hall (1976) and Trompenaars (1993) (f.e. Dafoulas and Macaulay, 2001; Zwikael *et al.*, 2005; Staples and Zhao, 2006). Especially Hofstede (1980; 1994) and Trompenaars (1993) have dominated the international debate on intercultural management with their models in which bipolar dimensions are used to analyze culture at a national level (Morden, 1999).

Hofstede (1980) used four value dimensions, i.e. (1) low and high power distance, (2) high and low uncertainty avoidance, (3) high and low individualism and (4) high and low masculinity. A dimension is an aspect from which one culture can be compared with another culture. The first value dimension is the level of acceptance by a country of the unequal distribution of power. The second value dimension refers to the extent in which people in a country feel threatened by ambiguous situations. The third value dimension is the level in which people look after themselves and neglect the needs of the country. The fourth value dimension refers to the level of dominant values such as assertiveness and materialism in a country. To analyse Asian values of long term versus short-term orientation, Hofstede (1980) also included a fifth value dimension; the Confucian dynamism. Hofstede arguments that these five value dimensions predict the cultural position of one country in relation to others.

Trompenaars (1993) used seven dimensions to describe a culture in a country: (1) universalism versus particularism, (2) individualism versus collectivism, (3) neutral versus affective relationships, (4) specific versus diffuse relationships and (5) achievement versus ascription, (6) attitudes to time and (7) attitudes to the environment. In the first set of dimensions universalism refers to the view that norms and values can be applied everywhere. Particularism in contrast, is said to prevail where the unique context and relationships are more important than universal abstract rules. The second set of dimensions refers to the level in which people regard themselves as a part of a group or society. Trompenaars (1993) follows the definition of Hofstede (1980) in this dimension. The third set of dimensions is related to the way in which emotions are generally expressed in a country; people can let themselves go and react emotionally or instead, they might tend to intellectualize their emotions and remain controlled in their responses. The fourth set of dimensions refers to the degree of involvement in which individuals are comfortable when dealing with other people. The fifth set of dimensions deals with how status and power are attributed in a country. In achievement-orientated countries the power and status of someone depends on the position of employment he has reached in the course of his career. In ascription-orientated cultures status is attributed to someone and in general independent of a specific task or function. The sixth dimension shows the way in which societies look at the concept of time. The seventh dimension deals with the society's attitude towards environment.

National cultural maps (f.e. Hofstede, 1980; Trompenaars, 1993) have helped to understand the reasons behind cultural differences among countries and to realize how an understanding of these differences is crucial in order to know what is appropriate management behaviour with regard to specific cultural contexts (Low, 2002). But we must remember that the objects of study in cross-cultural cooperation are human beings. Human beings cannot easy be pressed in simplistic models and schemes. Human beings can act, speak, think, have desires and drives, exist in space and time and, are simultaneously the object and subject of science (Chanlat, 1994). Jacob (2005: 517) rips the dominant cultural paradigm apart and argues that effective managers do not need to learn the country scores, but rather need to learn to detect what leadership style works in given culture and develop the necessary skills in order to work with this required leadership style. Interestingly, from personal experiences with consultancy firms working with multiple management models, I have learned that these consultants themselves aren't convinced about the usefulness of the country scores. Therefore, Jacob (2005: 516) asks "If there is no practical utility in organizing countries in clusters, why bother to do so?".

Apart from the debate on assumption of universal application of country clusters other critics have been given to multi value models (Jacobs (2005):

- A rather over-simplified and static perspective on the handling of cultural differences
- Focus on differences rather then similarities amongst people
- Pointing out the otherness of others is said to increase stereotyping and resentment towards the other
- A disputable assumption that cultural differences are stable, and measurable
- The absence of power issues and situational use of cultural differences
- The assumption that cultural differences can be overcome

Lowe *et al.* (2007), therefore, encourage cross-cultural researchers to employ *bricolage* in the context of local moralities, relationships and actionable outcomes. *Bricolage* is a concept used in anthropology to illustrate the way in which societies combine and recombine different symbols and cultural elements in order to come up with recurring structures (Lévi-Strauss, 2004). The objective is to further elaborate the understandings of social and cultural phenomena over theoretical or methodological 'purity' and paradigmatic struggle (Lowe et al., 2007: 244). People construct their social reality through their actions and in turn, this social reality prescribes the behavior of the people. People always construct, deconstruct and reconstruct their reality from both old and new experiences. Through this process culture is reproduced. As a consequence strategic behavior of people can transform social reality because culture is constantly being reproduced.

Towards New Perspectives on Managing Multiple Cultures

To move away from multiple value models the core task of cross cultural management should be:

> to facilitate and direct synergistic action and learning at interfaces where knowledge, values and experience are transferred into multicultural domains of implementation (Holden, 2002: 59).

The goal of cross cultural researchers is to understand how particular management practices work in a certain context. Söderberg and Holden (2002) and Sackmann and Philips (2004) propose a social constructionist approach on studying cross cultural management. Therefore, Jackson and Aycan (2006) focus at social interaction between employees of cultural divers background and study emerging local management models as well as new cultural practices. These patterns of meaning are produced and reproduced and negotiated in the course of social interaction (Holden, 2002). Such an interpretative perspective focuses at processes of meaning, sense making and social construction of culture by actors and comes to a '*verstehen*' of the constructed social reality (Weick, 1995). Projects are then considered to be the object and outcome of social interactions as much as any other form of organizing within a multiple context of socially interdependent networks (Hodgson and Cicmil, 2006). A social constructionist approach of cross cultural management in projects includes cultural heterogeneity, power issues, situational behavior and hybridization of cultural practices (Low, 2002; Söderberg and Holden, 2002; Jacob, 2005; Jackson and Aycan, 2006).

Peterson (2007) points out new directions for cross cultural research to overcome the misunderstandings in the cultural value perspective. Based upon an analysis of the anthropological roots of the value perspective he suggests to reconsider the concept of cultural boundaries and to include and better representation of the local culture. Indeed, cross cultural cooperation is increasingly based on boundary spanning networks as is the case with cross-cultural collaboration in project teams (Hasting, 1995). Organizations and nation-states aren't homogeneous entities (Söderberg and Holden, 2002). In the case of, for example, India it is quite clear that the nation-state can't be perceived as a homogeneous culture (Singh, 1990: 75; Anisya and Annamma, 1994; Chatterjee and Pearson, 2001; Fusilier and Durlabhji, 2001: 223). India is a complex mosaic of many languages, cultures and religions (Gopinath, 1998). Furthermore, 'traditional' Indian values are changing in interaction with Western values (Sinha and Sinha, 1990; Anisya and Annamma, 1994; Sahay *et al.*, 2003)

Jacob (2005) criticizes single and multiple value models for measuring average scores. Jacob argues that these statistical models do not take exceptions to the rule into account. Given the large variance in individual scores, the system of capturing country scores does not help producing any predictions. Genuinely, she raises the question that if managers in Mexico have an overall high average on high power distances, how will this knowledge help you when meeting individuals from Mexico? (Jacob, 2005).

In conclusion, new directions in cross cultural management can be found in a better representation of local culture, in exploring multiple levels of cultural differences and in studying the social interaction between employees of divers cultural background. In this chapter I want to explore the latter two concepts here.

Hybridization or Crossvergence

Crossvergence is all about fusing together management practices of two or more cultures, so that a practice relevant to a heterogeneous culture can be assembled (Jacob, 2005). Jacob (2005) adresses the relevance of hybrid cultures by explaining that many people grow up in various, possible overlapping cultural groups and may chose to make use of one approach in a given situation and another approach in a different situation. The scholar shows how crossvergence put emphasize on bringing together management practices of two or more cultures in order to create a more relevant management style. She praises crossvergence and hybridization as effective ways for managers to create "management practices that are efficient, while simultaneously aligned to the local culture" (Jacob, 2005). She denounces the Hofstedian canon for force-fitting countries into one management style, stating that they are either of one nature or of another. Such approaches do not leave room for understanding management style in its context.

New cultural practices emerge from social interaction between employees of cultural divers background (Jackson and Aycan, 2006). To study interaction, Shimoni and Bergmann (2006) developed a cultural hybridization approach which focuses at interactions, negotiations and mutual learning. In this approach, dichotomies of Western and local management are replaced by new hybrid work practices with sources of both local and Western culture (Shimoni and Bergmann, 2006). Examples of hybridization are found in Brannen and Salk (2000) who studied work practices in a German – Japanese strategic alliance and observed that Germans and Japanese had different attitudes toward working hours. New collaboration practices emerged as some of the German managers began to stay later at work while many of the Japanese worked fewer hours than they were accustomed to in Japan (Brannen and Salk, 2000). Another example is the study of Clausen (2007) who used multi-contextual analysis to describe the dynamics and complexity of sense-making processes at the interface of meaning exchange in the collaboration between Danish and Japanese managers. In the collaboration between a Danish company and its alliance partner in the Japanese market a 'negotiated' culture emerged. A third example is the study of Shimoni (2008) who discusses emerging management styles in the collaboration of managers from Thailand, Mexico and Israel and comes to the conclusion that new practices emerged. Finally, an example of hybridization can be found in Western and Indian work practices as consequence of offshoring. Kaker et.al. (2002) name corporations where Western and Indian management practices are mixed, hybrid firms. Indian management is a fusion of western models and indigenous practices without hardly any uniformity throughout India (Anisya and Annamma, 1994; Gopinath, 1998; Virmani, 2007). Sapra (1995) warned managers of India's corporate sector to change the work culture of their employees and bring in stringent quality control in the manufacture of their products. Similarly, he warned managers of multinational corporations to understand and appreciate daily Indian work practices and show due respect for Indian culture and customs (Sapra, 1995). Virmani (2007) calls this 'confusion' of the indigenous management caused by the need to adapt to different norms and practices to new and foreign concepts of management throughout history. These examples show the usefulness of a hybridization approach as a new direction for cross cultural studies.

Multiple Cultures in Projects

Kendra and Taplin (2004) note that a project culture consists of multiple fragmented subcultures. To study these subcultures Chanlat (1994) emphasizes the necessity of focusing on human behavior and at the same time exploring all levels of organizational life.

> The complexity of the problems that confront us both on national and an international scale, the importance of cultural elements, the emphasis on individual aspirations...all of these influences have, in effect led us to propose models of management that will henceforth be based on a true anthropology of organizations (Chanlat 1994: 160).

In order to grasp human reality within organizations five closely linked levels of organizational reality can be distinguished (Chanlat, 1994). (1) The first level is the individual level, in which Chanlat sees human reality as a subtle interaction of the biological, the psychic and the social. At this level individuals construct and deconstruct their own reality and cope with conflicts, tensions, uncertainties and ambiguities. (2) At the second level, the interactional level, the identity of the individual is formed in interaction with others. The interactions, both formal and informal, can appear between two different individuals or two different groups. (3) The third level, the organization level, focuses on the organizational cultures. (4) The fourth level, the society level, concerns national cultures. These national cultures evolved due to geography, history, political and economic forces, language and religion. (5) The fifth level, the world level, deals with transnational ideologies such as religion, globalization and liberalization.

In line with Chanlat, Alvesson and Berg (1992) distinguish six different levels involved with organizational culture: (1) national cultures, (2) regional and industrial cultures, (3) company culture, (4) professional culture, (5) department culture (6) and worker culture. Schneider and Barsoux (1997: 47) also distinguish six different cultural levels or spheres, which exert influence on business practice. Each sphere of influence has its own set of artifacts and behaviors, beliefs and values and, underlying assumptions. Schneider and Barsoux stress that it might not be useful to argue which sphere is more dominant because the spheres interact in complex ways. (1) The first sphere concerns the national cultures. National cultural differences have been discussed in the debate on intercultural management. (2) The second sphere concerns the regional and community cultures. Within national borders strong regional ties can maintain a strong sense of regional identity. Regional differences derive from history and language. (3) The third sphere focuses on the industry cultures. An industry culture is a subculture of a specific industry or sector. Industry cultures rise from the unique activities and problems encountered within industries For instance, the industry culture of the telecom sector differs from the health care sector. (4) The fourth sphere deals with organizational cultures. (5) The fifth sphere focuses on professional culture. The members of a occupational group share meanings they ascribe to work-related events and develop shared occupational ideologies. (6) The sixth sphere concerns the functional cultures. The nature of the task of the different functions such as finances, production, marketing and research and development results in different cultures.

The complexity and interdependence of local and global processes within megaprojects makes it necessary to conceptualize the different levels, without establishing a hierarchy between the different levels. A multi level analysis of cultural differences could be helpful to understand cultural dynamics in megaprojects. However, the analysis of cultures at different levels would not be sufficient without knowing what influences the success of cross-cultural cooperation.

Managing Multiple Cultures Successfully

Four different factors are related to successful cooperation in complex megaprojects (Van Marrewijk, 2004). The balance of power between the partners in an international project is the *first factor* that has influence on successful cross-cultural cooperation. Cross-cultural cooperation does not take place in a power free context. Power has to be understood here in a wide sense (Clegg, 1993). Power is defined by the size of the company, the financial resources, the access to political power, the access to technical knowledge and knowledge of the local market. A struggle for power can result from inequality in the balance of power or from rivalry, and can affect trust and cooperation in the project. Not the objective but the perceived inequality of power causes tensions in projects (Clegg, 1981). Therefore, Nicholson and Sahay (2001) included power and politics in their qualitative study of cross-cultural collaboration in a British - Indian software outsourcing. Ambivalent relationships and the opposing interests of partners can result in a politicization of a project with different hidden agendas (Faulkner, 1995) and incompatibility of strategic objectives (Cauley de la Sierra, 1995) .

The *second factor* affecting the success of cross-cultural cooperation in international megaprojects concerns the historically developed inequality and latent ethnic tensions between the home bases of the organizations. Over 350 years, former colonies have developed a rich tradition of resistance to the former Europeans colonials. These societies have developed a thorough knowledge of Dutch society and culture and are therefore generally familiar with cultural differences and thus know how to cope with them. The manifestation of ethnicity can obstruct cooperation within projects as national governments stressed the importance of national identity and unity. The longer and more intense the tradition of resistance, the more difficult cross-cultural cooperation can be (Van Marrewijk, 2004).

Studies on cross-cultural management are dominated by an essentialist conception of culture (Söderberg and Holden, 2002). A cultural identity is always a result of defining similarities and difference with other individuals and groups (Jenkins, 1997). Therefore, interaction is seen as a prerequisite for identification (Barth, 1969; Royce, 1982; Jenkins, 1997). Identity provides continuity, safety and stability for both individuals and groups. The identity of a person is constructed of distinct social identities. In distinct situations persons can arrange their social identities differently, which is called the hierarchy of social identities (Jenkins, 2004). Royce (1982) also states that cultural elements are created or invented by members of a group to distinguish themselves from another groups.

Koot (1997: 332) stressed that strategies of tolerance, harmony, interdependence and synergy are instruments of dominant Western companies and states that 'harmony is the catchword of those who want to maintain the status quo'. For non-dominant partners cooperation is more difficult as their risk of losing cultural identity is higher. Child and Faulkner (1988: 245) included the perspective of the partner organization and formulate a possible strategy in reaction to ethnocentric strategies. They label this as the 'breakdown strategy'; if one of the partners in the alliance is culturally dominating against the will the other partner.

> A condition in which the different groups in the alliance or joint venture are incapable of working with each other, and considerable tension and conflict will ensue so long as the alliance is kept in existence (Child and Faulkner, 1998: 248).

National identity and cultural differences can be understood as the result of social interaction that can change over time and is situational (Royce, 1982; Jenkins, 1997; Eller, 1999). In reaction to an ethnocentric strategy by a dominant organization, national identity and cultural differences can be used strategically by the non-dominant partner (Ailon-Souday and Kunda, 2003; Van Marrewijk, 2004). We can speak of 'ethnicisation' when the construction of an organizational identity based upon a notion of a shared national identity and shared cultural values is used strategically in projects (Van Marrewijk, 2004).

The *third factor* of influence concerns the formal and informal corporate cross-cultural strategies. Corporate strategies for coping with cultural differences are intertwined with organizational culture, and should therefore be understood in that context. Partners with cultural experience in the project's country are more successful in establishing a joint venture than partners without such experience (Van Oudenhoven and Van der Zee, 2002). Therefore, the use of experience and personal networks can be a decisive factor in acquiring a position in a foreign market.

Strategies for handling multiple cultures are culture-specific approaches (Chevrier 2003). Strategies in (Western) management literature show a great deal of similarity and can be divided into three groups (Adler, 1986; Schneider and Barsoux, 1997; Holden, 2002). The first group of strategies concerns ethnocentric strategies which support the cultural dominance of home base companies (see figure 1). Unity, control by the headquarters of the parent company, home base values and home base management models characterize this group of strategies. The second group of strategies consists of polycentric strategies which stress the importance of the culture of a host country. The acceptance of cultural diversity, the relative autonomy of local branches and the minimization of the cultural distance to the local market are all characteristics of this group of strategies. The final group consists of strategies which are combinations of the first two groups of strategies. Fung (1995), for instance, explained the origins of the ethnocentric and polycentric cross-cultural strategies from a culture-historical perspective and proposed the geocentric strategy as an attractive alternative to Western and local ethnocentrism. Strategies of tolerance, harmony, interdependence and synergy are instruments of dominant Western companies of those who want to maintain the status quo (Van Marrewijk, 2004). Like the concepts of the global multicentric strategy (Adler and Ghadar, 1993), the utilizing strategy (Schneider and Barsoux, 1997) and the synergy strategy (Adler, 1986), this concept is based on the assumption that cultural difference can be overcome or be constructively used for competitive advantage.

Adler (1986)	Adler and Ghadar (1993)	Fung (1995)	Schneider and Barsoux (1997)	Chevrier (2003)
Cultural Dominance	Ethnocentrism	Ethnocentrism	Ignoring	Trial and error
Cultural Compromise	Regio-centrism	Polycentrism	Minimization	Tolerance and self-control
Cultural Synergy	Multicentrism	Geocentrism	Utilization	Transnational

Figure 1. Strategies of cross-cultural management,

The *fourth factor* of influence concerns the formal and informal individual cross-cultural strategies. Individual strategies of managing multiple cultures proved to be important for successful cooperation. Project managers with cultural experience in the project's country are more successful in establishing a joint venture than partners without such experience (Van Oudenhoven and Van der Zee, 2002). The use of experience and personal networks can be a decisive factor in acquiring a position in a foreign market. Project employees use three different strategies to cope with cultural differences. (1) Adhering strictly to the culture of the home base and rejecting the host country's culture. (2) Getting thoroughly involved in the culture of the host country and rejecting the culture of the home base. (3) Establishing personal relations in the cultures of both the host country and the home base. A study of Chevrier (2003) showed that three kinds of cross-cultural practices emerged from project groups. Firstly, strategies were based upon individual tolerance and self-control. Secondly, a trial-and-error process was coupled with relationship development. Thirdly, capitalizing on transnational corporate or professional cultures. This is supported by a study of Mäkilouka (2004) who studied the leadership styles of project managers of multicultural teams. The large majority of project leaders appeared to be task oriented with cultural blindness, ethnocentrism, and in-group favoritism. The project managers that indicated a relationship orientation showed cultural sympathy and maintain team cohesion.

The discussion the the four factors related to successful cooperation show that the multi value models are insufficient for studying cross cultural management in complex megaprojects.

Towards an Interpretative Model of Studying Cross Cultural Management in Complex Projects

The exploration of an interpretative model starts with a discussion of the concept of culture. The integrative perspective on organization culture gave rise to much academic discussion (Martin, 2002). The concept of culture that has been used in the integrative perspective is far too simple. Van Maanen (1991) was among the first to stress different subcultures in organizations. The instrumental and functional character of culture with its emphasis on cultural systems is criticized. The interpretative perspective has increasingly received attention in organizational studies (Czarniawska-Joerges, 1992; Kunda, 1992; Barley and Kunda, 2001). In contrast to the general perception of organizations *having* a culture, organizations have to be perceived *as* cultures (Bate 1994). In this "root" metaphor, organizations are modern tribes with artefacts, practices, values, multiple cultures, power relations, conflicts, and abnormalities. To describe organizational culture as a phenomenon Martin (2002) uses three classifications (see Figure 2).

The first classification analyzes content themes which consists of espoused and inferred cultural values orientations. Espoused values are those who are communicated by the organization to employees and external audiences. The inferred values are those actual at the work shop in the day to day practices. The second classification maps the formal and informal practices such as (unwritten) social rules, activities, and behavior. Finally, the third classification analyzes cultural forms which describe the physical arrangements, stories, rituals, humor, myths, and heroes.

Based upon the 'root' metaphor of culture, Bate (1994) states that strategic change is *synonym* to cultural change. The transfer of a cultural system is an active interaction instead of a passive transfer as organizations are social worlds in which people construct their own cultural system in constant interaction between employees and the cultural system. Many strategies of changing organizational culture are based upon Lewin's (1958) suggestion that cultural change is a transformation from a stable situation to a new one. This perspective has been criticized for Lewin's assumption that organizations operate in a stable state, that he ignored power and politics, and that he had a top-down and management driven perspective.

According to Alvesson (1993) culture is much more a dynamic than a static concept as the project environment, management focus, and partners change during time.

Content themes		Cultural manifestations		
Espoused	Inferred			
		Practices	Formal	Structure
				Tasks and rules
				Control
			Informal	Social rules
				Activities
				Behaviour
				Perceptions
		Forms		Rituals, stories
				Jargon
				Humor
				Material artefacts

Martin, 2002

Figure 2. Classification of organizational culture.

Methodological Reflections

To study multiple cultures in complext projects qualitative fieldwork methods is needed. Anthropological fieldwork methods are becoming increasingly popular in management and organization studies (Czarniawska-Joerges, 1992; Schwartzman, 1993). Field research is a research strategy to describe, to interpret and to explain behavior, meaning and cultural products of persons involved in a general limited field by direct data collection of researchers who are physically present. The major invention of anthropologist is the “doing” of ethnographic fieldwork by means of participant observation (Bate, 1997). The aim is to give an emphatic understanding of the daily activities of the employees, to give the impression of having ‘been there’ and, to describe the connections of these employees with social, historical, cultural, political and economic processes from outside the organization (Bate, 1997). Organizations are perceived as a cultural phenomenon.

Three methodological instruments to guarantee the reliability of the research instruments and the internal validity have been used in this study (Hart *et al.*, 1996). Firstly, data-, researcher-, and methodological triangulation were applied. The methodological triangulation included biographical interviews, observations, participant observation, group interviews, and desk research. Eighty-five biographical interviews were distributed over the (former) management and work floor employees of Environ Megaproject and involved partners. Biographical interviews helped understanding of the development of value orientations and stimulated reflexivity on the part of those interviewed (Koot and Sabelis, 2002). To study the daily activities of employees the research instruments observation and participant observation have been used. Participant observation was executed for a year at the project’s headquarters, regional offices, and offices of the principal partners. Researcher triangulation was used as all interviews were conducted by two researchers, one taking notes, the other doing interviews.

Data triangulation included interviews with project managers, workfloor employees, employees of public and private partners. Secondly, field data was systematically handled and analyzed. During the research, four kinds of field notes were made: observational, theoretical, methodological, and reflective notes. The notes were direct worked out in interview reports. Thirdly, on two occasions a group of professional project managers related to Environ Megaproject reflected upon the research findings. Finally, findings were discussed with employees during lunch readings in the project offices and in meetings.

Apart from the evaluation research between 2003 en 2004, two studies have been done on the project control and international collaboration in 2005. These studies included twenty extra interviews on the topics, In 2007 desk study has been done to study the finalization of the Environ megaproject.

The Case of the Environ Megaproject

Environ Megaproject was one of the largest infrastructural projects in The Netherlands. The Environ Megaproject is a technological-complex project that uses non-proven technologies, involves participants from different industries and focuses on a result that is difficult to split in rational parts. The project included a large number of fly-overs, tunnels and bridges and is situated in dense populated areas. From start the project gave much debate in the parliament and society due to environmental questions of protecting landscape. Many technological complex problems had to be solved in order to dig tunnels in instable clay, to build bridges over wide rivers, to stabilize swampy grounds and to reduce environmental impact in dense populated areas. Furthermore, thousands of civilians living in the area affected by the Environ Megaproject were involved as well as nineteen local governments, three counties and twelve offices for water management. The many construction and engineering companies, governmental departments, pressure groups and other organizations increased the complexity for realization of Environ Megaproject.

The project was an independent project-organization under the supervision of the Ministry of Public Works of The Netherlands. The management team initiated, managed and executed all activities related to the realization of the project. All other related organizations had little authority in the project. In a conventional approach, mega projects developed in a number of distinct phases; alternatives study, feasibility study, safety study, environmental impact study, project appraisal and first decision by parliament (Flyvbjerg *et al.*, 2002). Then a state-owned enterprise is established to implement the project, application for required permits, finance, recruiting consultants for design and supervision, recruit contractors, and finally supervise and initiate operations (ibid). Environ Megaproject has not been developed in this linear time frame as project phases were overlapping and ambiguous.

The Environ Megaproject is a Public Private Partnership that started in the early 1990s and was finished in 2008 (see figure 3). The project was a Public Private Partnership in which the national government, construction firms, engineering and consultancy firms, investors and private companies participated. The contract was a Design, Build, Finance and Maintenance (DBFM) contract. It was partly pre-financed by the national government and partly by private banks and investors. The large DBFM contract was split into six sub-contracts in order to manage the megaproject properly. Each subcontract was managed in a sub-project by a project manager. These managers will be called sub-project managers here to make a distinction with central project management. The Ministry of Public Works reported to the Minister and controlled project budget in order to avoid cost overrun, time delays and changes in scope. Two departments of the Ministry of Public Works were responsible for the project. Steer was responsible for the initiation and decision-making phases, while Flow was in charge of the realization phase. Another important partner in the project was Straight, a centre of expertise for project management and infrastructure construction, which gave account to Steer.

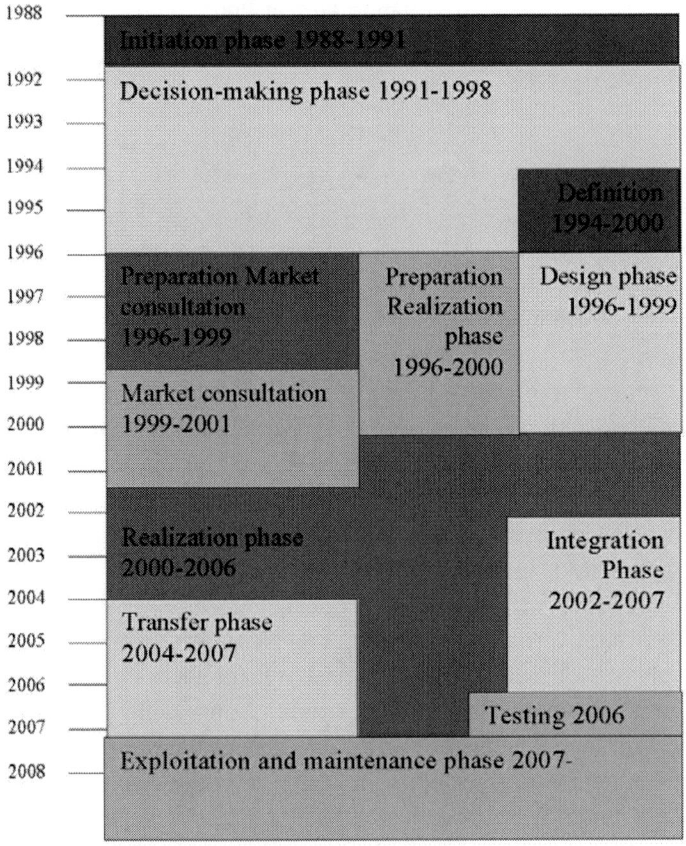

Figure 3. Different phases in the Environ Megaproject.

The project's context during the start was characterized by decision making processes, uncertainty, political discussion, and technological complexities. The political process of decision making dominated the start of the project. The government discussed the inclusion of private capital, and the different impacts. At the same time, the management team had to prepare for market contracting and for realization. Furthermore, the soil of The Netherlands, the many canals and rivers were a serious challenge for the engineers. Due to the exceptional size and the innovative character, the outcomes of the project were unsure.

Given the cultural diversity of public and private organizations, organizations of different national backgrounds, organizations from banking and construction sectors and the six subcontracts, the management of the Environ Megaproject had to manage cross-cultural themes at different levels during the project design, implementation and realization phases.

Industrial Level: Public and Private Organizational Discourses

The Environ Megaproject is a Public Private Partnership that is defined by the project's management as a cooperation between government and businesses, based upon clear contractual agreements in which:

- Contractual agreements on ownership of risks and costs
- Concerning societal as well as commercial goals
- Public and private partners expect to realize a better result against lower costs due to the contribution of specific knowledge.
- All partners stick to their own identity and responsibility

Different public and private partners are involved in the realization of the Environ Megaproject. The public partners are the Ministry of Public Works, Local governments, Flow, Steer and Straight (see figure 4). The private partners involve consultancy firms, drilling companies, construction companies and investment banks.

The project was considered by its public partners to be a project at a distance as public partners did not want to or were not able to support the project with employees and knowledge. Consequently, it was hard to find experienced, qualified employees. More than 95% of the employees working on the Environ Megaproject were hired from engineering consultancy firms rather than sourced from within the partner companies. Therefore, the experience and knowledge of Flow, Straight and the other partners were not included to a great extent in the Environ Megaproject. According to the responsible governmental partner the Ministry of Public Works:

> We have had sessions with the partners to discuss the cooperation model. But there wasn't a cooperative attitude (Interview with manager Ministry of Public Works).

Based upon the work of Lane (1994), Veenswijk (2003) works out a distinction between the public and the private discourse (see figure 5). These different sets can be conflicting in public private partnerships such as the Environ Megaproject. Cooperation within the Environ Megaproject was especially difficult for the public partners. They were used to be a commissioner rather than an equal partner. However, a Public Private Partnership is based upon equality of all partners. In Environ Megaproject there was a constant threat for public partners to fall back in the role of a traditional commissioner with all of the power instruments and discourse connected to it. This resulted in a serious conflict during the bidding of the Environ Megaproject. In a meeting to consult the market private partners showed no will of cooperation.

> In the meeting our project director announced another relation with the construction firms. He told that the Environ Megaproject was looking for a new way of cooperation. The directors of the construction firms were not enthusiastic and thought 'what an idiots'. Afterwards, during the drinks a former colleague asked me 'what are you guys up to? This must be a joke? (Interview with a project employee of Environ Megaproject)

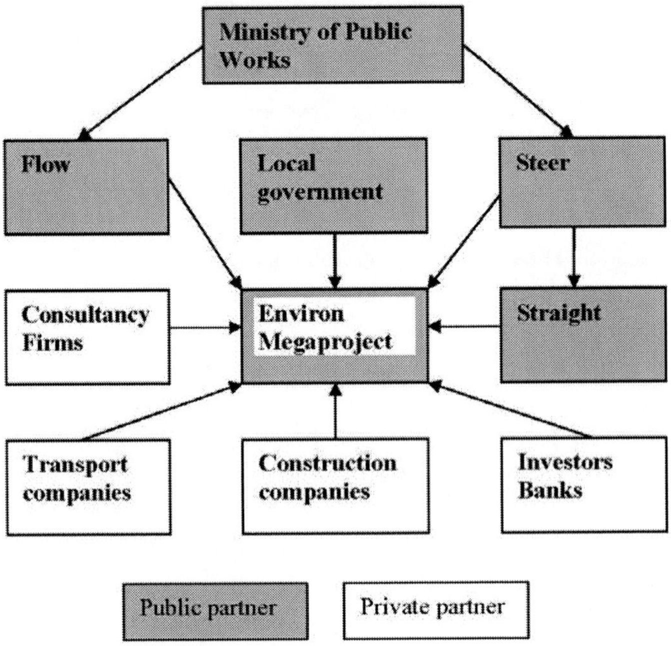

Figure 4. Public and private partners involved in the Environ Megaproject.

In 1999, during the market consultation phase, a serious crisis between public and private partners arose. Offerings were up to 60% higher than cost calculations done by the government. Confidence and trust between public and private partners was broken by the incident.

> We can state that September 1999 was the largest breach of confidence in the long historical relation between public commissioner and private client in the Netherlands (Interview with a member of the Board of Directors of a construction firm).

Although the construction of the Environ Megaproject is a temporary project, over time the project organization changed into a bureaucracy in which much time is spend on normal internal governmental processes. In contrast, the private partners had made a competitive tender with calculated risks. Both project management and private consortium wanted to prove that this Public Private Partnership was a success, a showcase for future megaprojects.

The project management of the Environ Megaproject was not successful in obtaining the commitment of partner organizations during realization phase. Extending a common project culture beyond the limits of project alliance partners' sovereignty is difficult; when stakeholders have to deal with the world of other organizations and individuals outside their sovereign realms, they lack authoritative resources to impose their will (Clegg *et al.*, 2002). The commitment of involved partners was seriously under pressure, due to different interpretations of the Environ Megaproject goals. According to our research, the project mission and goal were clearly formulated and relatively constant across time. However, the interpretation of partners differed across both time and setting. The meaning given to the formal project goals was dependent on organizational context and interests (see figure 6).

Public discourse	Private discourse
Rules	Results
Decent behaviour	Efficiency
Anticipation	Adjustment
Responsibility	Leadership
Accuracy	Innovation
Public	Private
Voice	Exit
Legitimacy	Outcome
Mission	Egocentrism
General interest	Profit

Veenswijk, 2003: 61.

Figure 5. Public versus private discourses.

Partner	Interest in Environ Megaproject
Straight	Operating the infrastructure Construction of the project Preservation of monopoly on infrastructure project
Steer	Possibility to reform the monopoly in the infrastructure sector
Environ Project management	Not a construction project, but an innovative concept of PM Fundamental change of the infrastructure sector Introducing new concepts of project management Independently realizing an innovative concept
Flow	Becoming a monopolist in the management of infrastructure projects
Private consortium	The return on investment Putting risks and costs to the State
Construction firms	Turn over and financial profit Putting risks and costs to the State
Consultancy firms	Delivering of employees, engineers and calculators

Figure 6. Different interests on the Environ Megaproject.

Financial partners have an important stake in DBFM contracts. Their interest is to have a profitable return on the investment done in a megaproject. While the investors are important partners in the cooperation, the contacts between the investors and the project organization was minimal during the execution of the project. The investors were interested in finishing the project within budget and time agreed, as their return on investment starts from that moment onwards.

After contracting the private parties tried to renegotiate contract changes. All changes have been studied for risk consequences and renegotiated with the pubic partner. These negotiations can delay the project. Private parties are concerned with the continuity of the corporations and try to maximalize their return of investments. In contrast, the government is responsible for societal goals, transparency, safety, participation, equality of rights and legitimacy. Fraud scandals in the construction sector have increased the public and political demands for transparency. New demands of local governments have resulted in extra fire safety measuring, anti vandalism, wild life protection and extra pedestrian protections. These changes had a large impact on influences for the private partners who had to calculate their risk profile over and over again.

Every aspect that was not foreseen at the beginning of the Environ Megaproject had to be renegotiated. As megaprojects last for many years it is nearly impossible to predict and include all changes and supplementary demands and with that, the risks that are connected to the project. All changes and extra demands have to be negotiated. Therefore, it is advised to have an extra risk fund reserved for the negotiation process.

National Level: Anglo-Saxon and Rhineland Models

In the Environ Megaproject organizations with different national cultural backgrounds operated. The Environ project management, Steer, Flow and Straight were all Dutch companies. A larger part of the construction firms were Dutch, only one construction company originated from France. One of the most important actors in the Environ Megaproject was a consortium responsible for the Design, Build, Finance and Maintenance contract. The industrial shareholders consisted of American, German and Dutch contractors held 51% of the consortium. The American project management firm Stars is one of the world's largest and most experienced project management firm. Their expertise is found in engineering, construction, maintenance and technical services. They operate in The Netherlands for many years but their American cultural background is still dominant. The German supplier is one of the leading technical industrial suppliers, with also many years of experience in The Netherlands. Furthermore, English and Hong Kong investor banks participated with a 49% stake in the consortium.

An important cultural tension could be observed in the cooperation between the Dutch project management and the American led consortium. This gave rise to the question regarding what the differences in the national cultures of America and the Netherlands in fact might be. In three of the cultural value dimensions of Hofstede (1980) the national cultures of the USA and the Netherlands are in fact quite similar. Only the value dimensions of masculinity differ in both cultures. Trompenaars' (1993) seven dimensions indicate that the American and Dutch cultures are similar. Both the Dutch and American cultures are universalistic, individualistic, achievement-orientated, specific-orientated and neutral cultures (see figure 7). These findings suggest an easy cooperation between American and public Dutch partners.

Value Dimension	Score of the Netherlands	Score of USA
Power distance	38 (No. 40 at the list)	40 (No. 38 at the list)
Individuality	80 (No. 4/5 at the list)	91 (No. 1 at the list)
Uncertainty avoidance	53 (No. 35 at the list)	46 (No. 43 at the list)
Masculinity	14 (No. 51 at the list)	62 (No. 15 at the list)

Hofstede 1980.

Figure 7. Cultural distance between Dutch and American cultures

Although there are large cultural similarities between the Dutch employees and employees of the American contractor in Hofstede's dimensions, in the daily practice of cooperation there were many conflicts. Central in these conflicts were misunderstandings with the Dutch project management on the behavior of the American led consortium.

> I don't understand why he acts the way he does. It must have to do with the American culture. (Interview with Dutch project manager).

At one point in the execution of the project a number of specialists are called by the project director for a meeting. Goal of this meeting is to better understand the motivations, behavior, culture and interests of the consortium in order to improve the efficiency of the project. The consortium partners are not invited as a strategy of how to cope cultural differences had to be developed by the project management.

> This is a multi cultural, multi actor complex organization. (Interview with Dutch Project director Environ Megaproject)

The contract between the public partner and private consortium is written in English. This is problematic in the daily work practices as many of the Dutch do not master the language as well as their private counterparts. Understanding each other's English is one of the largest problems noted by both Dutch and non-Dutch project employees. Pinto (2005) acknowledges that overcoming problems in the use of the English language are important hidden costs due to laborious collaboration and misunderstanding. According to English trained professionals the English vocabulary of Dutch employees is limited. To overcome ambiguity in the interpretation of the spoken English language informants prefer to use emails and written communication. This language problem was not included in the start of the project as many Dutch think of themselves as a fluent English speaker . However, this is not the case when collaborating with native English speakers of the American led consortium. Other experiences were asking questions and summarizing conversations. If one do not understand your colleague at the other side of the telephone line or in a chat session. Employees ask their colleagues to repeat their sentences and ask questions about the meaning given to words or expressions. Furthermore, employees summarized the content of the conversations in order to confirm their understanding of the discussed issues. As a result the concept of partnership never really developed.

The DBFM contract is based upon an Anglo-Saxon model. The private consortium was used to the Anglo-Saxon style in contrast to the Dutch partners that were used to work with the Rhineland model (see figure 8). The Anglo-Saxon model is based upon law developed from practices. Collaboration is based upon contracts that have to reduce uncertainty (Brouwer and Moerman, 2005). In the Rhineland model contracts are a foundation of collaboration, but in case of problems or opposing viewpoints these can be solved in a pragmatic way. Debating, discussing and trying to reach a consensus with all partners is therefore a core competence of the Dutch partners in this case. In contrast, the American management tried to get the best deal and hold on to that deal as long/firm as possible (Brouwer and Moerman, 2005). Differences between the Anglo Saxon and Rhineland models could be observed in the cooperation of mangers of public and private partners at the workfloor

> I was lucky that I could choose my own counterpart manager. We were very complementary to each other. What he did, I didn't do, and visa versa. That worked very well in the mutual tuning. I demanded that he was physically present one or two days a week; that worked out well. Other private managers were much less present at the work floor. That resulted in less commitment is my feeling. (Interview with Dutch manager of Environ Megaproject)

Characteristics	**Anglo-Saxon model**	**Rhineland model**
Law	Based upon practice	Based upon principles
Cooperation	Based upon contract	Based upon harmony and trust
Contextual change	Reducing uncertainty	Pragmatic
Conflict solution	Conflict orientation	Consensus, debate
Individual	Little attention	Central
Leadership	Directive	Coaching
Management	Central	Decentralize
Risk	Tunnel vision	Vague decision making
Metaphor	Coherent	Fragmented
Countries	USA, UK	Scandinavia, Germany
Interlaced in Dutch system	*****	********

Brouwer and Moerman, 2005.

Figure 8. Characteristics of Anglo Saxon and Rhineland model.

When a serious problem with the delivery of a safety system by the private partner arose, the private partner wanted the Dutch management to pay the extra costs to solve the problem. Dutch project management expected to find some point of consensus after a long negotiation process. Instead, the American project manager of the private consortium pointed at the original contract and refused to come to a setting of the conflict. After one year of negotiations, the Dutch project management finally had to pay for the extra safety system and the project was delayed with another couple of months.

The two models have fundamental oppositions which hindered smooth cooperation between partners involved. Employees and managers of Environ megaproject. Therefore, both Environ Megaproject and the private consortium needed to develop sensitivity towards cross-cultural theme's to prevent problems in the construction process.

Organizational Level: Fight over Power in the Project

The project's context during the start was characterized by decision making processes, uncertainty, political discussion, and technological complexities. The political process of decision making dominated the start of the project. The government discussed the inclusion of private capital, and the different impacts. At the same time, the management team had to prepare for market contracting and for realization. Furthermore, the soil of The Netherlands, the canals and rivers were a serious challenge for the engineers. Due to the exceptional size and the innovative character, the outcomes of the project were unsure.

In 1996, the Ministry of Public Works selected a project director with a clear vision, who could handle uncertainty, motivate people, and who could support the political decision making process. In the perception of the project director the Environ Megaproject was an innovative concept.

> The project is not only about construction. It is also a reorganization inside and outside the government. With this project we will show how construction industry and the Ministry of Public Works will work within the next ten years. (Interview with former project director)

In order to realize such an innovative concept, the project had to cooperate intensively with public and private partners, all with different organizational cultures. Flexibility and social political sensitiveness to the discussions in and changes from the political context were needed as the project scopes were not very clear. Therefore, according to the project management, the organization had to be problem oriented.

As the Environ Megaproject was an independent project, rather than one run from within the bureaucracy of the public sector or the hierarchy of a single company, bureaucratic control mechanisms were not very clear. The project organization experienced a lot of freedom in the initiation and decision making phase. Control of the project was at its most blurred in the year 1999, as the project was under the supervision of both Steer and Flow at the same time. Flow was responsible for parts of the project that could already be realized, while Flow was responsible for parts that were still in the process of decision making. The decision to split the Environ Megaproject in this way was opposed by the project management. These two departments, Flow and Steer, represented two different organizational cultures. The project management was afraid of loosing control over the project as these two departments had a poor record of cooperation. Splitting the project would give maximal flexibility to adapt to political developments in the context of public administration. The disadvantage of this situation turned out to be vague bureaucratic control and cooperation problems between Flow and Steer.

> Environ Megaproject had a vague structure, in which Flow, Straight, and consultancy firms all did something but in which the responsibilities were not clear. Straight wanted a structure with a clear commissionaire role. (Interview a with manager from Straight).

Due to their relatively independent situation, the management of the Environ Megaproject initiated, managed, and executed all activities related to the construction of infrastructure. All other partners, including Straight, had little authority in the project. That was remarkable; given that Straight is a centre of expertise for the construction of infrastructure. The Environ Megaproject needed Straight as a partner. To include the cooperation of Straight and other partners a steering committee was set up, in which all partners would participate to prepare for the realization phase. However, the partners did not agree on the organization of activities recommended for the project and the result was that the Environ Megaproject was in a state of conflict between nearly all partners, so serious, in fact, that these partners no longer wanted to cooperate. Informants stressed the lack of enthusiasm for the dominant and autonomous position of the project organization. Partners had no direct influence on decision making or control over the activities, as they would only be responsible for support in terms of people, knowledge, and experience. The partners preferred to opt for a matrix model in which they would have extended authority and would be responsible for specific parts of the project. They wanted to design infrastructure and manage a part of the project themselves.

> Our proposition was to give certain parts of the project to the different partners, and that these partners would give account to the project management. (Interview with a manager from Straight).

During the preparation of the realization phase, the management of the Environ Megaproject selected a non-classical model of project management to support the innovative character of the project. The model was based upon transparency, an orientation on organizational processes and a coaching leadership style in contrast to the more traditional model used in Flow, in which control and hierarchy dominated.

> Flow is a control organization, completely different than the life-cycle principle of our project. There was a dilemma of freedom versus control. I didn't want to control the project but to make it transparent in order that all involved organizations could easily follow the process. (Interview with project manager of Environ Megaproject).

Flow perceived the Environ Megaproject as a project full of risks, due to the lack of focus on control. When Flow took over the project during the realization phase they replaced the project manager. With the introduction of a more traditional project manager from Flow, conventional control and hierarchy were re-established. The management supervised compliance in terms of the formal rules and stipulated procedures, and was supported in this by the controller. Employees were now confronted with less freedom and a more bureaucratic organization.

Socialization mechanisms hardly worked in the Environ Megaproject. Only a few employees of the partner organizations participated in the project. During the preparation of the realization phase, the project organization's autonomy gave rise to much irritation and discussion within the ranks of Straight, Flow, and Steer.

> There was a strong identification with the project. Consequently they were not open and developed an attitude that put others off. They went their own way. (Interview with a manager from Steer).

The commitment of Flow and Straight to the Environ Megaproject was crucial for success. However, cooperation between both organizations did not work out well. Both Flow and Straight operate in distinct cultural settings, due to their technical specialties. Both organizations comprise closed communities of specialists. In the past, conflicts over control had occurred between Flow and Straight. Therefore, in the documents joint projects were called 'touchy works', for which a very detailed protocol of cooperation needed to be designed. In these protocols responsibilities were demarcated in discourses. To overcome the fighting over the question 'who is in charge of the project' a joint venture was proposed as solution.

> Given the size of the project, the complexity and the changes in such a project for innovative organization of construction as well as technologies, it is necessary to maximally utilize all the available knowledge. (Cooperation agreement 2000).

During the preparation for the realization, an advisory council was installed to coordinate the interfaces between both sub-projects. Both organizations were integral in their responsibility for the realization of the project. However, Straight's role was minimized by both Steer and the national government. European regulation prohibited Straight from gaining competitive advantage, compared to other competitors, in the realization of the project. From the perspective of Flow, Straight was unable to bring about the innovation necessary for this large project. Furthermore, Flow's administrative system of control was chosen as a platform for the Environ Megaproject system. According to Straight, these facts gave too much control to Flow. The equal cooperation envisaged turned out to be project control by Flow with (human) resources and knowledge inputs from Straight. This was not acceptable for Straight, and they threatened to resign from the project. When it became clear that the control was indeed in Flow's hands, Straight ended its cooperation and no longer supported the Environ Megaproject. The project manager of the Environ Megaproject reflected upon this period:

> My most impressive personal experience with this project was the clash of cultures and structures. (Interview with project manager of the Environ Megaproject).

During the realization phase the commitment between the Environ Megaproject and the partners Straight, Flow and Steer slowly restored. The project management acknowledged that it needed the cooperation of other partners in order to reach the objectives successfully. A number of public employees from Flow joined the project and changed the management style in the project to a more diplomatic style, avoiding conflicts, and focusing on cooperative behavior. The organizational culture of Environ Megaproject changed towards a centralized hierarchical organization with a focus on procedures, (financial) control, and human resources. Through knowledge management and the exchange of knowledge organizational networks were restored with Straight. Although the project management encounters old sores and antagonists, slowly more and more partners committed to the Environ Megaproject. In the next paragraph it is learned what interventions have been done to change the organization culture of Environ Megaproject and to restore collaboration.

Project Level: The Episodes of the Gideon's Gang and the Diplomats

Two distinct episodes could be analyzed in the development of the project culture of the Environ Megaproject; the episode of the Gideon's gang and the episode of the Diplomats. Both episode will be discussed below.

The Episode of the Gideon's Gang

With the selection of the visionary project director a new cultural episode started in 1996. This episode will be described according to Martin's (2002) classification of organization culture; content themes, practices and cultural forms.

Content Themes

The project developed a fighting spirit in which employees committed themselves to the project and to the belief in the innovative concept. The managers were innovators, strong in conceptual thinking, strong in development of new ideas, strong in the communication of enthusiasm to others and strong in overcoming difficulties of resistance in the (political) context. As a former staff member explained:

> We needed people with alternative minds, not traditional constructors. It was called the 'alternative disease', there was a constant change, and everything was overthrown. One has to be able to stand that. I needed people who were flexible. (Interview with former staff member of Environ Megaproject)

Employees experienced a strong sense of uniqueness. They were constructing something that was never done before, something that was not easy to realize. Entrepreneurship, flexibility, independency, responsibility, and creativity were perceived as important value orientations to make the project a success. Not the traditional control mechanism but transparency and freedom for managers and employees to solve problems were the main management tools. The former project director wanted, in contrast to traditional project management, to emphasize the innovative process management :

> I didn't want to control the project but to make it transparent so all involved partners could easily follow the process. (Interview with former project director).

At strategic positions a number of civil servants were employed as the management had to report to the Minister of Public Works.

> One always has to consider; do I tell the minister or do I solve the problem myself? One can't always put issues on the Minister's agenda, but if you don't tell and the issue turns out to be a major problem, you are in trouble. (Interview with project manager of Environ Megaproject)

Practices

The entrepreneurial, innovative approach broke with the project management traditions of the Ministry of Public Works. New employees and managers were not recruited from experienced partner organizations but from engineer consultancy firms and personal networks. It appeared to be difficult to recruit civil servants from the Ministry of Public Works and other public partners.

> It was difficult to find people as there was a large difference in culture between the department en the project. They were not entrepreneurial and had a negative image of Environ Megaproject. I was frequently told: 'my boss has told me not to cooperate with this project'. (Interview with former manager of Environ Megaproject)

To realize the innovative concept, Environ Megaproject was brought to the construction market in six different Design and Construct contracts. The project management choosed a decentralized management model, which they called "decentralism, unless" to manage the project during the realization phase. In this concept six managers, called project office managers, were fully responsible for realizing their 'own' Desing and Construct contract. Only issues that would influence the outcomes of other projects were discussed at the level of the project director. The entrepreneurial and independent managers were stimulated by the 'decentralism, unless' model to focus on the supervision of their contract. One of the critical aspects for the success of this decentralized management model was to uphold the integral focus of the project. Due to the late implementation of central scopes and the weak uphold of it, managers experienced little central control.

> Management was afraid to use power, confrontations were avoided. Management was afraid to be too direct; they were not good in discussing bad news. (Interview with employee of Environ Megaproject)

The management style that was practiced in this episode could be characterized as non-conformist, with a lot of freedom for personal development, creativity, freedom and innovation. Some experienced this style as rather chaotic. Such a style appealed the employee's personal responsibility and self dependency. Informal practices were rather chaotic.

> It was ok. You have got freedom of operation, you could be creative. 'Find your own way to reach your goal' was the motto. But it was also a hard time; you were thrown in the deep end. Eventually, you were managed. (Interview with former employee of Environ Megaproject)

Cultural Forms

The identity of independency and innovation was reflected in the logo of the Environ Megaproject that was completely different from the Ministry of Public Works. This logo appeared at all offices, papers, flags, articles, and public announcements. Each of the six project office managers had their own staff and was situated in a small field office next to 'their' part of the project and at distance of central head offices.

Due to the fact that few civil servants were willing to work at the project, a large amount of over 90% of contract employees was hired from engineering firms to realize the

Megaproject. These employees strongly identified themselves with the innovative project culture, which they also referred to as the 'Gideon's gang'. In the bible story, the Lord has chosen Gideon to head up the deliverance of Israel from the Midianites. God told Gideon that he needed only 300 out of his army of 30,000 men. It was better to have a small army of men who trusted God than to have a big army that included the fearful, because fear is contagious. Gideon's gang is a metaphor for a brave group of men that knows no fear and uses creative, innovative methods to reach their goals. Involved partners called the group of innovative managers in the Environ Megaproject the 'Gideon's gang'.

The Project Culture in the Episode of the Gideon's Gang

In conclusion, the project culture during the episode of the Gideon's gang could be described as a non-conformist and innovative culture (see figure 9). This project culture helped the project to be flexible, to adapt to changes and to design an innovative concept. Furthermore, it supported the project to avoid heavy protests of environmental groups and public. Finally, it stimulated employees to commit themselves to the project.

Project culture in Gideon's gang episode (1996-2001)		
Cultural forms	physical arrangements	• special logo, flag, paper, • headquarter apart from government
	rituals	• get-togethers, farewell parties • speakers corner speeches of the project director
	myths	• Gideon's gang • Environ Megaproject is a reorganization of the infrastructure market • heroes: engineers
Practices	rewards and punishment	• rewards for creative and innovative behavior • punishment for not sharing the vision, being conservative
	communication	• internal: informal, based on personal networks, • external: focused on public relation, information giving
	recruitment and selection	• competences: creative, dealing with chaos, entrepreneurial, independent, result oriented, young, non-bureaucratic, committed • using personal networks for recruitment • the best engineers together gives the best result • 90% external contracts
	management style	• visionary, creative, non conformist • chaotic, hectic, ad-hoc • technologic rational orientation: reduction of complex problems to technologic problems
Content themes	value orientations	• traditional project management will not bring the project to a success • innovation is necessary to be successful • Environ Megaproject is an unique project with unique solutions • budget is important but reliability of the solution is more important • protestant ethics; hard working, serve the public, be honest, lawfulness • strong loyalty to government, the minister has to be served at all times

Figure 9. Environ Megaproject culture during episode 1996-2001.

The Cultural Episode of the Diplomats

During the execution of the six Design and Construct contracts, some content themes in the project culture started to become, in Bate's (1994) terms, dysfunctional. The strong commitment, fighting spirit and non-traditional style of the project management caused irritation and non-cooperative behavior with the involved public and private partner organizations. This caused intensive negotiations on cooperation and finally the withdrawal of some of the partners. The former project director explained that the cultural cooperation with the partner organization was the most difficult part of his job:

> My most intensive personal experience with the project was the clash of cultures and structures. (Interview with former project director of Environ Megaproject)

The large freedom for project office managers to manage their 'own' contracts in combination with a weak central control mechanism resulted in laborious cross-functional cooperation within the project. The 'decentralism, unless' management model was not supported by the project culture.

> Our decentralized organization model increased problems. Nobody was really responsible for the interfaces between the projects; everybody was focused at his own project, hardly any integral management thinking could be detected. (Interview with staff member)

Increasingly, the Ministry of Public Works perceived the non-traditional project management as a risk. The chaotic and creative management style with emphasis on transparency and less attention for control reinforced the impression of the Ministry of Public Works. At the same time, cost overrun in the Environ Megaproject attracted public and political attention. The political parties asked the Minister for explanation and a parliamentary inquiry was started. As a result of these developments a new project director was selected to bring in more control and traditional project management experience.

Traditional planning and control, centralized management and cooperation with partner organization became central in the new management style. However, the managers in Environ Megaproject had experienced managerial freedom for a long time, and resisted the undesired cultural change.

> I encountered an icy atmosphere with no support for the new project director. Meetings were held behind his back and no information was given. (Interview with a former staff member of Environ Megaproject)

Consultants were hired to analyze internal cooperation, trust, and identification with integral project goals. The analysis showed distrust among the project office managers and a strong identification with their own contracts. The employees perceived the project from a technological rational perspective and were unable to connect their project to the integral scope of the project. The analysis resulted in a management meeting in which mutual images, fears and trust were discussed. A number of managers were replaced by other, more traditional project managers. The change process was supported with the implementation of a new control system creating a new position for a manager control. The change process was also reflected in fundamental changes in the organizational structure. Most important, a cultural diffusion started between the organization culture of the Ministry of Public Works and the project culture.

> More and more it is becoming a bureaucratic top-down organization with a very large overhead. It is just like the Ministry of Public Works. (Interview with an employee from Environ Megaproject)

The intervention was an, in terms of Bate (1994), aggressive strategy of cultural change. Turnover of management personnel occurs predominately during the execution phase of the project life cycle and is generally perceived as negatively affecting the performance of the project (Parker and Skitmore, 2005). Parker and Skitmore (2005) makes an exception; if a manager who is ineffective or not performing a turnover could increase performance. This was the case with the Environ Megaproject as the competences of the project director did not match the competences needed in the realization phase.

With the intervention of replacing the project director a new cultural episode started in 2001. This episode will be described according to Martin's (2002) classification of organization culture; value orientations, practices and cultural forms.

Content Themes

The newly introduced cultural values were based upon the organization culture of the Ministry of Public Works. In the Ministry of Public Works loyalty to the minister, political stability, control, hierarchy, and power were important value orientations. Conflict avoidance, risks avoidance, lawful action, accountability and cooperation were central concepts. These cultural values were introduced in Environ Megaproject. Important was the introduction of new control mechanism. Integral financial risks were calculated and managed through an intensive system of control. As all political and public attention was focused at the project, the communication with the Minister and the government had become a central topic.

> Political stability is a very important issue in the success of this project. We have to predict costs overruns and time delays". (Interview with controller of Environ Megaproject)

Practices

The cooperation between the Environ Megaproject and the public and private partners was intensified. The project management acknowledged that it needed the cooperation of other partners in order to reach the objectives successfully. A number of public servants from the Ministry of Public Works joined the project. Although the project management encounters old sores and antagonists, slowly more and more partners committed to the Environ Megaproject.

> We are now preparing for the next project phase in which our partner slowly gets more involved in order to take over the maintenance of the infrastructure. (Interview with employee)

The new project director used a more central and formal management style than the former project director. To obtain more control in the project, cooperation among the project office managers was stimulated and supported. Central staff units, such as ICT and communication, were professionalized and extended. Finally, formal procedures, bureaucratic control mechanism and a more strict management style were introduced:

> I stick to my principles and I get things done. I'm not afraid to offend someone if it is in the organization's interest. If I think something is a bad piece of work I will tell someone. (Interview with manager control)

To support the cultural change in the Eviron Megaproject, the six separate project offices were integrated into two offices; office North and office South. Cultural differences between the project office subcultures were managed to merge into two offices.

> We had a good atmosphere with a team spirit. A drink at Friday evening was important to us. When we merged with North we experienced differences. They were much more hierarchical. To reduce cultural differences, we exchanged people between the two offices. (Employee of former South office)

Cultural Forms

No longer the innovators but financial controllers and risk managers were the new heroes in the project. Company days and other activities were organized to stimulate the identification of the employees with the overall scope of the project. The logo of the Environ Megaproject was redesigned in such a way that it strongly reminded of the logo of Ministry of Public Works. Not all employees agreed with this change.

> Environ Megaproject has a very strong brand with the former logo. The Ministry of Public Works is associated with rules and dullness. The new project director has changed the house style. It has cost a lot of money, I have never understood. (Interview with an employee of the Communication Department.)

In conclusion, the project culture during the Diplomat's episode was characterized as diplomatic, centralized, cooperative and bureaucratic (see figure 10). This project culture helped the organization to control and to realize the integral project goals. It supported the control of financial resources and the prediction of cost overruns and time delays, necessary to keep out of political unrest. Finally, the culture was supportive for the cooperation with partner organization and transfer of knowledge.

Two clearly distinct episodes have been analyzed. In the episode of the Gideon's gang, the project developed an autonomous non-conformist culture with a strong fighting spirit. Bate (1994) calls this the functional growth episode. In this phase one can speak of a highly developed fighting spirit, where the central department and politics are regarded as the 'natural adversaries' of the project group. The project group derived its primary meaning from such concepts as authenticity, originality and, more generally speaking, 'operating against the beaten track'. The 'us-them'-thinking developed in this phase shows the seeds for the most important problems in the area of mutual trust and connection.

In the episode of the Diplomat's, strategies of cultural transformation can be observed resulting in the abandoning of the existing cultural form in the project. The old paradigm was overthrown by the new management. Bate (1994) calls this the aggressive approach, in which current culture is disturbed in order to create a new cultural system.

The dysfunctional decay started during the realization of the contracts. Disciplinary practices developed blind spots for the integral management of the project. Isolation from partners, lack of financial control, and cost overrun resulted in public and political pressure and forced the Ministry of Public Works to intervene and replace the project director. The connection between 'old' and 'new' management needs to be continuously reconfirmed by means of a series of external interventions, and to be shaped by means of discussion meetings, conferences and new 'rules of play'. It is precisely around these new rules of play that, owing to the lack of a shared form of sense-giving, conflicts arise, leading once again to a tightening of the rules, and consequentially creating a unique 'paper' reality, which can be described as a continuous siege of both the front- and backstage (Goffman, 1959). The process of cultural diffusion between the Ministry's organizational culture and the Environ Megaproject organization's culture resulted in a collapse of last culture.

> Our intended cultural change did not succeed. The bureaucracy has prevailed and the change agents has left the project. (Interview with former employee of Environ Megaproject)

Project culture in Diplomat's episode (2001-2004)		
Cultural forms	physical arrangements	• redesigning the logo to the ministry of Public Works • integration of regional offices
	Rituals	• celebrations of project goals: arrival tunnel boring machine, sinking of last compartment in river, completing bridges • company days
	myths	• the tunnel bore machine • Saint Barbara as the patron of tunnel workers • heroes: financial controllers, risks managers
Practices	rewards and punishment	• rewards for control and integral management, diplomatic behavior, calculating risks • punishment for isolating, focusing on own project, independent behavior and bad control
	communication	• internal: formal • external: risks have to be avoided or be communicated well in advance in order to prevent political unrest
	recruitment and selection	• competences: risk avoidance, diplomatic, empathetic, trustworthy, control • civil servants
	management style	• diplomatic, avoidance of conflicts, • cooperative, centralized, control, procedural • technologic rational orientation: reduction of complex problems to technologic problems
Content themes	value orientations	• control, hierarchy • traditional project management bring success • lawfulness, integrity, political stability • conflicts can be avoided • protestant ethics; hard working, serve the public, be honest, lawfulness • strong loyalty to government, the minister has to be served at all times • trusting personal networks

Figure 10. Environ Megaproject culture during episode 2001-2004.

The main implications of the research findings for project managers and project performing organizations are focused around the management of the project culture during the project's life cycle. Each project phase was characterized by a distinct project culture with different set of dominant value orientations (see Figures 10 and 11). The organization of the reflection on the project culture is the responsibility of the project management and includes an orientation on competences needed for managers and employees in a new phase. In too many cases, like the project manager of Environ Megaproject, capable managers are replaced too late.

Professional Level: Project Controllers versus Engineers

Professional jurisdiction, expressed by professional authority, autonomy and sovereignty, is the link between professionals and their work. Professions are distinguished from other occupations by having members who have the exclusive right to determine how their professional knowledge is passed on and to whom; who can legitimately do its work; how the work should be done; if and when it should be evaluated and by whom (Abbott, 1988; Amabile *et al.*, 2001). This professional jurisdiction is attractive to occupational groups because of material and status rewards associated with it. These are usually assumed to include prerogatives such as autonomy, power over clients and financial resources.

Professional jurisdictions are central to the understanding of professions and professional identity (Abbott, 1988). When studying the rise and fall of professional jurisdictions, of most interest are the professional boundaries where occupational groups battle for authority over types of goods and services delivered. These battles are often followed by the division of the jurisdiction into *functionally* interdependent parts known as (a) 'division(s) of labour'. Jurisdiction is not fixed in time, and are a reality to be taken into account in the process of change (Abbott, 1988). Hence, professions are not static, their roles and boundaries change as opportunities in the environment emerge (Abbott, 1988). With a deeper understanding of jurisdiction we are able to see how professional groups reconceptualise the nature of their professional and occupational groups in new environments and how they carefully plan their future role in the labour market.

The relation between professionals and the organization in which they operate, is one of power and organisational control. Traditionally-autonomous professions have to continually cope with the bureaucratisation of organisations as the latter attempt to have greater managerial control over work processes. New occupations seek professional status in their own domain within these organisations. In the process of claiming rights to perform certain tasks, members of occupations naturally tend to emphasise what makes them alike yet different from other workers. As members of occupations seek autonomy and control over their work in organisations, they discuss the issue of professional jurisdiction with other professional groups. The grounds of the argument may differ, but most rely on the assumption that '*we* know what this organisation needs and *they* do not' (Parker, 2000: 209). However, organisational managers can (and do) succeed in reducing occupational autonomy by the standardisation and routinization of work, and they do this via the creation of policies and procedures and supporting organisational structures. As a result, organizational managers and members of a professional subculture fight over control and power.

Occupational subcultures have their origins in the distinct ways of solving practical problems. In modern society, people are identified by their occupation and are given exclusive rights to perform and control the tasks related to their respective occupations. People with the same occupation develop shared ways of coping with shared task-related demands and uncertainties (Roberts, 2000). These ideas are work-related and originate for a larger part from outside the organization through means of occupational socialization at schools and universities during education and training.

Occupational subcultures have six characteristics (Trice and Beyer, 1993). (1) Members identify themselves with their profession. Self-definition of members determines the boundaries of occupations. (2) The members of the occupational group use each other as points of reference. They seek support for and confirmation of the meanings they ascribe to work-related events. In this manner, they develop shared occupational ideologies. (3) Members of an occupational culture use stories, language, myths, taboos and rituals to cope with the emotional demands related to their work. (4) Members derive favorable self-images and social identities from their respective occupations. Three features contribute towards favorable self-image: the personality traits which arise in the face of danger, fundamental esoteric skills and socially valuable services. (5) Members tend to mix their work life with their private life. In some professions, such as those of fishermen, international consultants and police officers, little space is left for a separation between private and work life. (6) Occupational cultures stimulate ethnocentrism, particularly when they become communities. Outsiders are treated with suspicion and "our way" of doing things becomes the right and only way.

Professional culture in the Environ Megaproject could be observed with the drilling engineers and workman in the rituals of project starting. Before the start of a large drilling project the employees of the French construction firm held a mass in the entrance of the tunnel. A Dutch catholic priest was asked to hold this mass and blessed the saint of the drillers called St. Barbara. As drilling is a dangerous profession and the drilling bore was exceptional large in this project, more than 100 employees showed their respect to uncontrollable risks. St. Barbara, with a helmet on, is placed at a depth of twenty meters in a small chapel near the tunnel entrance. Ever since medieval times St. Barbara protects people against unsuspected calamities.

The professional culture of engineers conflicted with the professional culture of project controllers.

> In water engineering there is the concept of flow but once every time one needs force. Flow is your guide but force is needed for intervention. That is the role of controllers in this project. Within the ministry of Public Works too many engineers use the concept of flow. (Interview with project director)

Engineers are trained to solve problems, they derive a existential pleasure from solving technological problems (Florman, 1996). To solve these problems they tend to leave out social, financial and ethical peripheral (in the engineers eyes) phenomenon (Davis, 1998). As a result of this way of thinking, the implementation of technical solutions is sometimes problematic. Therefore, engineers in megaprojects are increasingly working in inter-professional teams (Amabile *et al.*, 2001). Knowledge regarding professional jurisdiction helps to understand cooperation in interdisciplinary work teams. Furthermore, processes of organizational change and resistance can be better understood when the role of professionals and the protection of jurisdictions are taken in account.

Given the development of the project culture, engineers in the Gideon's episode were not keen to work with project controllers. The controllers were perceived as traditionalists, who would block the innovative project management style developed in Environ Megaproject. The project control director decided to professionalize the staff departments such as juridical business, finance, control, planning and risk management. Registration of working hours, a central database for documents and an internal computer network helped to centralize and increase central control in the project. By strengthening these departments within the project organization the project controllers were better equipped to function as the 'conscience' of the megaproject. The cooperation between the project controllers and the project management has increased the quality of project management and the predictability of financial risks.

Among the critical success factors of collaboration between engineers and project controllers named by respondents was the equal appreciation of technology and procedures. The quality was no longer tested by engineers but determined by processes. Both engineers and project controllers hold on to their own professional culture, but worked on good relations between both professions. Professionals orientated at the technological content generally dominate Megaprojects. These professionals were less occupied with following procedures. Project controllers confronted engineers upon their behavior not following the procedures. They confront the engineers with questions and process appointments. Project controllers don't have many kindred spirits in the organization as they are focused at processes and human behavior. According to respondents, project controllers can be satisfied when the quality of processes are good, the audit certification has been given and the quarterly figures have been received well. As a result of mutual understanding of each needs, the cooperation between engineers and project controllers increased significantly.

Departmental Level: Project Director versus Sub-project Managers

The Environ Megaproject was organized as a decentralized project organization. The central management concept was called 'decentralism unless' which signified that department managers got (financial) objectives of central project management. These department managers are very independent and have a large freedom in finding ways to realize their objectives. The second important management concept was the concept of integral management. In this concept the sub-project managers are responsible for the contract, finance, administration, housing, Human Resource Management, environment and information. The sub-project manager is controlled by the project director. They have to work within the central scope of the project. In short, the 'decentralism unless' concept was based upon:

- Independent management of parts of the project by sub-project managers
- Geographical splitting of the project in six smaller projects
- Settlement of departments near the construction workplace
- Departments in direct contact with local environment

- Central frame for controlling the project objectives

This management philosophy was based upon earlier experiences with large projects and worked well in a decentralized technical firm. To successfully execute the 'decentralism unless' model rigorous enforcement of clear frame and a powerful board of directors who know what is going on at the work floor was needed. Furthermore, clear frames and bonus system stimulate managers to reach their sub-project objectives as well as organizational objectives had to be included. Finally, a strong management that give space to entrepreneurship of sub-project managers was needed for a successful execution.

According to the interviewees it was difficult for the project management of Environ Megaproject to embed these competences in the project culture. The sub-project managers were selected for their innovation, creativity, entrepreneurship, dealing with uncertainty, flexibility and their leadership. The independent managers were passionate with their sub-project. In the daily practice of the execution of the project, department managers developed independent 'kingdoms'. Although the agreement was that a financial setback in sub-project would have to be compensated in another part, slowly the department managers found their own ways of solving problems. They increasingly concentrated on their own sub-projects and less at the integral project. One member of the project staff explains:

> There is a healthy distrust to the staff. Department managers put everything for their own sub-project. Central frames are not well known. Central doesn't hear the stories from the field because of the distrust to the staff. (Interview with staff member Environ Megaproject)

In 2001, central frames were implemented in the sub-projects. According to many employees of Environ megaproject this was far too late. The department managers are already working for more than a year with construction firms. Many of the department managers interpreted the 'decentralism unless' model as 'decentralism' because they experiences little control from the central project management.

> The 'frames' of central weren't consistent and changed too much. This resulted in a relaxed attitude in the organization. (Interview with department manager)

The strict control of frames and integral project objectives is absence at the start of the project. Rather a more visionary, innovative, flexible, creative style of management is executed. A project employee tells:

> The management style didn't fit with the model, there was an avoidance of conflicts, and the management was afraid to use power (Interview with project employee)

Slowly, internal adjustments between the different sub-projects were becoming more problematic. The sub-projects have become more and more independent and autonomous. The sub-project interests prevail over the Environ Megaproject's interests, resulting in sub-optimalization of the project. Tensions arose between the sub-projects.

> The informal agreement was; you don't pass their boundaries, they don't pass mine. Stay out of each other's terrain. (Interview with department manager)

> The model increased the coordination problems. Nobody felt really responsible for managing the interfaces between the sub-projects. Everyone argued from their own sub-project. There was hardly any integral thinking. (Interview with staff member)

The 'decentralism unless' concept has resulted in a number of negative effects such as difficult project control, difficult risk management, and problems with interfaces between the sub-projects. Every sub-project manager adapted central management instruments to local circumstances. As a consequence, these instruments, such as HRM and communication, are difficult to be exchanged between the different sub-projects.

The new project director is confronted with the difficult task to redefine central framework and to control a strict observance of this framework. The sub-project managers however, were used to work independent and autonomous for a long time. There was no need for strict control and central management. This resulted in the earlier discussed intervention and replacement of the sub-project? managers. Finally, a director control was created to increase the control within the Environ Megaproject. The management slowly changes through interventions of the director control

> I get things done, I'm not afraid to confront people if this is for the benefit of the organization. If something is really bad I will tell them. (Interview project director control)

In conclusion, the cooperation between sub-projects came under pressure due to the failure of central project management to strictly implement the management model 'decentralism unless'. Sub-project managers had become independent entrepreneurs not taking care of other sub-projects. Interventions by the central management included the firing of a number of sub-project managers, introduction of central information systems, and the implementation of project controllers. Finally, a new director control was installed to increase the internal control.

Conclusion

This chapter focused at new understandings of cross cultural management in megaprojects. Contemporary studies on megaprojects are primarily focused at contracting, policy making and economic failures in terms of time, scope and budget (f.e. Flyvbjerg *et al.*, 2002; f.e. Flyvbjerg *et al.*, 2003). There is little empirical information about the daily cross-cultural management practices within megaprojects that give an better understanding of how and why many megaprojects fail to deliver in time, scope and budget. Project management literature that focuses at cross-cultural cooperation takes a rather static and integrative perspective of culture (Kendra and Taplin, 2004; Mäkilouka, 2004; Staples and Zhao, 2006). These studies are based upon multi value models which indicated the cultural 'distance' between nations (Morden, 1999). This integrative perspective of culture has met a larger part of academic critics (f.e. Low, 2002; Söderberg and Holden, 2002; Sackmann and Phillips, 2004; Jacob, 2005; Jackson and Aycan, 2006; Sackmann and Friesl, 2007). The management of multiple cultures appeared to be an interesting new direction (Jacob, 2005). This perspective includes differences in industrial culture, regional culture, organization culture, professional culture, and departmental culture in cross cultural studies (Söderberg and Holden, 2002; Jacob, 2005).

To apply the management of multiple cultures model in an empirical setting the case of the Environ Megaproject has been studied. Applying multi value models to this complex Public Private Patrnership would have learned us about different (national) cultural background of involved partners and their cultural 'distances', but would have given little or no information about the power struggles, cultural heterogeneity, cultural dynamics, situational behavior, professional conflicts and daily work practices within the Environ Megaproject. Findings based upon the multiple cultures model and presented in the case of Environ Megaproject gave a much better understanding of cross-cultural cooperation in megaprojects.

Cross cultural cooperation in the Environ Megaproject has to be understood in the context of power and history. The lack of hybrization of practices between public partners Flow and Straight at organizational level is directly related to their struggle for power over the project. The two organizations have a long history of negotiations and difficult cooperation. No new cultural practices could be negotiated which resulted in the break down of the cooperation. This has seriously hindered the project's mutual learning as valuable knowledge of Straight employees was absent in the Environ Megaproject.

Cultural differences could be observed at sector level in the social interaction of public and private partners of the Environ megaproject. At national level the Anglo-Saxon and Rhineland models dominated cultural practices. While the Dutch project management wanted to come to a consensus over the safety system conflict the American led consortium were after a confrontation. Conflicts over Anglo-Saxon and Rhineland management models resulted in unexpected time delay and budget expansion. At project level new management practices were introduced by the Diplomats resulting in conflicting cooperation between adapts of the Gideon gang and the Diplomats. New practices were developed in managing cultural transitions in the new project phase At departmental level the cooperation between central management and sub-project management was central. Finally, professional cooperation between project controllers and engineers resulted in new practices of challenging.

The critical management debate on project management needs more attention of both practitioners and academics (Hodgson and Cicmil, 2006). The interpretive perspective on daily activities in megaprojects better helps to understand the cultural dynamics that result in budget overrun, time delays and scope changes.

Acknowledgment

I want to thank Karen Smits for comments and suggestions on this chapter

References

Abbott, A. D. (1988), *The system of professions: an essay on the division of expert labour,* University of Chicago Press, Chicago.

Adler, N. (1986), *International dimensions of organisational behaviour,* Kent publishers, Boston.

Adler, N. and Ghadar, F. (1993), "A strategic phase approach to international human resource management", Wong-Rieger, D. and Rieger, F. (Eds.), *International Management Research: Looking to the Future,* Walter De Gruyter, Berlin, pp. 136-161.

Ailon-Souday, G. and Kunda, G. (2003), "The Local Selves of Global Workers: The Social Construction of National Identity in the Face of Organizational Globalization", *Organization Studies,* Vol. 24, No. 7, pp. 1073-1096.

Alvesson, M. (1993), *Cultural Perspectives on Organizations,* Cambridge University Press, Cambridge.

Alvesson, M. and Berg, P. (1992), *Organisational culture and Organisational Symbolism,* Walter de Gruyter, Berlin.

Amabile, M., Patterson, C., Mueller, J. and Wojcik, T. (2001), "Academic-practitioner collaboration in management research: A case of cross professional collaboration", *Academy of Management Journal,* Vol. 44, No. 2, pp. 418-431.

Anisya, T. S. and Annamma, P. (1994), "India; Management in an ancient and modern civilization", *International Studies of Management and Organization,* Vol. 24, No. 1, pp. 91-105.

Barley, S. R. and Kunda, G. (2001), "Bringing Work Back In", *Organization Science,* Vol. 12, No. 1, pp. 76-95.

Barth, F. (1969) Ethnic groups and boundaries: the social organisation of cultural difference. Allen and Unwin.

Bate, P. (1994), *Strategies for Cultural Change,* Butterworth Heinemann, Oxford.

Bate, P. (1997), "Whatever Happened to Organizational Anthropology? A Review of the Field of Organizational Ethnography and Anthropological Studies", *Human Relations,* Vol. 50, No. 9, pp. 1147-1171.

Brannen, J. V. and Salk, J. E. (2000), "Partnering across borders: Negotiating organizational culture in a German-Japan joint venture", *Human Relations,* Vol. 53, No. 4, pp. 451–487.

Brouwer, J. J. and Moerman, P. (2005), *Angelsaksen versus Rijnlanders. Zoektocht naar overeenkomsten en verschillen in Europees en Amerikaans denken,* Garant, Amsterdam

Cauley De La Sierra, M. (1995), *Managing Global Alliances. Key Steps for Successful Collaboration,* Addison-Wesley Publishing Company, Wokingham.

Chanlat, J. F. (1994), "Towards an anthropology of organisations", Hassard, J. and Parker, M. (Eds.), *Towards a New Theory of Organisations,* Routledge, London, pp. 155-190.

Chatterjee, S. R. and Pearson, C. A. L. (2001), "Perceived societal values of Indian managers; some empirical evidence of responses to economic reform", *International Journal of Social Economics,* Vol. 28, No. 4, pp. 368-379.

Chevrier, S. (2003), "Cross-cultural management in multinational project groups", *Journal of World Business,* Vol. 38, No. pp. 141-149.

Child, J. and Faulkner, D. (1988), *Strategies of cooperation, managing aliances: networks and joint ventures,* Oxford Press, Oxford.

Clausen, L. (2007), "Corporate Communication Challenges: A 'Negotiated' Culture Perspective", *International Journal of Cross Cultural Management,* Vol. 7, No. 3, pp. 317-332.

Clegg, S. (1981), "Organization and Control", *Administrative Science Quarterly,* Vol. 26, No. 4, pp. 545-562.

Clegg, S. (1993), "Narrative, Power and Social Theory", Mumby, D. K. (Ed.), *Narrative and Social Control: Critical Perspectives,* Sage, Newbury Park, pp. 16-45.

Clegg, S. R., Pitsis, T. S., Rura-Polley, T. and Marosszeky, M. (2002), "Governmentality Matters: Designing an Alliance Culture of Inter-organizational Collaboration for Managing Projects", *Organization Studies,* Vol. 23, No. 3, pp. 317-338.

Czarniawska-Joerges, B. (1992), *Exploring Complex Organizations. A Cultural Perspective,* Sage Publications, Inc., London.

Dafoulas, G. and Macaulay, L. (2001), "Investigating Cultural Differences in Virtual Software Teams", *The Electronic Journal on Information Systems in Developing Countries,* Vol. 7, No. 4, pp. 1-14.

Davis, M. (1998), *Thinking like an Engineer. Studies in the Ethics of a Profession,* Oxford University Press, Oxford.

Eller, J. (1999), *From culture to ethnicity to conflict: an anthropological perspective on international ethnic conflict,* the University of Michigan Press., Michigan.

Faulkner, D. (1995), *International Strategic Alliances. Co-operating to Compete,* McGraw-Hill Book Company, London.

Florman, S. C. (1996), *The Existential Pleasure of Engineering,* St. Martin's Griffin, New York..

Flyvbjerg, B., Bruzelius, N. and Rothengatter, W. (2003), *Megaprojects and Risk. An anatomy of Ambition,* University Press, Cambridge.

Flyvbjerg, B., Skamris Holm, M. and Buhl, S. (2002), "Underestimating Costs in Public Works Projects. Error or Lie?" *Journal of the American Planning Association,* Vol. 68, No. 3, pp. 279-295.

Fung, R. (1995) Organisational strategies for cross-cultural co-operation: management of personnel in international joint ventures in Hong Kong and China. Rotterdam, Erasmus University.

Fusilier, M. and Durlabhji, S. (2001), "Cultural values of Indian managers: An exploration through unstructured interviews", *International Journal of Value-based Management,* Vol. 14, No. 3, pp. 223-236.

Goffman, E. (1959), *The Presentation of Self in Everyday Life,* Doubleday, New York.

Gopinath, C. (1998), "Alternative approaches to indigenous management in India", *Management International Review,* Vol. 38, No. 3, pp. 257-275.

Hall, E. T. (1976), *Beyond Culture.,* Double day/Anchor Books (1981), New York.

Hannerz, U. (1992), *Cultural Complexity. Studies in the Social Organization of Meaning,* Columbia University Press, New York.

Hart, H. T., Boeije, H. and Hox, J. (1996), *Onderzoeksmethoden,* Boom, Amsterdam.

Hasting, C. (1995), "Building the culture of organisational networking", *International Journal of Project Managment,* Vol. 13, No. 259-263, pp.

Henrie, M. and Sousa-Poza, A. (2005), "Project Management: a Cultural Review", *Project Management Institute,* Vol. 36, No. 1, pp. 5-14.

Hodgson, D. E. and Cicmil, S. (Eds.) (2006), *Making Projects Critical,* Palgrave McMillan, New York.

Hofstede, G. (1980), *Culture's Consequences: International Differences in Work-Related Values.,* Sage Publications., London.

Hofstede, G. (1994), *Cultures and Organizations: Intercultural Cooperation and its Importance for Survival,* HarperCollins Publishers, London.

Holden, N. (2002), *Cross-cultural management. A knowledge management Perspective,* Printence Hall, Essex.

Jackson, T. and Aycan, Z. (2006), "Editorial: From cultural values to cross cultural interfaces", *International Journal of Cross Cultural Management,* Vol. 6, No. 1, pp. 5-13.

Jacob, N. (2005), "Cross-cultural investigations: emerging concepts", *Journal of Organisational Change Management,* Vol. 18, No. 5, pp. 514-528.

Jenkins, R. (1997), *Rethinking Ethnicity; Arguments and Explorations,* Sage Publications, London.

Jenkins, R. (2004), *Social Identity,* Routledge, London.

Kakar, S., Kakar, S., Ketsdevries, M. F. R. and Vrignaud, P. (2002), "Leadership in Indian Organizations from a Comparative Perspective", *International Journal of Cross Cultural Management,* Vol. 2, No. 2, pp. 239 - 250.

Kendra, K. and Taplin, T. (2004), "Project Success: A Cultural Framework", *Project Management Journal,* Vol. 35, No. 1, pp. 30-45.

Koot, W. (1997), "Strategic Utilization of Ethnicity in Contemporary Organizations", Sackmann, S. (Ed.), *Cultural Complexity in Organisations; Inherent Contrast and Contradictions,* Sage Publications, California, pp. 315 – 339.

Koot, W. and Sabelis, I. (2002), *Beyond Complexity. Paradoxes and Coping Strategies in Mangerial Life,* Rozenbergh Publishers, Amsterdam.

Kunda, G. (1992), *Engineering culture : control and commitment in a high-tech corporation,* Temple University Press, Philadelphia, Pa.

Lane, J. E. (1994), "Will public management dirve out public administration?" *Asian Journal of Public Administration,* Vol. 16, No. 2, pp. 139 - 151.

Lévi-Strauss, C. (2004), *Het trieste der tropen,* Uitgeverij Atlas, Amsterdam / Antwerpen.

Lewin, K. (1958), "Group decision and social change", Maccoby, E., Newcomb, T. and Harley, E. (Eds.), *Readings in social psychology,* 197-211, New York, pp.

Low, S. (2002), "The Cultural Shcadows of Cross Cultural Research: Images of Culture", *Culture and Organisation,* Vol. 8, No. 1, pp. 21-34.

Lowe, S., Moore, F. and Carr, A. N. (2007), "Paradigmapping Studies of Culture and Organization ", *International Journal of Cross Cultural Management,* Vol. 7, No. pp. 237-251.

Mäkilouka, M. (2004), "Coping with multicultural projects: the leadership styles of Finnish project managers", *International Journal of Project Management,* Vol. 22, No. pp. 387-396.

Martin, J. (2002), *Organizational culture : mapping the terrain,* Sage Publications, Thousand Oaks, Calif., etc.

Morden, T. (1999), "Models of National Culture – Management Review", *Cross Cultural Management,* Vol. 6, No. 1, pp. 19-44.

Nicholson, B. and Sahay, S. (2001), "Some political and cultural issues in the globalisation of software development: case experience from Britain and India", *Information and Organization,* Vol. 11, No. pp. 25-43.

Oudenhoven Van, J. P. and Zee, V. D., K.I. (2002), "Successful International Cooperation: The Influence of Cultural Similarity, Strategic Difference and International Experience", *Applied Psychologie: an International Review,* Vol. 51, No. 4, pp. 633-653.

Parker, M. (2000), *Organizational culture and identity: unity and division at work,* Sage, London.

Parker, S. and Skitmore, M. (2005), "Project management turnover: causes and effects on project performance", *International Journal of Project Management,* Vol. 23, No. pp. 205-214.

Parker. S and Skitmore, M. (2005), "Project management turnover: causes and effects on project performance", *International Journal of Project Management,* Vol. 23, No. pp. 205-214.

Peterson, M. F. (2007), "The Heritage of Cross Cultural Management Research: Implications for the Hofstede Chair in Cultural Diversity ", *International Journal of Cross Cultural Management,* Vol. 7, No. 3, pp. 359-377.

Pinto, J. A. M. (2005), "Swimming Against the Tide: The Hidden Costs of Offshoring", *The CPA Journal,* Vol. 75, No. pp. 9-11.

Roberts, S. (2000), "Development of a positive professional identity: Liberating oneself from the oppressor within", *Advances in Nursing Science,* Vol. 22, No. 4, pp. 71-82.

Royce, A. P. (1982), *Ethnic Identity; Strategies of Diversity,* Indiana University Press, Bloomington.

Sackmann, S. A. and Friesl, M. (2007), "Exploring cultural impacts on knowledge sharing behavior in project teams – results from a simulation study", *Journal of Knowledge Management,* Vol. 11, No. 6, pp. 142-156.

Sackmann, S. A. and Phillips, M. E. (2004), "Contextual Influences on Cultural Research: Shifting Assumptions for New Workplace Realities", *International Journal of Cross Cultural Management,* Vol. 4, No. 3, pp. 370-390.

Sahay, S., Nicholson, B. and Krishna, S. (2003), *Global IT Outsourcing: Software Development Across Borders,* University Press, Cambrigde.

Sapra, C. L. (1995), "Managing the Indian economy in the cross-cultural context", *Cross Cultural Management: An International Journal,* Vol. 2, No. 2, pp. 38 - 41

Schneider, S. and Barsoux, J.-L. (1997), *Managing across-cultures,* Prentice Hall, London.

Schwartzman, H. B. (1993), *Ethnography in Organizations,* Sage Publications, Inc., Newbury Park.

Shimoni, B. (2008), "Separation, emulation and competition: Hybridization styles of management cultures in Thailand, Mexico and Israel", *Journal of Organizational Change Management,* Vol. 21, No. 1, pp. 107 - 119.

Shimoni, B. and Bergmann, H. (2006), "Managing in a Changing World: From Multiculturalism to Hybridization - The Production of Hybrid Management Cultures in Israel, Thailand and Mexico", *Academy of Management Perspectives,* Vol. 20, No. 3, pp. 76-89.

Singh, J. P. (1990), "Managerial culture and work related values in India", *Organization Studies,* Vol. 11, No. 1, pp. 75-106.

Sinha, J. B. P. and Sinha, D. (1990), "Role of social values in Indian organizations", *International Journal of Psychology,* Vol. 25, No. pp. 705-714.

Söderberg, A. and Holden, N. (2002), "Rethinking Cross Cultural Management in a Globalizing Business World", *International Journal of Cross Cultural Management,* Vol. 2, No. 1, pp. 103-121.

Söderberg, A. and Vaara, E. (Eds.) (2003), *Merging across borders—people, cultures and politics,* Copenhagen Business School Press, Copenhagen

Spekman, R. E., Isabella, L. A., Macavoy, T. C. and Forbes Iii, T. B. (1996), "Creating strategic alliances which endure", *Long Range Planning,* Vol. 29, No. 3, pp. 346-357

Staples, D. S. and Zhao, L. (2006), "The Effects of Cultural Diversity in Virtual Teams Versus Face-to-Face Teams", *Group Decision and Negotiation,* Vol. 15, No. pp. 389-406.

Trice, M. T. and Beyer, J. M. (1993), *The Cultures of Work Organizations,* Englewood Cliffs, New Yersey. .

Trompenaars, F. (1993), *Riding the waves of Culture. Understanding cultural diversity in business,* The Economist Books Ltd., London.

Van Maanen, J. (1991), "The Smile Factory: Work at Disneyland", Frost, P. J., Moore, L.F., Louis, M.R., Lundberg, C.C. And J. Martin (Eds) (Ed.), *Reframing Organizational Culture,* SAGE Publications, Inc., Newbury Park, pp. 58 - 76.

Van Marrewijk, A. (2004), "The Management of Strategic Alliances: Cultural Resistance. Comparing the Cases of a Dutch Telecom Operator in the Netherlands Antilles and Indonesia", *Culture and Organization,* Vol. 10, No. 4, pp. 303-314.

Van Marrewijk, A. and Veenswijk, M. (2006), *The Culture of Project Management. Understanding Daily Life in Complex Megaprojects,* Prentence Hall Publishers / Financial Times, London.

Van Marrewijk, A. H., Clegg, S., Pitsis, T. and Veenswijk, M. (2008), "Managing Public-Private Megaprojects: Paradoxes, Complexity and Project Design", *International Journal of Project Management,* Vol. 26, No. 6, pp. 591-600.

Van Oudenhoven, J. P. and Van Der Zee, K. I. (2002), "Successful International Cooperation: The Influence of Cultural Similarity, Strategic Difference and International Experience", *Applied Psychologie: an International Review,* Vol. 51, No. 4, pp. 633-653.

Veenswijk, M. (2003), *Public entrepreneurs, private adventures : organizational change and identity in the new organization,* Rozenberg Publishers, Amsterdam.

Virmani, B. R. (2007), *The Challenges of Indian Management,* Response Books, New Delhi.

Weick, K. E. (1995), *Sensemaking in Organizations,* Sage, London.

Yanow, D. and Schwartz-Shea, P. (Eds.) (2006), *Interpretation and Method: Empirical Research Methods and the Interpretative Turn,* M E Sharpe, Armonk, New York.

Yin, R. (2003), *Case Study Research Design and Methods,* Sage Publications, London.

Zwikael, O., Shimuzu, K. and Globerson, S. (2005), "Cultural differences in project management capabilities: A field study", *International Journal of Project Management,* Vol. 23, No. pp. 454-462.

In: Progress in Management Engineering
Editors: L.P. Gragg and J.M. Cassell, pp. 43-68
ISBN: 978-1-60741-310-3

Chapter 2

Project Change Management System: An Information Technology Based System

Faisal Manzoor Arain
Construction Project Management,
School of Construction,
Southern Alberta Institute of Technology, Calgary, Canada

Abstract

In a perfect world, changes will be confined to the planning stages. However, late changes often occur during project processes, and frequently cause serious disruption to the project. The need to make changes in a project is a matter of practical reality. Even the most thoughtfully planned project may necessitate changes due to various factors. The fundamental idea of any change management system is to anticipate, recognize, evaluate, resolve, control, document, and learn from past changes in ways that support the overall viability of the project. Learning from past changes is imperative because the professionals can then improve and apply their experience in the future. Primarily, the chapter proposes six principles of project change management. Based on these principles, a theoretical model for project change management system (PCMS) is developed. The theoretical model consists of six fundamental stages linked to two main components, i.e., a knowledgebase and a controls selection shell for making more informed decisions for effective project change management. Further, the framework for developing an information technology based project change management system is also discussed. This chapter argues that the information technology can be effectively used for providing an excellent opportunity for the professionals to learn from similar past projects and to better control project changes. Finally, the chapter briefly presents an information technology based project change management system (PCMS) for the management of changes in building projects. The PCMS consists of two main components, i.e., a knowledgebase and a controls selection shell for selecting appropriate controls. The PCMS is able to assist project managers by providing accurate and timely information for decision making, and a user-friendly system for analyzing and selecting the controls for change orders for projects. The PCMS will enable the project team to take advantage of beneficial changes when the opportunity arises without an inordinate fear of the negative impacts. By having a systematic way to manage changes, the efficiency of project work and the likelihood of project success should increase. The chapter would assist professionals in developing an effective change management system. The system would be helpful for them to take proactive measures for reducing changes in projects. Furthermore, with further generic enhancement and modification, the PCMS will also be useful for the management of changes in other types of projects, thus helping to raise the overall level of productivity in the industry. Hence, the system developed and the findings from this study would also be valuable for all project management professionals in general.

Introduction

In a perfect world, changes will be confined to the planning stages. However, late changes often occur during construction, and frequently cause serious disruption to the project (Cameron, *et al.*, 2004). Great concern has been expressed in recent years regarding the adverse impact of changes to the building projects. The need to make changes in a project is a matter of practical reality. Even the most thoughtfully planned project may necessitate changes due to various factors (Ibbs, *et al.*, 2001). Developments in the social and technological aspects of life may foster the need for renovation or extension of existing buildings. The construction of building poses risks. Changes during the design and construction processes are to be expected. Arain and Low (2005a) identified the design phase as the most likely area on which to focus to reduce the changes in future building projects. If one were to seriously consider ways to reduce problems on site, an obvious place to begin with is to focus on what the project team can do to eliminate these problems at the design phase (Arain, 2005a; Arain and Low, 2005b).

Considering the hectic working environment of building projects, decisions are being made under pressure and cost and time invariably dominate the decision making process (O'Brien, 1998). Most forms of contract for building projects allow a process for changes (Arain and Low, 2005b). Even though there may be a process in place to deal with these late changes, cost and time invariably dominate the decision making process. If the changes affect the design, it will impact on the construction process and, quite possibly, operation and maintenance as well (Cameron, *et al.*, 2004). To overcome the problems associated with changes to a project, the project team must be able to effectively analyze the changes and its immediate and downstream effects (CII, 1994; Arain and Low, 2007a). To manage a change means being able to anticipate its effects and to control, or at least monitor the associated cost and schedule impact (Hester, *et al.*, 1991). An effective analysis of changes and change orders requires a comprehensive understanding of the root causes of changes and their potential downstream effects.

In project management, changes in projects can cause substantial adjustments to the contract duration time, total direct and indirect cost, or both (Ibbs, *et al.*, 1998; Gray and Hughes, 2001; Ibbs, *et al.*, 2001). Every building project involves a multi-player environment and represents a collaborative effort among specialists from various independent disciplines (Arain, *et al.*, 2004). Because changes are common in projects, it is critical for project managers to confront, embrace, adapt and use changes to impact positively the situations they face and to recognize changes as such (Ibbs, 1997). The changes can be minimized when the problem is studied collectively as early as possible, since the problems can be identified and beneficial changes can be made (CII, 1994; Arain and Low, 2007a). The changes can be deleterious in any project, if not considered collectively by all participants. From the outset, project controls should take advantage of lessons learned from past similar projects (Ibbs, *et al.*, 2001).

The integration of construction knowledge and experience at the early design phase provides the best opportunity to improve overall project performance in the construction industry (Arain, *et al.*, 2004). To realize this integration, it is not only essential to provide a structured and systematic way to aid the transfer and utilization of construction knowledge and experience during the early design decision making process, but also to organize these knowledge and experience in a manageable format so that they can be inputted effectively and efficiently into the process.

Decision making is a significant characteristic that occur in each phase of a project. In almost every stage, decision making is necessary. Often, these decisions will, or can affect the other tasks that will take place. To achieve an effective decision making process, project managers and the other personnel of one project need to have a general understanding of other related or similar past projects (CII, 1994a). This underscores the importance of having a good communication and documentation system for better and prompt decision making during various project phases. If professionals have a knowledge-base established on past similar projects, it would assist the professional team to plan effectively before starting a project, during the design phase as well as during the construction phase to minimize and control changes and their effects. The current technological progress does not allow the complete computerization of all the managerial functions or the creation of a tool capable of carrying out automatically all the required management decisions. To insure the success of this important management function, it is believed that human involvement in this process remains essential. Thus the Decision Support System (DSS) approach for this kind of application seems to be the most natural idea (Miresco and Pomerol, 1995).

Information technology has become strongly established as a supporting tool for many professional tasks in recent years (Arain and Low, 2005c). Computerized decision support systems can be used by project participants to help make more informed decisions regarding the management of changes in projects by providing access to useful, organized and timely information (Miresco and Pomerol, 1995; Mokhtar, *et al.,* 2000). As mentioned earlier, project strategies and philosophies should take advantage of lessons learned from past similar projects from the inception. It signifies the importance of an organized knowledge-base of similar past projects. The importance of a knowledge-base for better project control was recommended by many researchers (Miresco and Pomerol, 1995; Mokhtar, *et al.*, 2000; Gray and Hughes, 2001; Ibbs, *et al.*, 2001; Arain and Low, 2005c).

A knowledge-based decision support system is a system that can undertake intelligent tasks in a specific domain that is normally performed by highly skilled people (Miresco and Pomerol, 1995). Typically, the success of such a system relies on the ability to represent the knowledge for a particular subject. Computerized decision support systems can be used by project participants to help make more informed decisions regarding the management of change orders in projects by providing access to useful, organized and timely information. The chapter presents a theoretical model for Project Change Management System (PCMS) for better management of changes in building projects. The system would assist the professionals in learning from past projects for reducing potential changes in the building projects.

The PCMS system was developed based on knowledge acquired from building projects carried our in Singapore; it provides the best opportunity to address the contemporary issues relevant to the management of changes in building projects. The PCMS would assist professionals in taking proactive measures for reducing potential changes in building projects. The PCMS includes a knowledge base that presents a comprehensive scenario of the causes of changes, their relevant effects and potential controls that would be helpful in decision making at the early stage of the changes occurring. The PCMS would assist project management teams in responding to changes effectively in order to minimize their adverse impact to the project. Furthermore, the PCMS will enable the project team to take advantage of beneficial changes when the opportunity arises without an inordinate fear of the negative impacts.

Management of Changes in Building Projects

The issue of managing changes has received much attention in the literature. Despite many articles and much discussion in practice and academic literature, the issue of learning from the past projects for making timely and more informed decisions for effective management of changes was not much explored in the literature. Many researchers have proposed theoretical models for managing changes. Krone (1991) presented a change order process that promoted efficient administrative processing and addressed the daily demands of changes in the construction process. The contractual analysis technique (CAT) found that early notification and submission of proposals helped to maintain management control and avoided impact claims. The CAT laid the foundation for future contract change clauses in construction management. The proposed process was limited to administrative processing and addressing the daily demands of changes in the construction process. Stocks and Singh (1999) presented the functional analysis concept design (FACD) methodology to reduce the number of change orders in construction projects. They found that FACD was a viable method that could reduce construction costs overall. Harrington, *et al.* (2000) presented a theoretical model for the management of change (MOC) in the organizational context. The model presented a structured process consisting of seven phases, namely, clarify the project, announce the project, conduct the diagnosis, develop an implementation plan, execute the plan, monitor progress and problems, and evaluate the final results. They suggested that the MOC structure can be applied outside the organization to any project change management.

A theoretical model was proposed by Gray and Hughes (2001) for controlling and managing changes. The central idea of the proposed model was to recognize, evaluate, resolve and implement changes in a structured and effective way. CII (1994) and Ibbs, *et al.* (2001) proposed a project change management system (CMS) that was founded on five principles. The five principles included: promote a balance change culture, recognize change, evaluate change, implement change, and improve from lessons learned. The change management system was a two-level process model, with principles as the foundation, and management processes to implement those principles. The proposed system lacked the basic principle and process of implementing controls for future changes in the building projects.

The basic principles of change management that are presented in this chapter were adapted from the research works by CII (1994) and Ibbs, *et al.* (2001).

Principles of Change Management

The fundamental idea of any change management system is to anticipate, recognize, evaluate, resolve, control, document, and learn from past changes in ways that support the overall viability of the project. Learning from the changes is imperative, because the professionals can improve and apply their experience in the future. This would help the professionals in taking proactive measures for reducing potential changes.

This chapter presents six basic principles of change management proposed. As shown in Figure 1, the six basic principles include identify change for promoting a balanced change culture, recognize change, diagnosis of change, implement change, implement controlling strategies, and learning from past experiences. Each of these principles works hand-in-hand with the others.

The decision-makers seek guidance from past decisions, like learning from the past experiences. The Adaption-Innovation Theory (AIT), proposed by Kirton (1976), defined and measured two styles of decision making: adaption and innovation. Kirton (1984) further explained that adaptors characteristically produced a sufficiency of ideas, based closely on, but stretching, existing agreed definitions of the problem and likely solutions. Kirton (1984) argued that the decisions made by adaptors were precise, timely, reliable and sound.

The first principle of change management is to identify changes. As shown in Figure 1, in this principle, referring to past projects for early recognition of a problem is very important, because it will assist in identifying the issue at the early stage. Furthermore, this will also assist in encouraging beneficial changes and discouraging detrimental changes. Beneficial changes are those that actually help to reduce cost, schedule, or degree of difficulty in the project. Detrimental changes are those that reduce owner value or have a negative impact on a project.

The second principle of change management is to recognize changes. In this principle, communication, documentation and awareness about trending are very important, because these would assist in identifying changes prior to their actual occurrence. The third principle of change management is to diagnose the change. As shown in Figure 1, nature evaluation, trending, and impact evaluation are very important aspects. This is because these would assist in determining whether the management team should accept and implement the proposed change.

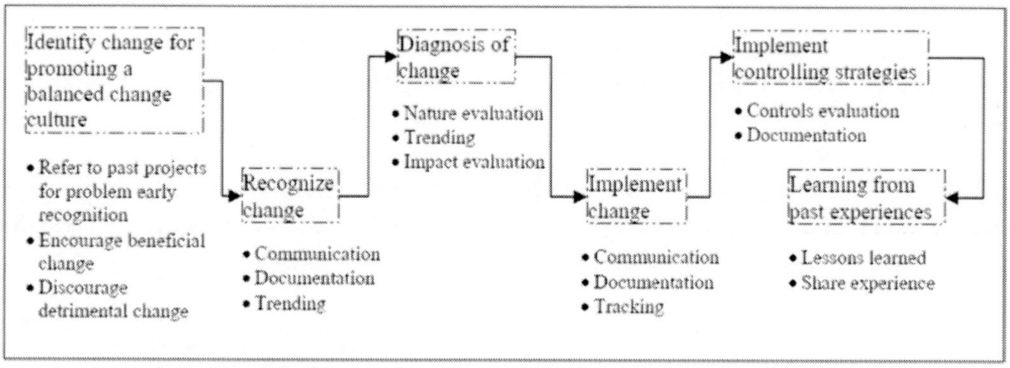

Source: adapted from Ibbs et al., 2001.

Figure 1. Fundamental principles of change management.

Implementing change is the fourth principle of change management. After evaluating the change, implementing change is an important step. As shown in Figure 1, in this principle, communication, documentation and tracking are very important. This is because these would assist in implementing change through communicating information between team members and developing database through documenting and tracking of the change implemented. Implementing controls for changes is the fifth principle of effective change management. It is a very important step, since this is the main reason to have the change management system. As shown in Figure 1, evaluating and documenting controls are very important, because evaluating suggested controls would assist in selecting effective controls for changes, and documenting the controls would assist in learning lessons from the change.

The sixth principle of change management is to learn from past experiences. In this principle, learning lessons and sharing experiences are very important because the main idea is to evaluate mistakes made so that errors can be systematically corrected. Such analysis should be shared between team members so that everyone will have a chance to understand the root causes of the changes and to control problems in a proactive way.

Model for Project Change Management System (PCMS)

Based on these principles, a theoretical model for project change management system (PCMS) is developed. The model consists of six fundamental stages linked to two main components, i.e., a knowledge-base and a controls selection shell for making more informed decisions for effective management of changes. The database will be developed through collecting data from source documents of past projects, questionnaire survey, literature review and in-depth interview sessions with the professionals who were involved in the building projects. The knowledge-base will be developed through initial sieving and organization of data from the database. The controls selection shell would provide support in decision making through a structured process consisting of building the hierarchy between the main criteria and the suggested controls, rating the controls, and analyzing the controls for selection through multiple analytical techniques.

The knowledge-base should be capable of displaying changes and their relevant details, a variety of filtered knowledge, and various analyses of the knowledge available. This would eventually lead the decision makers to the suggested controls for changes and assist in selecting the most appropriate controls.

As shown in Figure 2, the need for a change can originate from the client, user, design consultant, project manager and contractor. Considering the underlying principles of change management and the theoretical framework discussed earlier, the first step of the theoretical model for project change management is to identify changes for promoting a balanced change culture. Once the change is proposed, the proposal will be analyzed through a knowledge-base (level 1) for initial decision support to recognize the change at an early stage for encouraging beneficial changes and preventing detrimental changes. If options are required for certain changes, then the request for a proposal will be made. However, the proposals will be analyzed generally through a knowledge-base that will assist in establishing the first principle of change management.

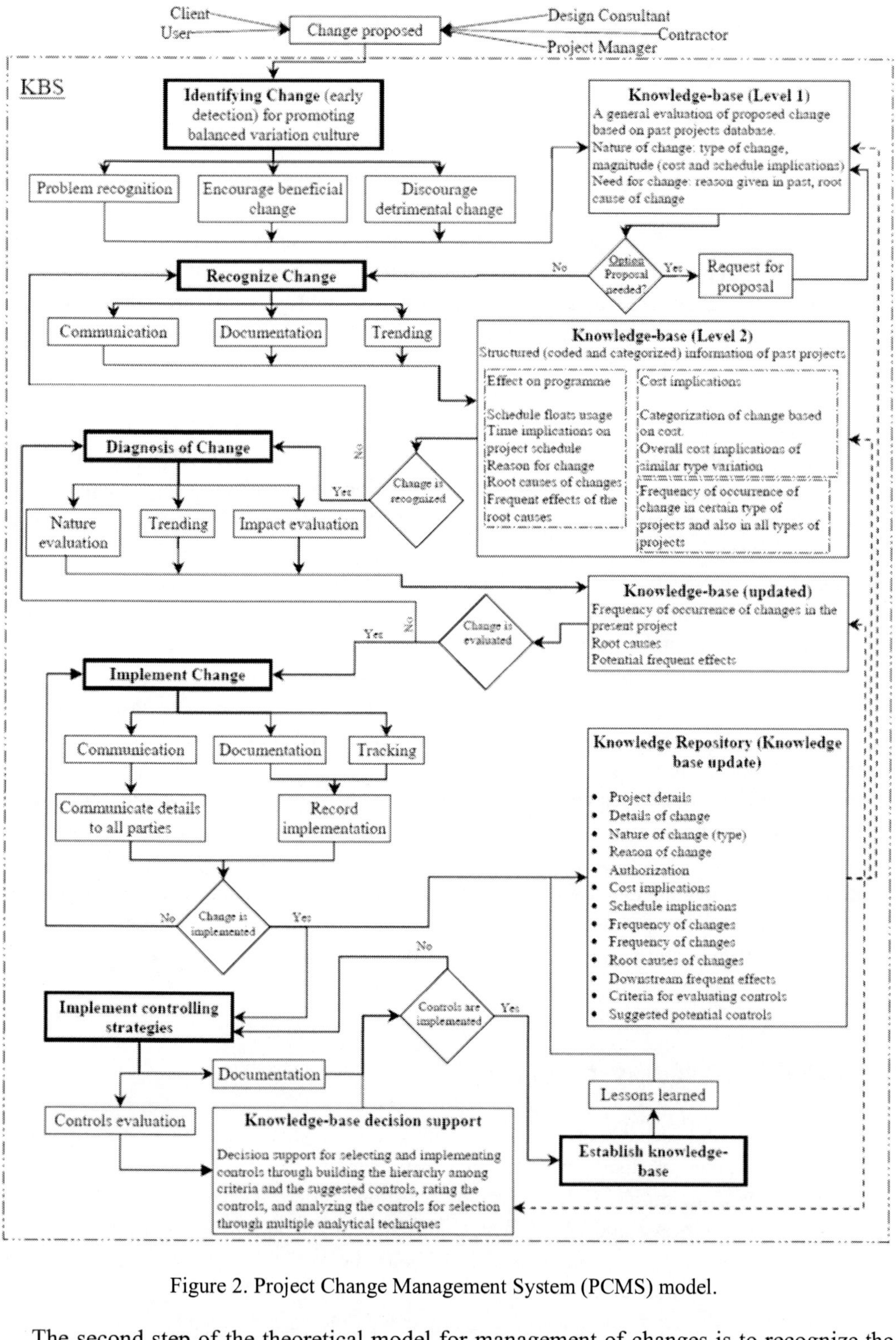

Figure 2. Project Change Management System (PCMS) model.

The second step of the theoretical model for management of changes is to recognize the change. Therefore, it is important that an environment be created that allows team members to openly communicate with one another. In this stage, team members are encouraged to discuss and to identify potential changes (Ibbs *et al.,* 2001; Arain and Low, 2006a). Identifying changes prior to their actual occurrence can help the team to manage changes better and earlier in the project life cycle. As shown in Figure2, the knowledge-base (level 2) provides structured information of past projects that would assist in effective communication between team members. The codes and categorized information relating to the effects on programme, cost implications, and frequency of occurrence of changes would eventually assist in recognizing changes at the early stage of their occurrence.

After the team recognizes the change, the diagnosis of change is carried out through the knowledge-base (updated). The knowledge-base (updated) contains information about the frequency of changes in the present project, their root causes, and potential effects. This information assists the management team in evaluating the change. The purpose of the evaluation is to determine whether the management team should accept and implement the proposed change.

After the evaluation phase, the team selects the alternatives and communicates the details of the change to all affected parties. Better team communication will allow for the timely implementation of the change selected. Documentation of the change implemented is an integral part of the implementation phase. The documentation contributes to the knowledge repository as shown in Figure 2.

After the implementation phase, selecting and implementing controls for changes are very important as shown in Figure 2. The knowledge-base eventually leads the decision makers to the suggested controls for changes and assists them in selecting the most appropriate controls. The controls selection shell would provide decision support through a structured process consisting of building the hierarchy between the main criteria and the suggested controls, rating the controls, and analyzing the controls for selection through multiple analytical techniques.

After selecting and implementing the controls for changes, establishing and updating the knowledge-base is the last yet most important phase of the theoretical model for management of change orders (Arain and Low 2006a). The knowledge-base will improve with every new building project, since the essence of the model is to provide timely and accurate information for the decision making process. The knowledge-base established may assist project managers by providing accurate and timely information for decision making, and a user-friendly system for analyzing and selecting the controls for changes.

Knowledge-Based System (KBS)

The fundamental idea of any strategic management system is to anticipate, recognize, evaluate, resolve, control, document, and learn from past experiences in ways that support the overall viability of the project (Ibbs, *et al.*, 2001; Arain, 2005b; Arain and Low, 2005c). The professionals can improve and apply their experience in the future projects hence learning from the changes is imperative. This would help the professionals in taking proactive measures for reducing potential changes.

A knowledge-based system was a system that could undertake intelligent tasks in a specific domain that was normally performed by highly skilled people (Miresco and Pomerol, 1995). Typically, the success of such a system relied on the ability to represent the knowledge for a particular subject (Mokhtar, *et al.*, 2000). Computerized decision support systems can be used by project participants to help make more informed decisions regarding the management of change orders in projects by providing access to useful, organized and timely information.

It is important to understand that the KBS for the management of project changes was not designed to make decisions for users, but rather it provided pertinent information in an efficient and easy-to-access format that allows users to make more informed decisions.

As mentioned earlier, the issue of managing changes has received much attention in the literature. In spite of many articles and much discussion in practice and academic literature, the issue of learning from the past projects for making timely and more informed decisions for effective management of changes was not much explored in the literature (Arain, 2005b; Arain and Low, 2006b). Many researchers have proposed principles and theoretical models for managing changes (Mokhtar, *et al.*, 2000; Ibbs, *et al.*, 2001; Arain and Low, 2005c). This chapter presents a project change management system (PCMS) containing a KBS for managing changes in building projects, which has not been studied and developed before. Hence, the study is a unique contribution to the body of knowledge about KBS towards the management of changes in construction. It is important to understand that the KBS for the management of changes is not designed to make decisions for users, but rather it provides pertinent information in an efficient and easy-to-access format that allows users to make more informed decisions.

The KBS consists of two main components, i.e., a knowledge-base and a controls selection shell for selecting appropriate controls (Arain and Low, 2007b). The database is developed by collecting data from the source documents of 80 building projects, questionnaire survey, literature review and in-depth interviews with the professionals who were involved in these projects. The knowledge-base was developed through initial sieving and organization of the data from the database. The knowledge-base was divided into three main segments, namely, macro layer, micro layer and effects/controls layer. The system contains one macro layer that consists of the major information gathered from source documents, and 80 micro layers that consist of detailed information pertinent to changes for each project. Overall the system contains 155 layers of information. The segment that contained information pertinent to possible effects and controls of the causes of changes for building projects was integrated with the controls selection shell. The shell contains 53 layers based on each of the causes of changes and their most effective controls. The controls selection shell provided decision support through a structured process consisting of building the hierarchy between the main criteria and the suggested controls, rating the controls, and analyzing the controls for selection through multiple analytical techniques.

The KBS is developed in the MS Excel environment using numerous macros for developing the user-interface that carry out stipulated functions. These are incorporated within a controls selection shell. The graphical user interface (GUI) assists users in interacting with the system on every level of the KBS. In addition, the GUI and inference engine will maintain the compatibility between layers and the decision shell. The KBS provides an extremely fast response to the queries. The KBS is capable of displaying changes and their relevant in-depth details, a variety of filtered knowledge, and various analyses of the knowledge available. The KBS is able to assist project managers by providing accurate and timely information for decision making, and a user-friendly system for analyzing and selecting the controls for changes in building projects.

The detailed information that is available on various layers of the KBS is briefly discussed below. The information and various filters that can be applied to the knowledge-base developed may assist the professionals in learning from past projects for enhancing management of changes in building projects.

Macro Layer of the KBS

As mentioned earlier, the macro layer is the first segment of the knowledge-base. It consists of the major information gathered from source documents of 80 building projects and through interview sessions with the professionals. As shown in Figures 3a, 3b and 3c, the macro layer contains the major information about the building projects completed, i.e., project name, program phase, work scope, type, date of commencement, project duration, date of completion, actual completion, schedule completion status, schedule difference, contract final sum, contingency sum percent, contingency sum, contingency sum used, total number of change orders, total cost of change orders, total time implication, total number of changes, frequency of change orders, frequency of changes, main contractors and consultants.

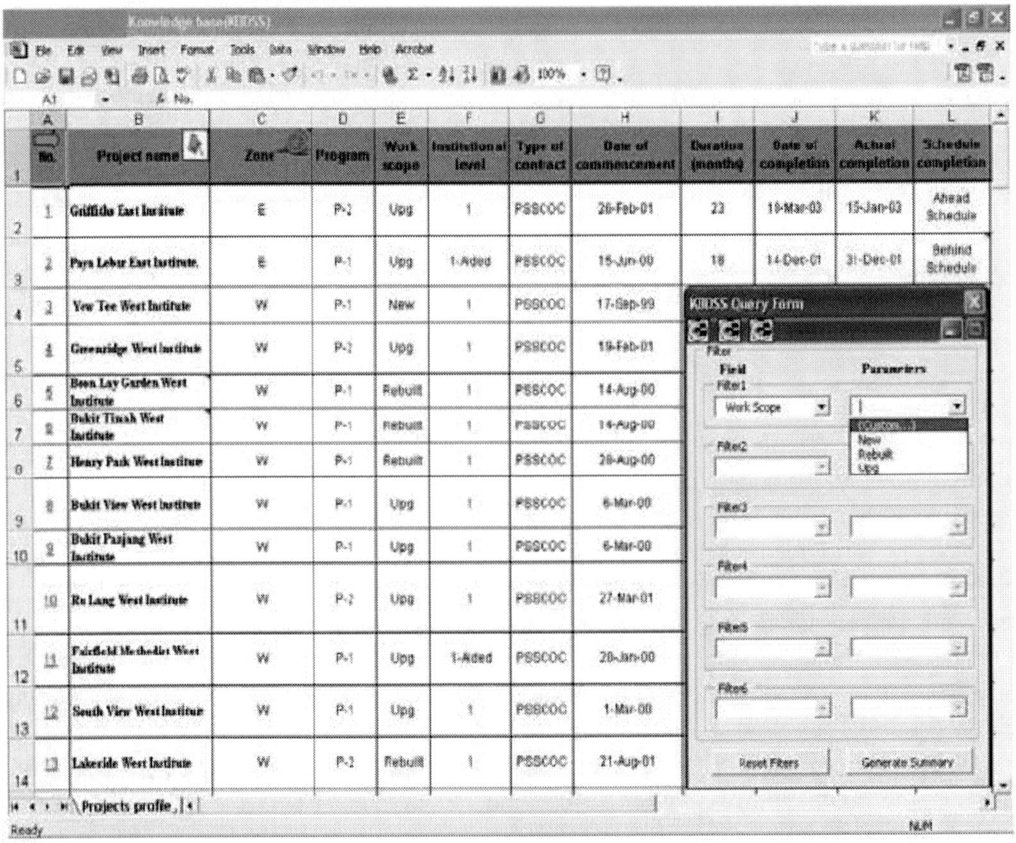

Figure 3a. Macro layer of the knowledge-base that consists of the major information regarding building projects.

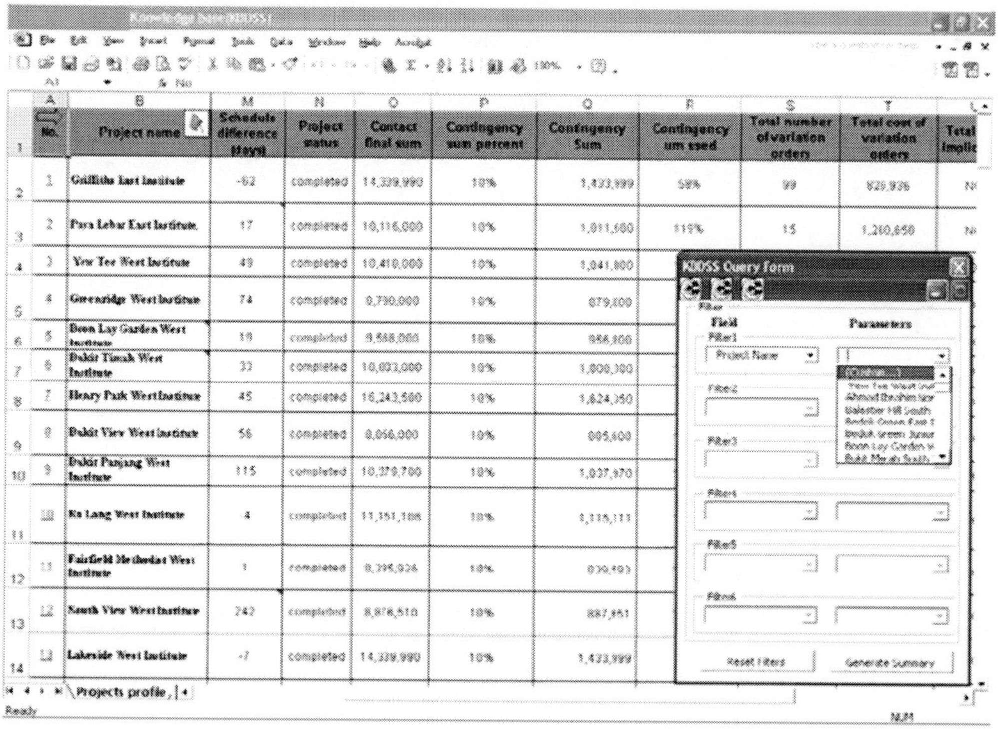

Figure 3b. Macro layer of the knowledge-base (cont'd).

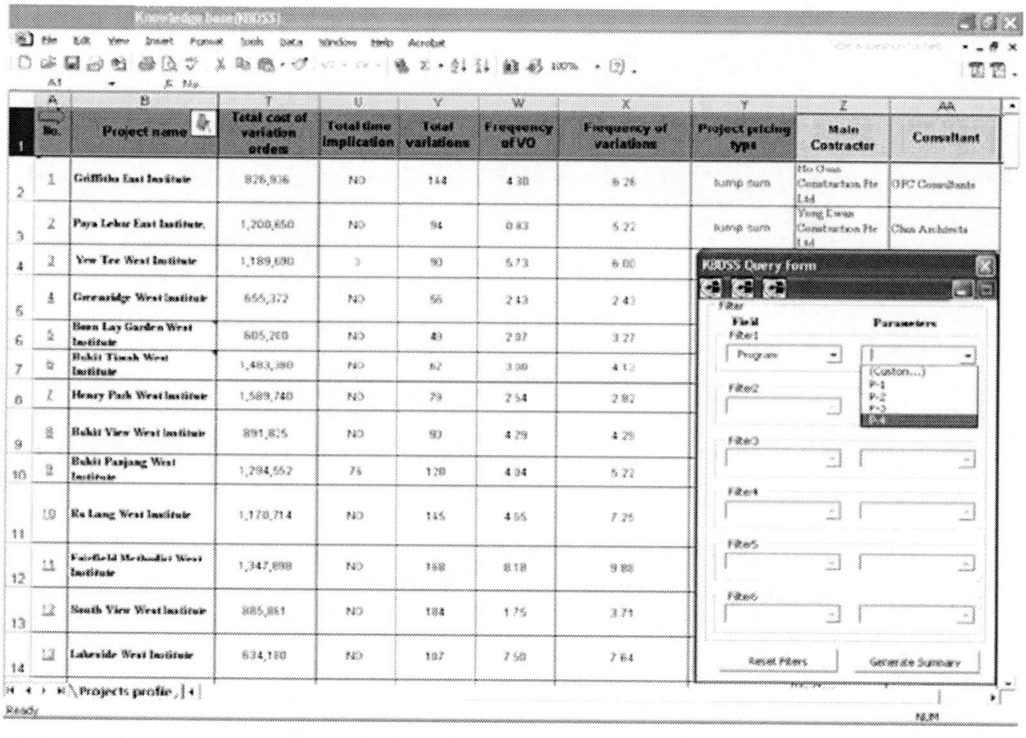

Figure 3c. Macro layer of the knowledge-base (cont'd).

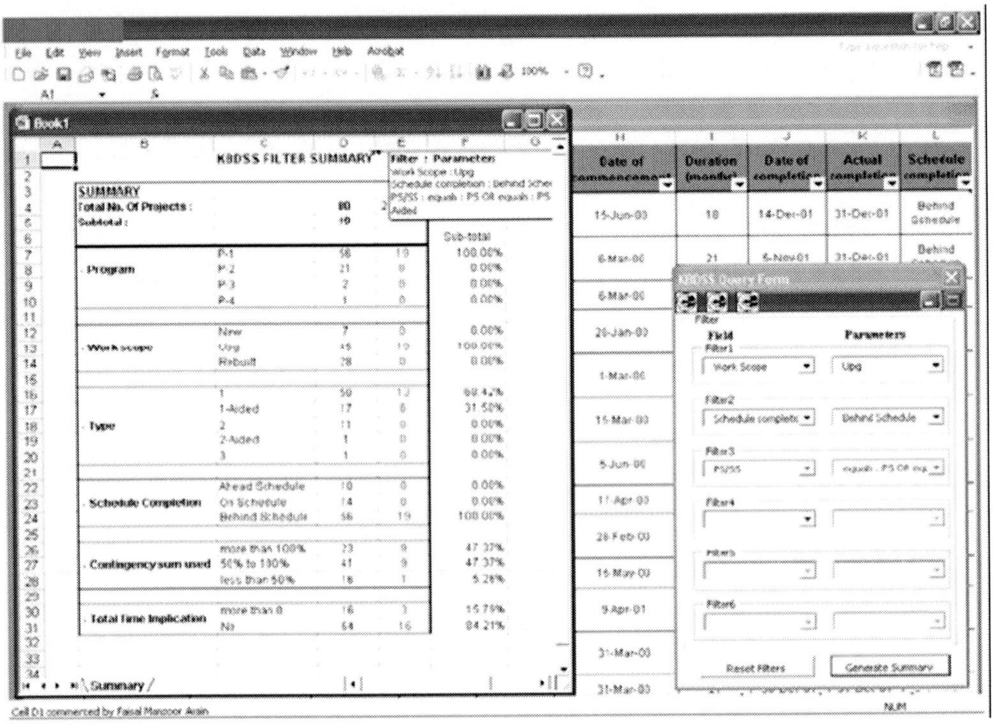

Figure 4. Summary section displaying the results of the filters applied on the macro layer.

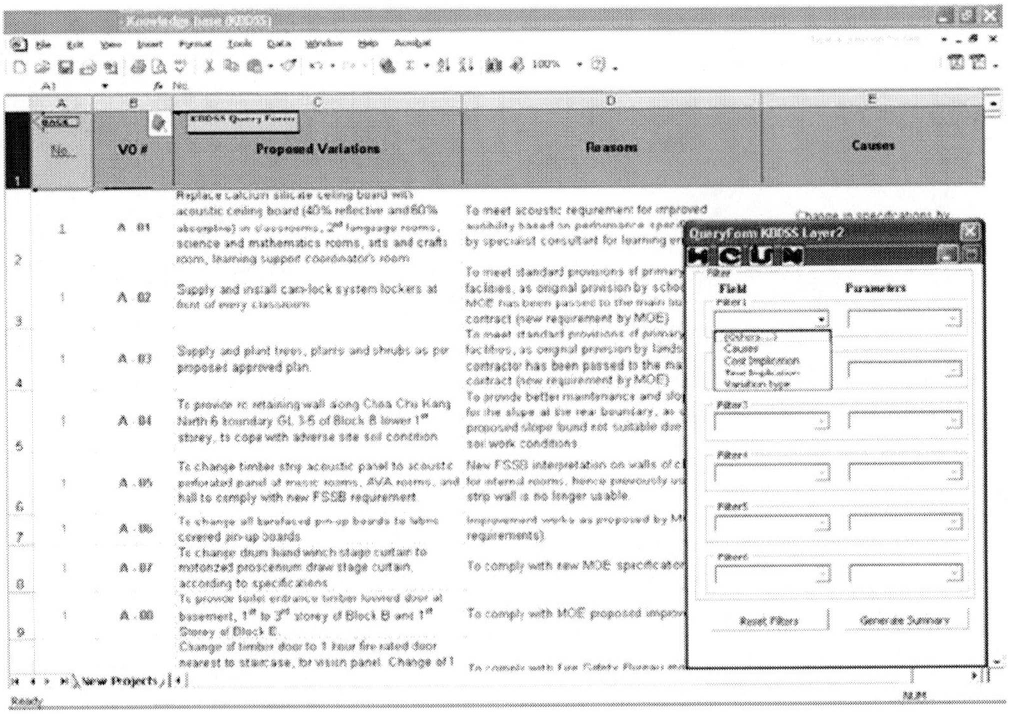

Figure 5a. Micro layer of the knowledge-base that contains the detailed information regarding change orders for the building project.

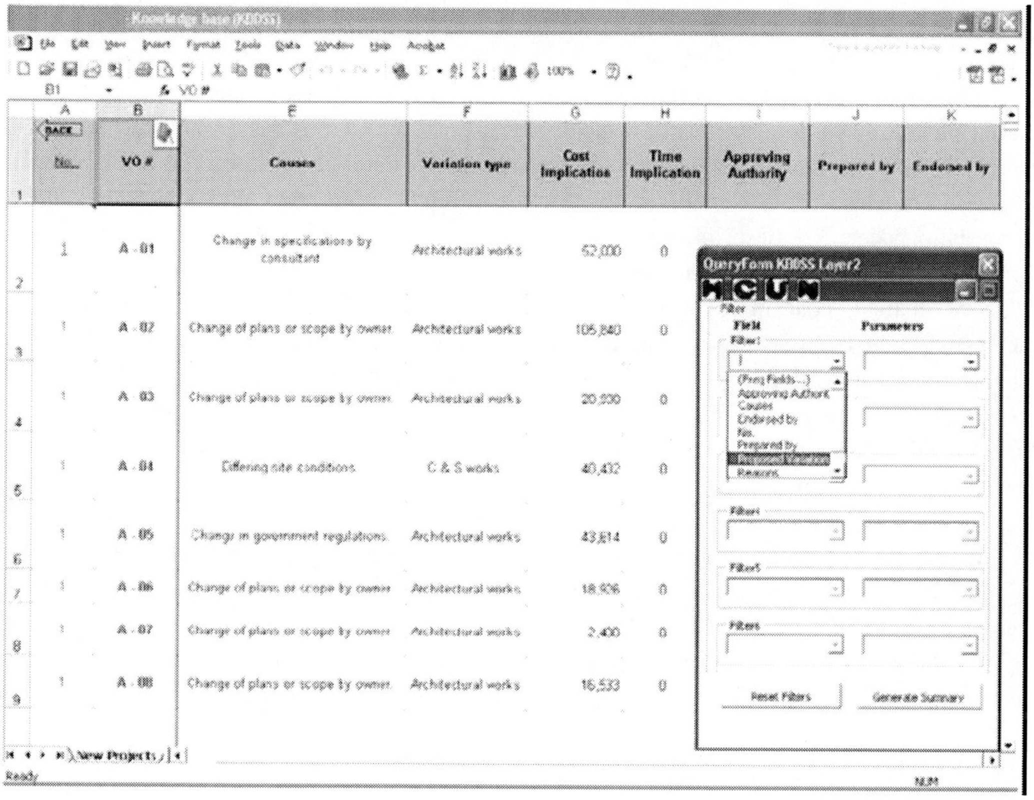

Figure 5b. Micro layer of the knowledge-base that contains the detailed information regarding change orders for the building project.

A variety of filters are provided on the macro layer that assists in sieving information by certain rules. The user would be able to apply multiple filters for analyzing the information by certain rules, for instance, the user would be able to view the information about the building projects that were completed behind schedule and among these projects, the projects with the highest frequency of change orders, highest contingency sum used, highest number of changes, etc. This analysis assists the user in identifying the nature and frequency of changes in certain type of building projects.

The inference engine provides a comprehensive summary of the information available on the macro layer as shown in Figure 4. Furthermore, the inference engine also computes the percentages for each category displayed in Figure 4. This assists the user in analyzing and identifying the nature and frequency of change orders in certain type of building projects. The information available on the macro layer would assist the professionals in identifying the potential tendency of encountering more changes in certain type of building projects. By applying multiple filters that are provided on the macro layer, the professionals would be able to evaluate the overall project variance performance. These analyses at the design stage would assist the professionals in developing better designs with due diligence.

Micro Layer of the KBS

The micro layer is the second segment of the knowledge-base that contains 80 sub-layers based on the 80 building projects respectively. As shown in Figures 5a and 5b, the micro layer contains the detailed information regarding changes and change orders for the building project. The detailed information includes the change order code that assists in sieving information, detailed description of particular change collected from source documents, reason for carrying out the particular change provided by the consultant, root cause of change, type of change, cost implication, time implication, approving authority, and endorsing authority. Here, the information regarding the description of particular change, reason, type of change, cost implication, time implication, approving authority, and endorsing authority were obtained from the source documents of the 80 building projects. The root causes were determined based on the description of changes, reasons given by the consultants, and the project source documents and were verified later through the in-depth interview sessions with the professionals who were involved in these projects.

In addition to computing the abovementioned information, the inference engine also computes and enumerates the number of changes according to various types of changes as shown in Figure 6. The inference engine also assists in computing the actual contingency sum by deducting the cost of changes requested and funded by the institution or other sources. This may assist in identifying the actual usage of contingency sum based on the project cost.

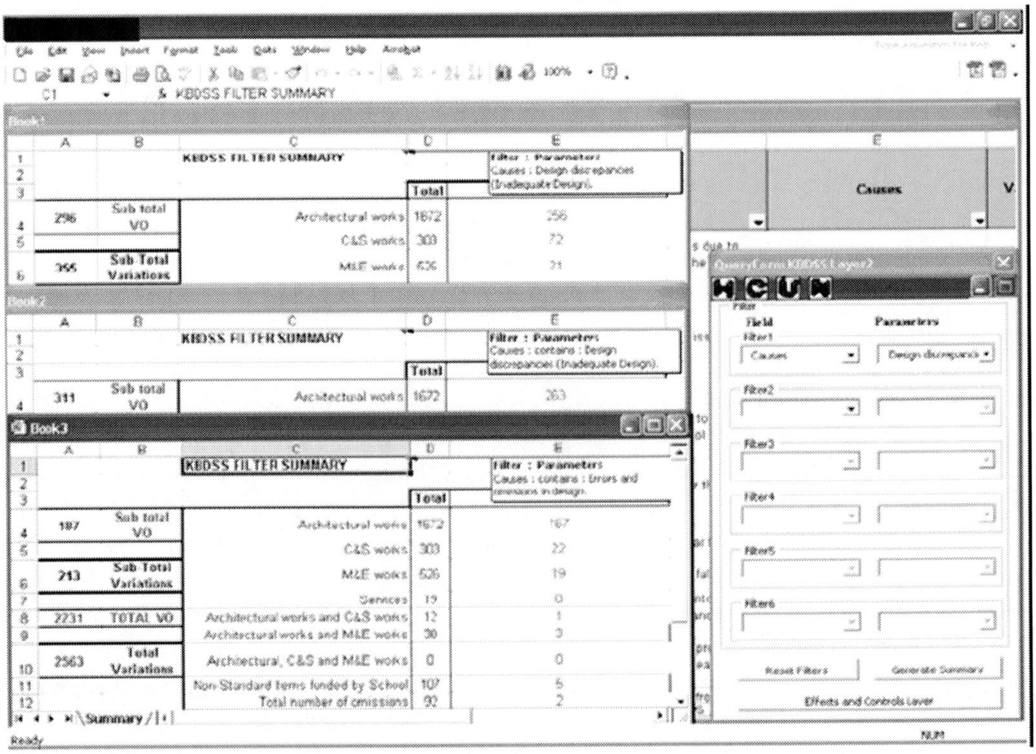

Figure 6. Multiple summary sections displaying the results of the filters applied on the micro layer, and the KBS query form showing the effects and controls layer tab that connects the micro layer with the effect and controls layer of the knowledge-base.

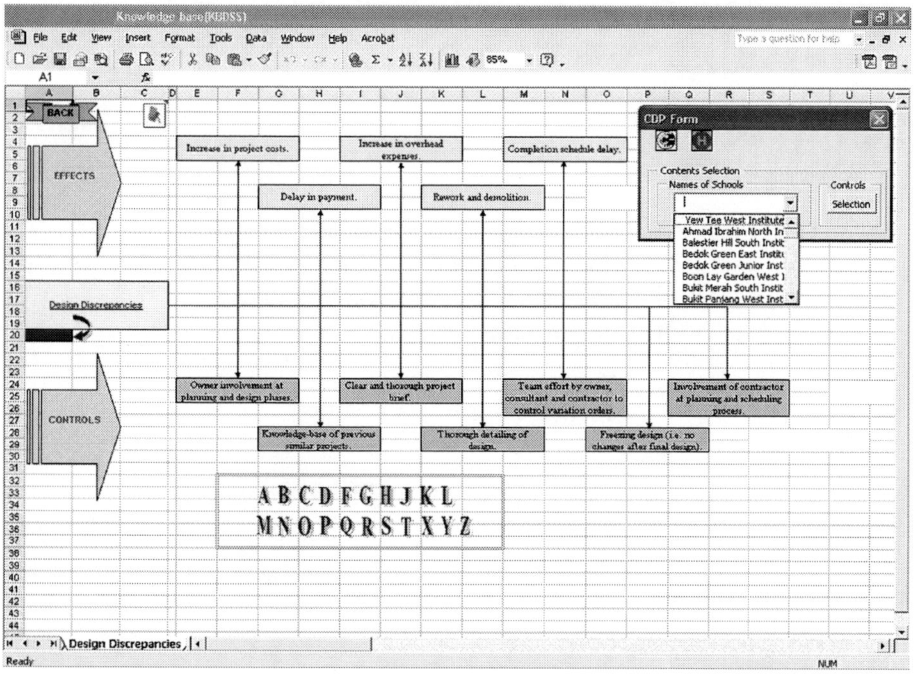

Figure 7. Effects and controls layer of the knowledge-base that pinpoints the most important effects and most effective controls for each cause of changes.

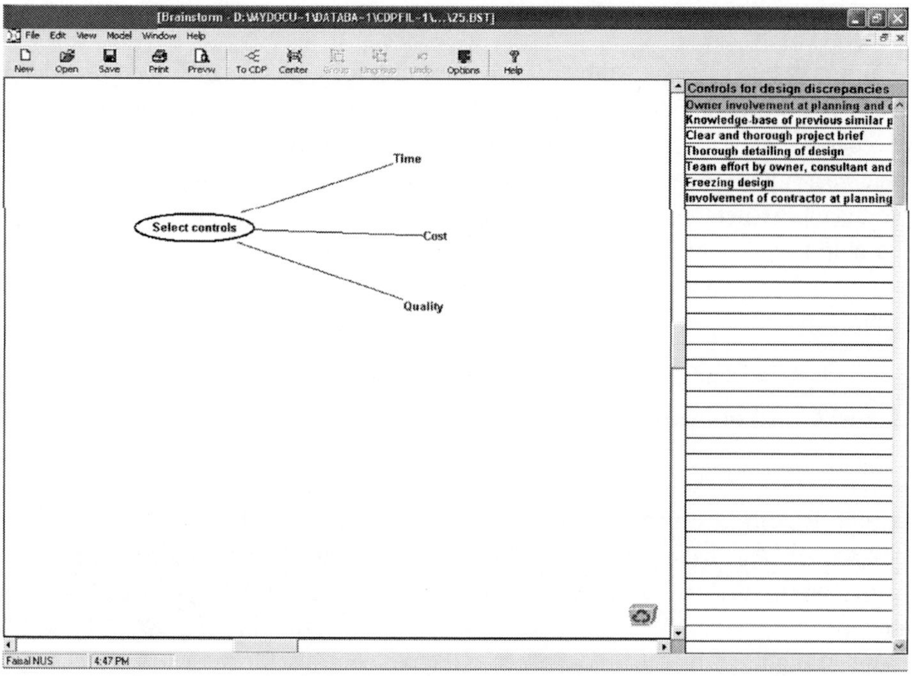

Figure 8. Main panel of controls selection shell that contains the goal, main criteria and the most effective controls for changes (focusing on Time, Cost and Quality).

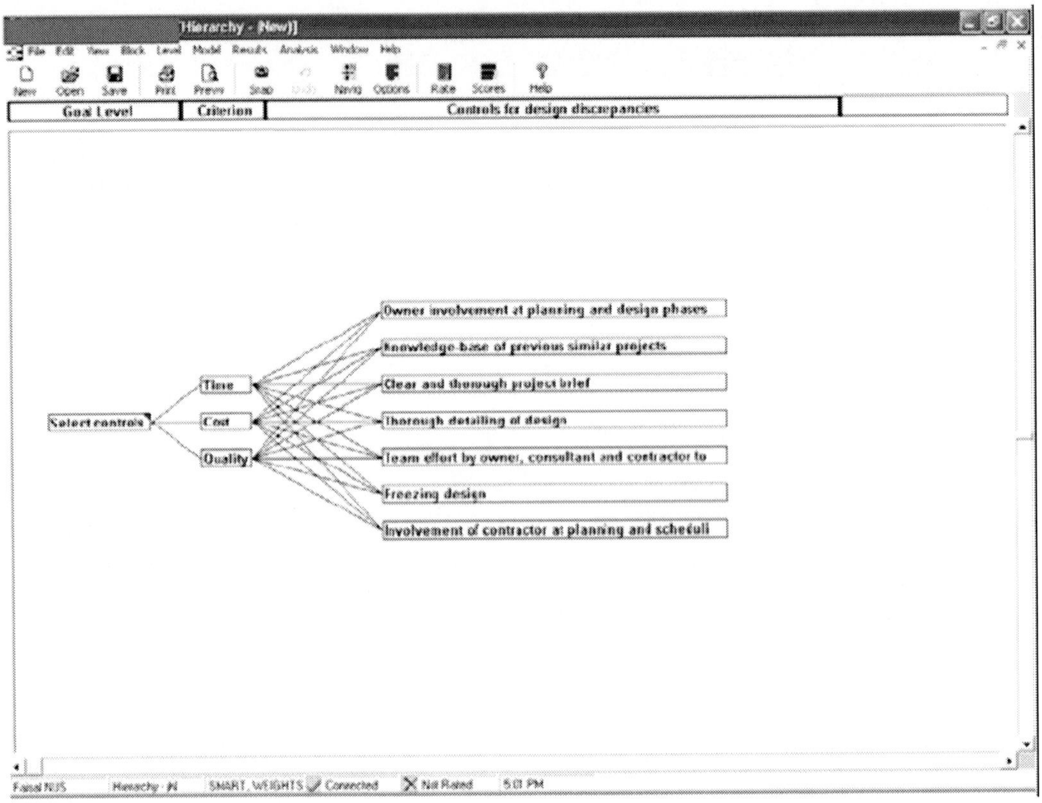

Figure 9. Building the hierarchy among the goal, main criteria and controls for changes.

The information can be sieved by certain rules through a variety of filters provided in the micro layer. The professionals would be able to apply multiple filters for finding out the most frequent causes of changes, most frequent types of changes, and changes with most significant cost implication and time implication. The multiple summaries that can be generated by apply filters and using the query form is presented in Figure 6. The professionals would be able to analyze the most potential changes in building projects. The information available on the micro layers would assist in pinpointing the root causes of changes in the past building projects.

Effects and Controls Layer of the KBS

The third layer of the KBS contains 53 sub-layers based on the potential causes of changes and 10 sub-layers of most important causes combined. The 53 causes can be modified in the event that new ones are discovered or emerged over time. The numerous filters provided in the macro, micro, and effects and controls layers will be updated automatically with every new project added. As shown in Figure 7, the graphical presentation of the 5 most important effects and 5 most effective controls for the cause of changes was presented. An understanding of the effects of changes would be helpful for the professionals in assessing changes. A clearer view of the impacts on the projects will enable the project team to take advantage of beneficial changes when the opportunity arises. Eventually, a clearer and comprehensive view of the potential effects of changes will result in informed decisions for effective strategic management of changes. It is suggested that changes can be reduced with due diligence during the design stages. Furthermore, the suggested controls would assist professionals in taking proactive measures for reducing changes in building projects. As mentioned earlier about the design stage, it is recommended that the controls be implemented as early as possible. As shown in Figure 7, the controls selection tab is provided in the CDP form. This feature assisted in linking the knowledge-base with the controls selection shell. This is required because the professionals may not be able to implement all the suggested controls. Therefore, the shell assists them in selecting the most appropriate controls based on their own criterions.

Controls Selection Shell

The controls selection shell is integrated with the knowledge-base to assist the user in selecting the appropriate controls of changes. As mentioned in the previous section, the 5 most effective controls for the cause of changes were presented on the effects and controls layer, and the layer was linked with the controls selection shell. The controls selection shell provides decision support through a structured process consisting of building the hierarchy among the main criterions and the suggested controls, rating the controls, and analyzing the controls for selection through multiple analytical techniques, for instance, the analytical hierarchy process, multi-attribute rating technique, and direct trade-offs. The controls selection shell contained four layers that were based on the structured process of decision making, namely, control selection criterions, building the hierarchy between criterions and controls, rating the controls, selecting the best controls based on the given criterions.

As shown in Figure 8, this layer of the controls selection shell contains the suggested controls for the cause of change selected in the controls and effects layer of the KBS. Hence, the controls selection shell contains 53 layers based on the each cause of changes and their most effective controls. Here the goal was to select the controlling strategies and the main criterions were time, cost and quality. In this layer, the professionals may add any suggested controls that are considered to be important. Furthermore, the professionals may specify their own contemporary criterions for selecting the controls. The provision of the facility for adding more controls and criterions would assist them in evaluating the suggested controls according to the project stages and needs. This may assist them in selecting and implementing the appropriate controls at appropriate time.

The main objective of this layer is to generate the hierarchy between the main criterions and the suggested controls for changes. The shell generates hierarchy among the goal, the criterions and the suggested controls as shown in Figure 9. The hierarchy assists in rating all the suggested controls.

The rating process includes four main activities i.e., choosing a rating method, selecting rating scale views, assigning rating scales and entering weights or scores. This layer will provide multiple techniques for rating. This layer essentially provides analytical hierarchy process (AHP) and simple multiple attribute rating technique (SMART) as rating techniques. This is because the decision will be based on purely qualitative assessments of the suggested controls. Skibniewski and Chao (1992) suggested that the AHP was an effective rating and evaluating technique for evaluation of advanced construction technologies. There are three rating methods available, i.e., direct comparison, full pair-wise comparison, and abbreviated pair-wise comparison. The direct method is the default rating method and is used for entering weights for this decision process. As shown in Figure 10, the first step for rating the controls was to assign weight to the criterions, i.e., time, cost and quality. The professionals should rate each criterion based on the project phases. This is because during the early stages of the construction projects, normally the implementation cost of suggested controls is not significant. More emphasis should be given on the available resources at the present stage of the construction projects.

The second step was to rate the suggested controls with respect to quality. This was because quality was rated critical as shown in Figure 11. The rating priority is based on hierarchy of the main criterions rated earlier in the first step. Here the professionals should assign more weight to the controls that may enhance the project quality. The third step was to rate the suggested controls with respect to time. Here the professional should rate the controls, which may require less time for implementation, as high. The user rated all the suggested controls and assigned weights to each alternative (control) as shown in Figure 12. Lastly, the fourth step was to rate the suggested controls with respect to cost. Here the professionals should select more weights for the controls that are not costly. The user rated all the suggested controls and assigned weights to each alternative (control) as shown in Figure 13. Overall, the rating of the suggested controls may vary according to the project phases. For instance, the controls may be implemented only in the design phase or in the construction phase of the building projects. Hence, the KBS would assist the professionals in selecting the appropriate controls for changes according to the present stage of the building project.

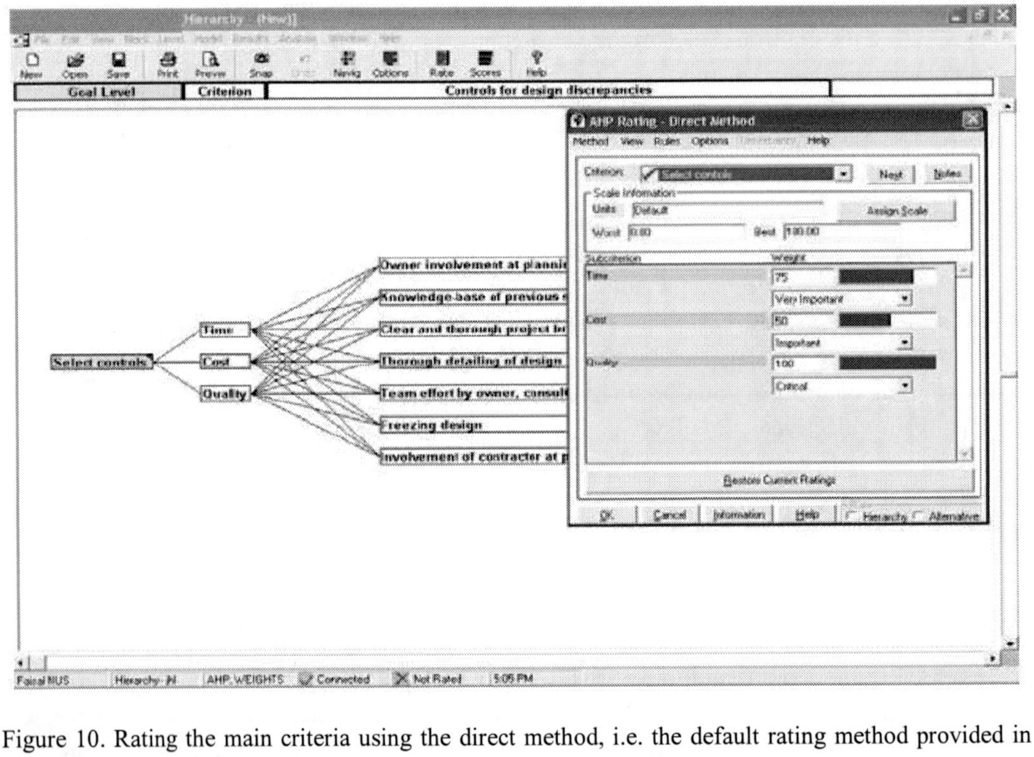

Figure 10. Rating the main criteria using the direct method, i.e. the default rating method provided in the KBS.

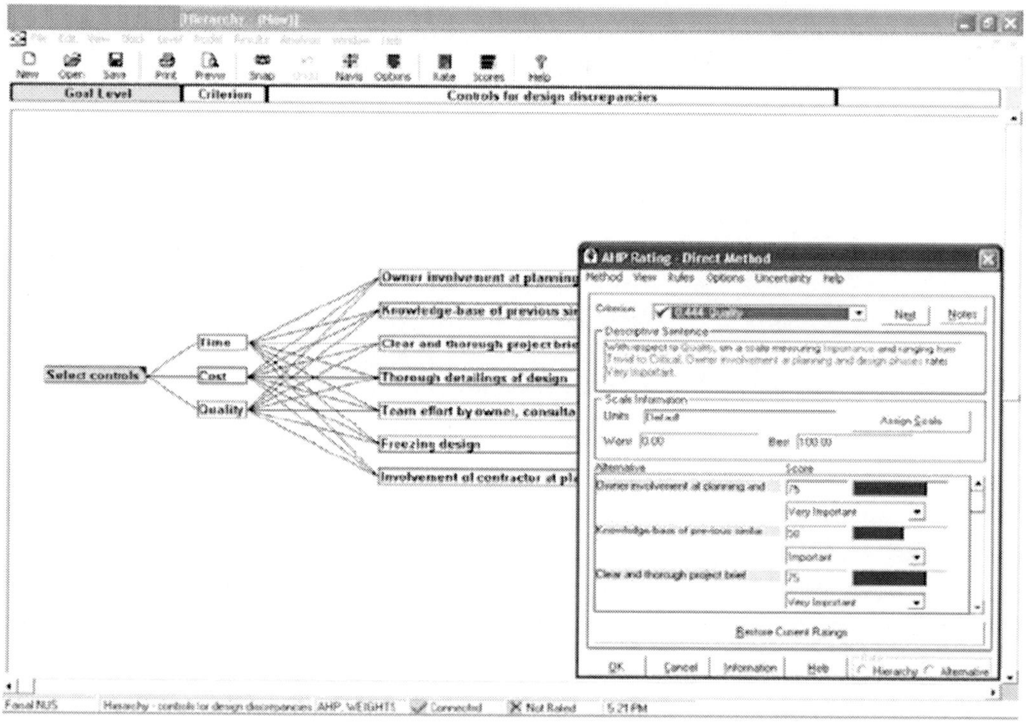

Figure 11. Rating the controls for changes with respect to quality.

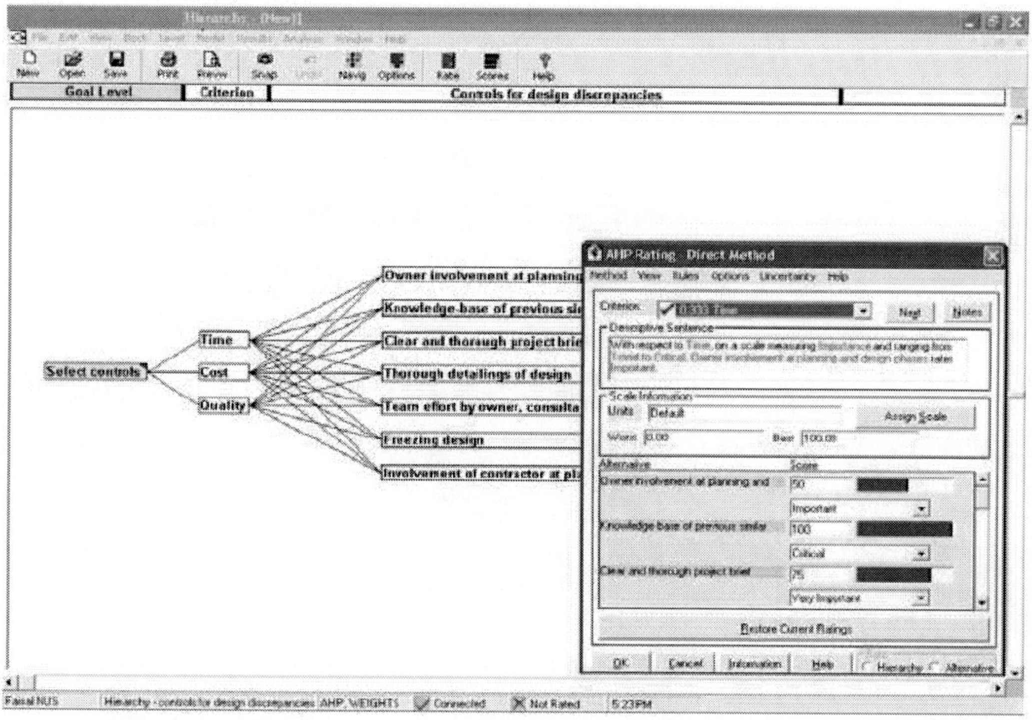

Figure 12. Rating the controls for changes with respect to time.

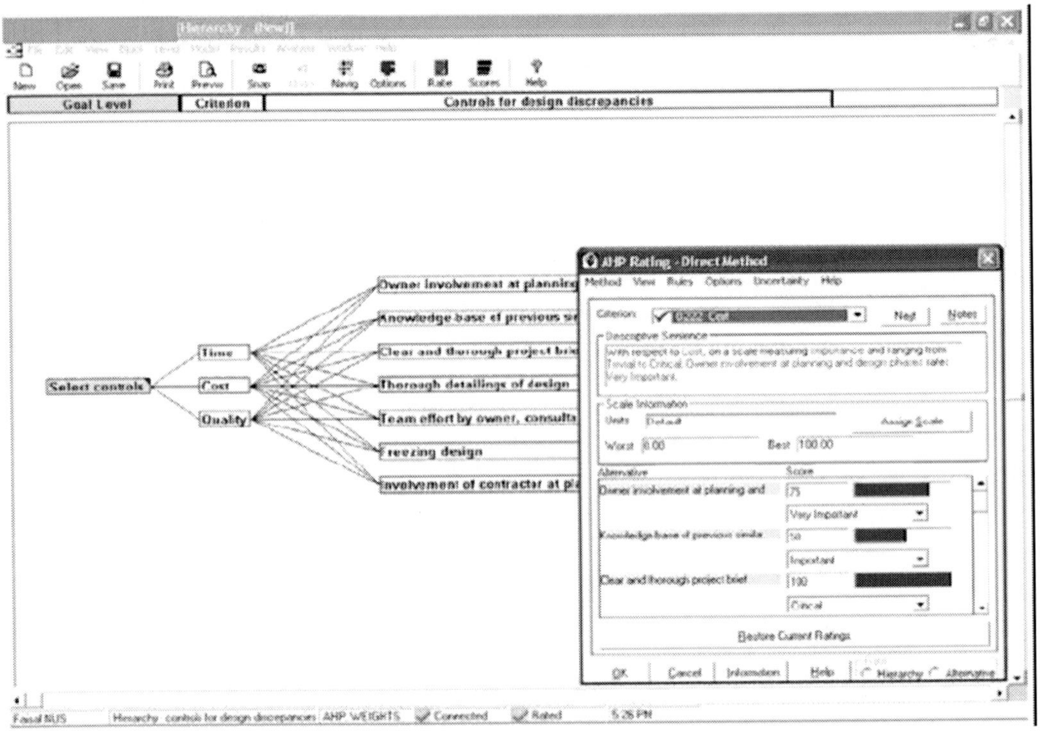

Figure 13. Rating the controls for changes with respect to cost.

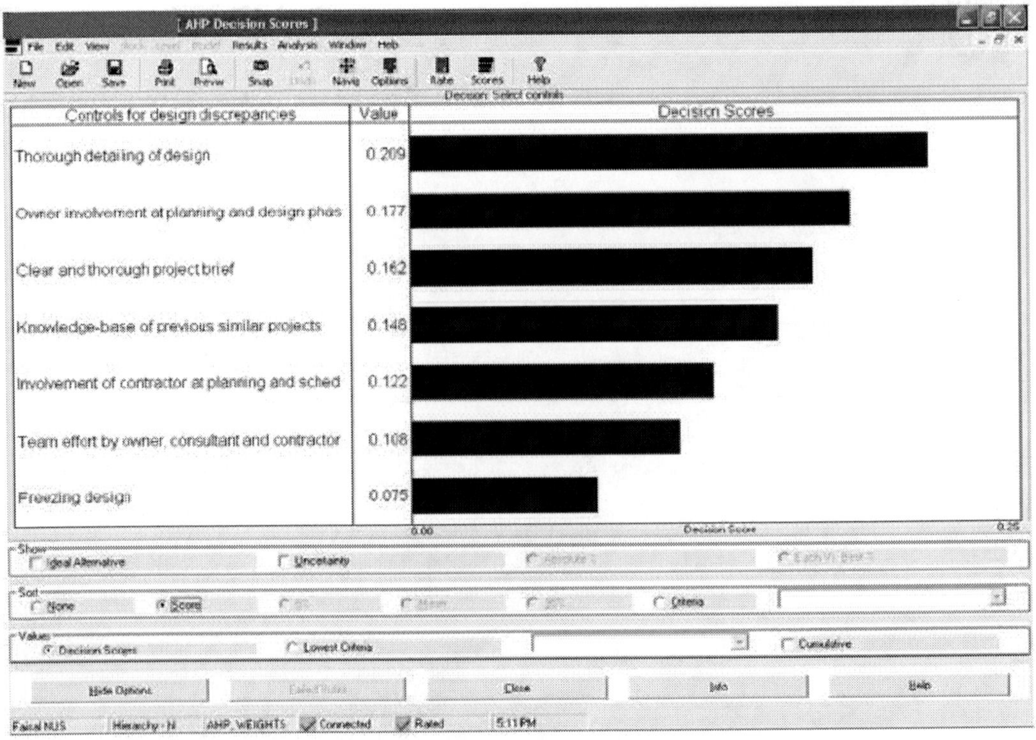

Figure 14. The controls for changes sorted according to the decision scores.

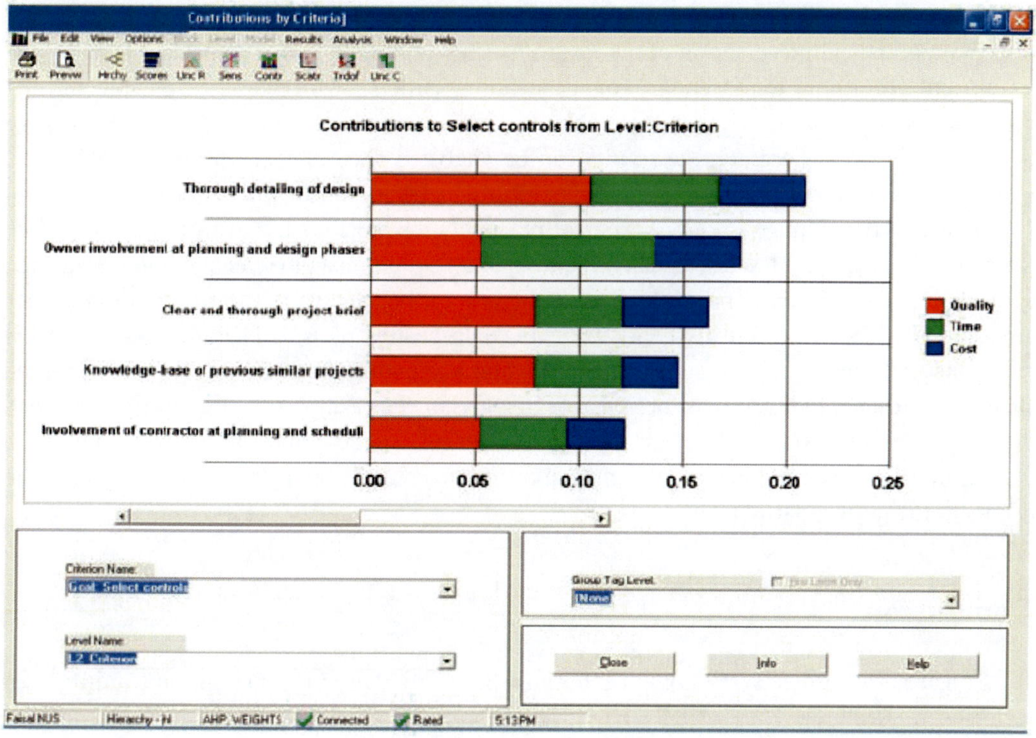

Figure 15. The suggested controls sorted according to contributions by criteria.

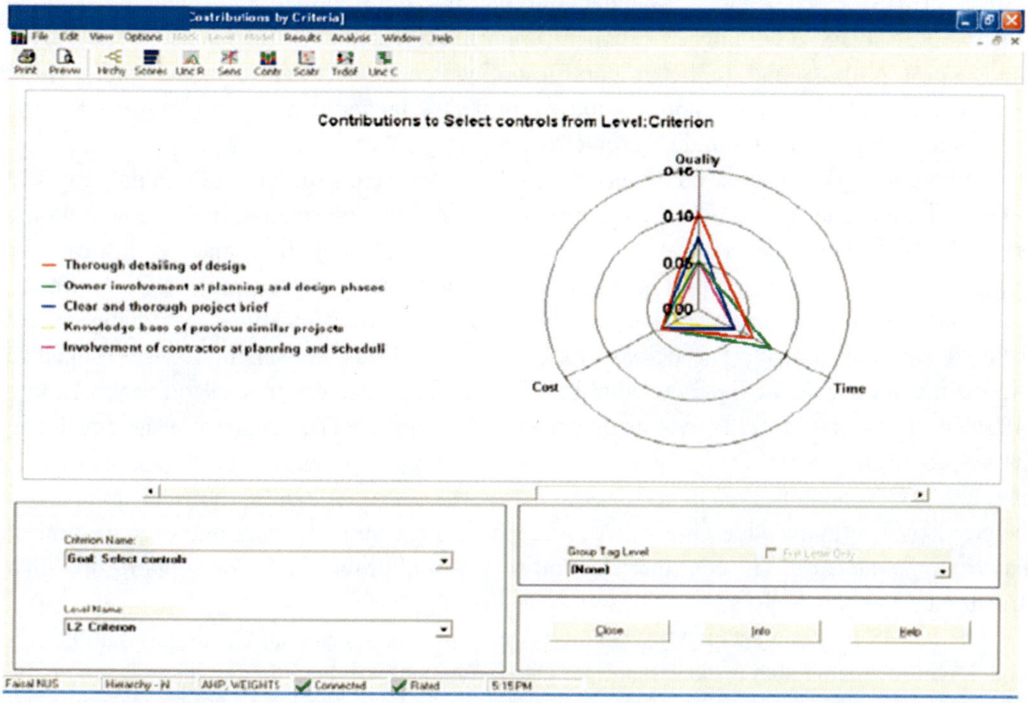

Figure 16. The results according to the contribution by criteria in radar form (web).

The controls selection shell calculates the decision scores based on the rating process and displays a graphical presentation of the results as shown in Figure 14. The decision scores can be sorted according to ascending or descending orders, which assist in viewing the comprehensive scenario. The professionals can easily select the best controls based on the decision scores. Furthermore, the results can be analyzed according to various contributions by criterions as shown in Figure 15. The graphical presentation of the results in radar form (web) is shown in Figure 16. The graphical presentations enhance the user-friendly interface that assist in analyzing the issues conveniently. The professionals may analyze the suggested controls by selecting any one of the criterions. For further analysis, various analysis modes are also provided, i.e., sensitivity by weights, data scatter plots, and trade-offs of lowest criterions. All these modes assist in analyzing and presenting the decision. Furthermore, the shell also presents various other options for displaying the results, i.e., decision score sheet, pie charts, stacked bars, stacked horizontal bars, and trend. The graphical presentations of the results not only assist in selecting the most appropriate controls but also help in presenting the results to the project participants.

Conclusion

Construction projects are complex because they involve many human and non-human factors and variables. They usually have long duration, various uncertainties, and complex relationships among the participants. Primarily, the study proposed six principles of change management. Based on these principles, a theoretical model for project change management system (PCMS) was developed. This paper argued that the information technology could be effectively used for providing an excellent opportunity for the professionals to learn from similar past projects and to better control project changes. Finally, the chapter briefly presented an information technology based project change management system for the management of changes in building projects.

Although every construction project has its own specific condition, professionals can still obtain certain useful information from past experience. This information will enable building professionals to better ensure that their project goes smoothly without making unwarranted mistakes, and it should be helpful to improving the performance of the project. Furthermore, it is imperative to realize which changes will produce significantly more cost change effect for a construction project. The PCMS model consisted of six fundamental stages linked to two main components, i.e., a knowledge-base and a controls selection shell for making more informed decisions for effective management of changes. The database was developed through collecting data from source documents of past projects, questionnaire survey, literature review and in-depth interview sessions with the professionals who were involved in the projects. The knowledge-base was developed through initial sieving and organization of data from the database. The controls selection shell would provide decision support through a structured process.

The PCMS model presented a structured format for management of changes. The PCMS model would enable the project team to take advantage of beneficial changes when the opportunity arises without an inordinate fear of the negative impacts. By having a systematic way to manage changes, the efficiency of project work and the likelihood of project success should increase. The PCMS model emphasized on sharing the lessons learned from existing projects with project teams of future projects. The lessons learned should be identified throughout the project life cycle and communicated to current and future project participants.

The PCMS provides an excellent opportunity to the professionals to learn from past experiences (Arain, 2005b). It is important to note that this system for the management of changes is not designed to make decisions for users, but rather it provides pertinent information in an efficient and easy-to-access format that allows users to make more informed decisions and judgments. Although this system does not try to take over the role of the human experts or force them to accept the output of the system, it provides more relevant evidence and facts to facilitate the human experts in making well-informed final decisions (Arain, 2005b). The PCMS should be applied in the early stages (design stages) of the building projects.

The PCMS contains the knowledge based system (KBS) that is a unique system developed specially for the effective strategic management of changes in building projects for the first time (Arain, 2005b). The KBS would assist professionals in analyzing changes and selecting the most appropriate controls for minimizing changes in building projects. The PCMS model discussed in this chapter is valuable for all the professionals involved with developing the building projects. The initial use of the system for management of project changes resulted in reducing changes by 30 – 35% in building projects in Singapore. Presently, the system is being utilized by the governmental organization (the developer) for developing educational building projects in Singapore.

Knowledge acquisition was the major component for developing this system. The PCMS is developed based on the data collected from the 80 building projects. The KBS consists of two main components, i.e., a knowledge-base and a controls selection shell for selecting appropriate controls. The database is developed by collecting data from the source documents of these 80 building projects, questionnaire survey, literature review and in-depth interviews with the professionals who were involved in these projects. The KBS provides a fast response to queries relating to the causes, effects and controls for changes. The KBS is capable of displaying changes and their relevant in-depth details, a variety of filtered knowledge, and various analyses of the knowledge available (Arain, 2005b). This would eventually lead the decision maker to the suggested controls for specific changes and assist the decision maker to select the most appropriate controls for managing the changes timely.

The system is dynamic and designed to accommodate information pertinent to changes in ongoing projects that provides a platform for the organization to continuously learn and develop based on current building projects. It has an extremely user-friendly interface. The documentation process would take place in the workflow with minimum extra effort as the system also assist in filling in pertinent information. It retains the learning points in the knowledge base. This facilitates multiple reuse of knowledge in a team environment. The knowledge base acts as an authoritative reference for decision making as the learning points have been improved through processing by experts. Also, by constantly adding new learning points to the knowledge base as more projects are analyzed, the knowledge base is updated.

In PCMS, the knowledge consolidation process of the past experience will allow such knowledge to reside within an organization rather than residing within individual staff that may leave over time. Furthermore, as the KBS systematically consolidates all the decisions that have been made for numerous projects over time so that individuals, especially new staff would be able to learn from the collective experience and knowledge of everyone. Hence, the PCMS has a great potential for training new staff members. The new staff will be able to explore the details of all previous actions and decisions taken by other staff involved with the building projects. This would assist them in learning from past decisions and making more informed decisions for effective management of changes.

The PCMS through its KBS will help to enhance productivity and cost savings in that: (1) timely information is available for decision makers/project managers to make more informed decisions; (2) the undesirable effects (such as delays and disputes) of changes may be avoided as the decision makers/project managers would be prompted to guard against these effects; (3) the knowledge base and pertinent information displayed by the KBS will provide useful lessons for decision makers/project managers to exercise more informed judgments in deciding where cost savings may be achieved in future building projects; and (4) the KBS provides a useful tool for training new staff members (new professionals) whose work scope include building projects.

The PCMS model would assist building professionals in developing an effective project change management system. The system would be helpful for them to take proactive measures for reducing changes. The system efficiently assists the professionals in learning from past experiences. It is recommended that the system should ideally be used during the design stages of construction projects. Furthermore, with further generic enhancement and modification, the KBS will also be useful for the management of changes in any types of building projects, thus helping to raise the overall level of productivity in the construction industry. Hence, this system would also be valuable for all building professionals in general.

Acknowledgements

The author sincerely acknowledges Professor Dr. Low Sui Pheng, Head Department of Building, School of Design and Environment, National University of Singapore for his guidance and support for carrying out the present study.

References

Arain, F. M., Assaf, S. and Low, S. P. (2004). Causes of discrepancies between Design and Construction. *Architectural Science Review*, **47**, 237-249.

Arain, F.M. (2005a). Potential barriers in management of refurbishment projects. *Journal of Independent Studies and Research*, **3,** 22-31.

Arain, F. M. (2005b). Strategic management of change orders for institutional buildings: Leveraging on information technology. *Proceedings of the PMI Global Congress 2005 North America,* Toronto, Canada, BNS04 - Donald S. Barrie Award Winning Paper, 01-17.

Arain, F. M. and Low, S. P. (2005a). Lesson learned from past projects for effective management of change orders for Educational Building Projects. *Proceedings of the MICRA 4th Annual Conference,* Kuala Lumpur, Malaysia, 10-1 to 10-18.

Arain, F. M. and Low, S. P. (2005b). The potential effects of change orders on institutional building projects. *Facilities*, **23**, 496-510.

Arain, F. M. and Low, S. P. (2005c). Knowledge-based decision support system framework for management of changes in educational buildings. *Proceedings of the MICRA 4th Annual Conference,* Kuala Lumpur, Malaysia, 1-12 to 1-24.

Arain, F. M. and Low, S. P. (2006a). A framework for developing a knowledge-based decision support system for management of changes in institutional buildings, *Journal of Information Technology in Construction (ITCon),* **11**, Special Issue Decision Support Systems for Infrastructure Management, 285-310.

Arain, F. M. and Low, S. P. (2006b). Value Management through a Knowledge-based Decision Support System for managing changes in educational building projects. *International Journal of Construction Management*, **6**, 81 – 96.

Arain F. M. and Low, S. P. (2007a). Strategic Management of Changes in Educational Building Projects: A Timeline-Based Checklist Approach. *Proceedings of the 5th International Conference on Construction Project Management* (ICCPM/ICCEM 2007), Singapore, 1 – 15.

Arain, F. M. and Low, S. P. (2007b). Leveraging on Information Technology for Effective Management of Changes in Educational Building Projects: A KBDSS approach. *Centre for Education in the Built Environment, Working Paper* **10**, 1 – 88.

Cameron, I., Duff, R. and Hare, B. (2004). *Integrated Gateways: Planning out Health and Safety Risk.* Research Report 263, Glasgow Caledonian University, UK.

CII (1994). *Project Change Management.* Special Publication 43-1, Construction Industry Institute, University of Texas at Austin, TX.

CII (1994a). *Pre-project Planning: Beginning a Project the Right Way.* Publication 39-1, Construction Industry Institute, University of Texas at Austin, TX.

Gray, C. and Hughes, W. (2001). *Building Design Management.* Butterworth-Heinemann, Oxford, UK.

Harrington, H. J., Conner, D. R. and Horney, N. L. (2000). *Project Change Management.* McGraw Hill, New York.

Hester, W., Kuprenas, J. A. and Chang T. C. (1991). *Construction Changes and Change Orders.* CII Source Document 66, University of California- Berkeley.

Ibbs, C. W. (1997). Change's impact on construction productivity. *Journal of Construction Engineering and Management,* **123**, 89-97.

Ibbs, C. W., Lee, S. A. and Li, M. I. (1998). Fast tracking's impact on project change. *Project Management Journal,* **29**, 35-41.

Ibbs, C. W., Wong, C. K. and Kwak, Y. H. (2001). Project change management system. *Journal of Management in Engineering*, **17**, 159-165.

Kirton, M. (1976). Adaptors and innovators: a description and measure. *Journal of Applied Psychology,* **61**, 622-629.

Kirton, M. (1984). Adaptors and innovators: why new initiatives get blocked. *Long Range Planning,* **17**, 137-143.

Krone, S.J. (1991). *Decreasing The impact of Changes: Ripple Effect, Scope Changes, Change Orders.* unpublished PhD Dissertation, The George Washington University, USA.

Miresco, E. T. and Pomerol, J. C. (1995). A knowledge-based decision support system for construction project management. *Proceedings of the Sixth International Conference on Computing in Civil and Building Engineering*, (**2**), 1501-1507.

Mokhtar, A., Bedard, C. and Fazio, P. (2000). Collaborative planning and scheduling of interrelated design changes. *Journal of Architectural Engineering*, **6**, 66-75.

O'Brien, J.J. (1998). *Construction Change Orders*. McGraw Hill, New York.

Skibniewski M. and Chao L. C. (1992). Evaluation of advanced construction technologies with analytical hierarchy process. *Journal of construction engineering and management*, **118**, 577-593.

Stocks, S. N. and Singh, A. (1999). Studies on the impact of functional analysis concept design on reduction in change orders. *Construction Management and Economics*, **17**, 251-267.

In: Progress in Management Engineering
Editors: L.P. Gragg and J.M. Cassell, pp. 69-86
ISBN: 978-1-60741-310-3

Chapter 3

COUPLING MECHANISMS IN THE MANAGEMENT OF DEVIATIONS: PROJECT-AS-PRACTICE OBSERVATIONS

*Markus Hällgren**
Umeå School of Business, Umeå, Sweden

Abstract

Traditionally projects are considered means for getting things done, simultaneously striving for efficient and accurate methods – that is, doing more in less time. A consequence, not often discussed, is that doing more things in less time with a closer focus on cost, will inevitably lead to a more complex and tightly connected project execution system which is more sensitive to deviations. Following a "project-as-practice" perspective this paper explores and analyses how deviations are managed. The findings suggest that even though the company under consideration manages about 120 projects per year deviations cannot be avoided. The deviations were found initially to decouple (a process of creating loosely coupled activities) from the overall project process and later on to recouple (a process of tightly coupling activities) when the deviation was resolved. The paper suggests that the management of deviations is dynamic and changing and that the concept of coupling is a fruitful way of exploring the process.

Introduction

Traditionally, projects are considered means for getting things done, an issue that is highlighted in the contemporary project management literature (c.f. Ekstedt *et al.*, 1999; Engwall and Westling, 2004) where a key ingredient in the development of project management tools and techniques is using methods that are more time and cost efficient. (Lindkvist *et al.*, 1998). Thus, the need for simultaneous speed and accuracy is emphasized, creating a demand for more elaborate planning and control methods. A consequence, not

* E-mail address: markus.hallgren@usbe.umu.se. www.markushaellgren.com. . (phone) + 46 90 786 58 85, (fax) + 46 90 786 66 74. 901 87 Umeå, Sweden

often discussed, is that doing more things in less time with a closer focus on cost will inevitably lead to a more complex and tightly connected project execution system (compare Perrow, 1984), which is more sensitive to deviations. To put it differently, planned activities in a project are doomed to be interrupted by deviations having various impacts (Hällgren and Maaninen - Olsson, 2005). Surprisingly, , as a research and practitioner community, we do not know how unexpected events (deviations) are managed in practice – away from theoretical constructs and suggestions on how to handle them (Cicmil, 2006). Equally disturbing is that we do not know much about the mechanisms and the deviation management logics. Therefore, the question that will be given focus in this paper is:

How are deviations managed in practice in tightly coupled projects?

Some organizations experience very few major deviations despite functioning in tightly coupled and complex environments (Roberts, 1990b; Weick, 2004). Interestingly, these organizations recognize that they rely on both tight and loose coupling in their activities – contrary to what is described in the literature. In the contemporary literature on projects and coupling processes, projects are generally described as loosely coupled organizations with planned, tightly coupled activities (Christensen and Kreiner, 1991; Dubois and Gadde, 2002). Although these insights are useful and valid, they neglect the micro-activities - the practice - of managing projects. Therefore the question of coupling is elaborated upon through a project-as-practice approach which implies that the activities of the practitioners are taken seriously (Johnson *et al.*, 2003). In this context, this means how practitioners make use of their abilities to manage the unexpected in tightly coupled power plant projects. (c.f. Whittington, 2003; Jarzabkowski, 2005; Whittington, 2006). The aim of this paper is to examine the logic of coupling processes with regard to deviations. This will shed some light on what makes projects able to handle a simultaneous need for flexibility and stability which for a long time have interested scholars of organizational theory (Burns and Stalker, 1961; Thompson, 1967; Weick, 1976). The conclusion ties the analytical pieces together and shows that the sum is larger than its parts. The paper is structured accordingly.

Background

Organizing Verbs

The struggle to stamp out the nouns and replace them with verbs is attributed to Karl Weick (1979:44). He suggests than any organizing process is achieved by ecological change, enactment, selection and retention. Enactment is about bracketing parts of the overall organizing process, or dividing it into events which assembled make up the "organizing"; selection about sorting out meaningful understandings and finally preserving them (Weick, 1979; Weick *et al.*, 2005:414). This bracketing process (which projects are argued to be a result of (Lundin and Söderholm, 1995)) is *not an outcome* of organizing but a sensitizing device in the shape of practices influencing what is, or is not bracketed (Weick, 1979:166). Using the word "practices" is intentional, it highlights the connection to the practice idea (Weick uses "previous events") and that is useful for the next part of the argument. The overall process which Weick describes as "enactment, selection and retention" could also be described as how the situated praxis influences the shape of the practitioners' practices and how these concepts merge in episodes of organizing. (There is more about the Practice Turn in for example, Whittington (2006) or Schatzki *et al* (2001)).

Hendry and Seidl (2003), drawing on Luhmann (1995) (who have the same interest in social systems as Weick) argue that the overall processes of an organization can be divided into streams of interrelated episodes of communication (as part of organizing). From a more general organizing viewpoint these episodes can be meetings, planning sessions, informal talks, going-away days (Whittington, 2006). Essentially the episodes are parts of the processes which are singled out from other organizing processes and are comparable with the enactment part of Weick's model.

When an episode is singled out (as part of previous experiences, and compared once again) a certain situated praxis follows, (again compared to the selection part of the model) and later on are either retained or not into the overall behavior (compared to practices). Thus what we have is a process of bracketing, acting and re-connecting. Hendry and Seidl suggest that it can be described as a process of coupling which we need to understand better in order to understand practice. However, as Weick (1976) and Orton and Weick (1990) argue, to say that something is loosely coupled (when it is given an identity and requires a response) is only the beginning as what is really interesting happens *within* the process., – that is, what happens when such a mechanism kicks in. This would support the ideas of Hendry and Seidl, but also the more general views from an "as-practice" perspective.

Loosely Coupled Systems

The core of the organizing argument is one of loosely coupled systems (Weick, 1979:236) The concept of coupling was formally introduced to organization theory by Professor Karl Weick in (1976) drawing of arguments from Glassman (1973). In the literature, loose coupling is commonly argued to be a barrier to change and learning and a facilitator of innovation on an organizational level (e.g. Dubois and Gadde, 2002). In projects, this is recognized not only in how projects are used for producing novel solutions when it is not necessary for the entire organization to change, but also the problem of learning between projects. Initially, loose coupling was seen as an enabler and a feature which explained the organizing of local practices, in turn admitting novel solutions in a limited context (Weick, 1976; Snook, 2002). The strength of coupling is that it allows the researcher to examine the rationality and indeterminacy of organizations simultaneously, as well as organizations which are open and closed at the same time (c.f. Orton and Weick, 1990:205; Scott, 2003:271-272). "[Loosely] coupled events are responsive, *but* [...] each event also preserves its own identity and some evidence of its physical or logical separateness. [...] Their attachment may be circumscribed, infrequent, weak in its mutual effects, unimportant, and/or slow to respond. [...] [Loose coupling] also carries connotations of impermanence, dissolvability, and tacitness all of which are potentially crucial properties of the "glue" that holds organizations together." (Weick, 1976:3) (For a comprehensive exposition of the subject and the literature see Orton and Weick (1990) On the other hand, tightly coupled activities are those that are directly responsive -ones in which a change spreads rapidly like rings on the water such as in a concurrent engineering project where a plan commonly specifies what each party is expected to do and the activities are both tightly coupled and dependent on each other.

Upset with how the coupling concept was being used, Orton and Weick subsequently made a critical literature review of the various applications. They started their critique with the claim that the concept "is widely used and diversely understood"(Orton and Weick, 1990:203). Five voices of loose coupling emerged; Causation, Typology, Direct Effects, Compensations and Organizational Outcomes. (see Orton and Weick, 1990) The voices all struggle with the issue of choosing between a dialectical and a uni-dimensional approach to coupling. The dialectical approach is the good one according to Orton and Weick. Being dialectic means that tight and loose coupling is present at the same time. Being uni-dimensional on the other hand is bad. It is bad because it drifts away from the original meaning in the sense of putting tight and loose coupling at the end of a scale without acknowledging their parallelism and mutual influence. To aid understanding Orton and Weick essentially broke down the coupling concept into distinctiveness and responsiveness describing them thus: "If there is neither responsiveness nor distinctiveness, the system is not really a system, and it can be defined as a non-coupled system [1]. If there is responsiveness without distinctiveness, the system is tightly coupled [2]. If there is distinctiveness without responsiveness, the system is decoupled [3]. If there is both distinctiveness and responsiveness, the system is loosely coupled [4]." The idea is that these concepts are dialectic, thus not mutually exhaustive and uni-dimensional. (Orton and Weick, 1990:205) The differences are illustrated in exhibit 1 below, the numbers corresponds to the numbers above.

"If a person selectively attends to the openness, independence and indeterminate links among some elements, he or she will describe what amounts to a decoupled system. That characterization, too, is incomplete and inaccurate because parts of the system remain coupled and closed". That said, the research does not become less powerful but it could benefit from a more uni-dimensional approach, possibly meaning that the micro-perspective of "The Practice Turn" (Schatzki *et al.*, 2001; Jarzabkowski *et al.*, 2007; Whittington, 2007) could be a fruitful development as suggested by Hallet and Ventresca (2006) in association with institutional theory development.

From an institutional perspective, construction industry projects are loosely coupled to their organizations. The projects are loosely coupled because there is a focus on the single project rather than the continuous business, a need for local adjustments at the building site, utilization of standardized parts, competitive tendering, market-based exchange, and self-determination; features which are met by the loose coupling mechanisms of localized adaptation, buffering, sensing, generation of variation and self-determination. (Dubois and Gadde, 2002) Even though project literature commonly treats coupling on an institutional level (c.f. Christensen and Kreiner, 1991; Kreiner, 1995; Lindkvist et al., 1998; Dubois and Gadde, 2002) the concept of coupling could also be brought down to a task management level where the prerequisites of the hierarchically higher levels are created, e.g. knowledge processes (Brusoni *et al.*, 2001) and where it contributes to the understanding of how different parts of an organization interact (Andersen, 2006:22).

		Responsiveness	
		No	**Yes**
Distinctiveness	**Yes**	[3] Decoupled system	[4] Loosely coupled system
	No	[1] Non-coupled system	[2] Tightly coupled system

Compare Orton and Weick, 1990.

Exhibit 1. Types of coupling.

Further down, within the project, the activities seems to be tightly coupled when examining them on the surface (Hällgren, 2007) even though they appear to be loosely coupled from the level above (Dubois and Gadde, 2002). The reason is that It would not be efficient (efficiency being a hallmark of projects) to carry around slack; there is commonly one accepted way of reaching the goal (the plan); a strong reliance on coordination; the project is centralized, not very delegated there is causal dependence between activities and so forth - all hallmarks of tight coupling (Weick, 1976:5-6). Loose coupling is generally considered to be a kind of local practice which is not necessarily according to the initial intentions of the planner (compare Snook, 2002). Thus it is fruitful to depart on a journey exploring the intricate meanings of coupling in a project context.

To put it differently, the micro-processes found on the task level commonly influence the levels above, or they are sedimented throughout the organization (Giddens, 1979:55). The problem with coupling is that the possibility that what seems to be loosely coupled from one perspective is not necessarily so when the context and the situation is examined from another point of view. (Weick, 1976:9-11) The importance of "perspective" is thus once again brought to our attention. In addition, what is coupled in one way at one occurrence is not necessarily so in a similar future situation. Coupling is thus a changing process which needs to be captured in the actions of the practitioners. These transitions from loose coupling to tight coupling are facilitated by the unexpected as this forces the organization to change its logic of activities. Thus deviations are essential to an understanding of the micro-patterns of loosely coupled projects.

Project Management in the Face of Deviations

Traditionally projects are considered means for getting things done. (Ekstedt et al., 1999). The reason that projects are efficient is that they are limited which makes it possible to initiate action and to pay attention to deviations. From this perspective the unexpected is regarded as unnatural (Tsoukas and Chia, 2002:567) - something that should be avoided and managed accordingly - although it is recognized that there different types of projects (c.f. Turner and

Cochrane, 1993; Shenhar and Dvir, 1996; Crawford and Pollack, 2004) and that some will experience more deviations than others.

Following the line of argument deviations are defined as unexpected events which are identified (given distinctiveness) and requiring attention (given responsiveness). which conforms with the idea that it is not a deviation until someone has said so in a social setting beyond the individuals' own reasoning (Weick, 1979; Engeström, 2000). The first step is thus that someone recognizes a deviation and announces it, or in other words, that someone `brackets and labels the event′ (Weick, 1979; Weick *et al.*, 2005). These deviations obviously constitute an important part of the project manager's work. Loose coupling of the organization (Dubois and Gadde, 2002) or the activities (Weick, 1976) is however seen as a feature which limits the impact of deviations and has therefore been given extra scrutiny (c.f. Weick and Roberts, 1993).

In project management literature this local practice is a neglected field of research as most literature focuses on plans, methods and tools or what Clegg et al (2005) call "synoptic" accounts. Such accounts are generally brief and general However, some studies – most unintentionally, some intentionally - take a project-as-practice approach. For example, Simon (2006) studied the role of the project manager in the computer industry based on the actions performed; Nilsson (2005) found that a project manager's day in the software industry was highly fragmented and meeting-intensive; Hällgren and Maaninen-Olsson (2005) researched deviations and their handling in a rolling-mill project. Pitsis*, et al* (2003) on the other hand, studied project management meetings during the "Sydney 2000 Olympic infrastructure project" where a chase for "future perfect" experienced several deviations and obstacles but still reached the goal. The above-mentioned studies reveal a local practice that departs from the make-believe constructs of the traditional literature. Thus,: "*[G]etting more involved with analytical and mostly rational theoretical models of projects will only provide more make-believe statements on project management issues.*" (Blomquist *et al.*, 2006:3-4)

Planning for Deviations

Whether they reflect local practice or not, plans are assumed to be means for specifying what tasks should be done, by whom and when, in endeavors that are complex in the sense that many interrelated activities are conducted in parallel, involving many people. Plans are thus instrumental. Diesel power plants which consist of literally thousands of activities and hundreds of people and apply a severely compressed concurrent engineering approach are a case in point. (Alsakini *et al.*, 2004; Lindahl, 2005). By means of plans, otherwise complex activities are broken down to allow more tightly coupled and efficient systems (systems in the sense of interacting activities (Simon, 1996:183-185)). This has made plans one of the most important features of project management (Dvir and Lechler, 2004).

Tools for planning are rational and straightforward. Through knowledge of the goal and the context and the application of past knowledge, they try to find the stabilizing features of the development. Whenever an event that challenges the plan occurs, it is expected that it receive a structured response (c.f. Nicholas, 2001; Gray and Larson, 2006). These are change and risk management procedures which commonly rely on quantitative measurements, reports, negotiations and formal decisions (Barkley, 2004). Such procedures serve to keep the plan and the tasks tightly coupled to the overall goal.

On a detailed level a plan describes the execution of the task, even the task broken down to a single afternoon. On this level a plan can be sequential, that is when the tasks come after each other. The tasks can also be arranged, as in this case, in a parallel pattern. If intentionally and heavily used this is addressed by concurrent engineering or fast tracking. That is, the delivery time in the project is deliberately shortened as much as possible which decreases the amount of available slack in the project and increases the complexity (defined as the number of activities (Simon, 1996)) and tight coupling (responsiveness but not distinctiveness) between tasks (Weick, 1976; Orton and Weick, 1990) and need for coordination and adjustment (Thompson, 1967:55). Applying concurrent engineering thus puts an extra challenge to the neglected project execution (Love *et al.*, 1998; Kloppenborg and Opfer, 2002) when it produces more deviations that require attention.

Method

Recent developments in academia have taken on the task of capturing the practice of the practitioners and bringing back what is relevant not only to education but also academia as well as providing the practitioners themselves with the possibility of reflecting upon their own actions.

The research is based on Geertz's (1973:6) argument that in order to build strong theories and understanding of an organizing phenomenon, there is a need to focus on the actions of the individuals in the specific context. The ethnographic approach is believed to capture these actions. As Geertz (1973:6) put it "If you want to understand what science is, you should look first to what its practitioners do: anthropologists do ethnography." It is important to note however that there are other methodological means that may achieve practice-based results-interviews, for example, but also quantitative methods – but for the question and aimed posed observations are the obvious choice.

The empirical findings, on which this paper reports, comprise a twelve week long ethnographic observational study of two project teams in the independent power solutions business. In addition, triangulating the findings of the observations, there are 60 interviews, reports, contracts, minutes-of-meetings, more than 5000 emails, two week-long on-site project implementation visits and innumerable informal discussions related to the projects and their context.

So far, project management and temporary organizing have focused primarily on tools and methods but they also adopt a process approach to the developments (Söderlund, 2004). The practice approach is believed to provide additional unique insights to the findings by giving them content. Practice is thus examined as the actions of the practitioners which could be further elaborated and transformed into, or become part of, a process. This brings us to a question of what practice is. Practice can be defined in, and by, three broad concepts. These are the Praxis (situated actions) of the Practitioners (the men and women in project management) and their Practices (tools, methods, norms, values) (see Jarzabkowski, 2000; Johnson et al., 2003 for a review). These three concepts cannot be separated but rather they co-exist as they shape each other. The strength of the practice approach can also be found in this co-existence. Capturing the praxis of the practitioners will, if implemented correctly, bring organizing mechanisms to the surface rather than make-believe statements found in many "theories", tools and methods. Thus, the practice turn provides the foundation upon which statements can be elaborated.

This particular research has been analyzed in a number of steps using the project-as-practice perspective and Nvivo 8.0 software. First of all, let us be clear on the fact that the choice of the particular deviations that are addressed in the paper is important. The deviations that are addressed all had significant impact upon the projects where they occurred. They also show intricate relations among individuals, actions and organizational entities. More importantly, the deviations are representative of the other 12 found in table 1. Secondly, during the course of the research, there were a number of weeks for reflection which provided the possibility of thinking about the developments (compare Yin, 1994). Thirdly, directly after the observations, the three chosen projects were written down as summary descriptions and sent for validation by the project managers. The fourth step was to write articles and present it in various forms and occasions (presentations, conferences etc.). This provided theoretical, empirical and analytical tools for further development. Fifthly, in this specific paper, the knowledge that has been accumulated during the course of the research breaks down the deviations into behaviors and analysis that address the issues in a broader sense. Sixthly, the analysis was brought together to form a somewhat unified theory of deviation related practice. The last stage was possible by using the multiple perspectives that emerged earlier in the research and is believed to answer to the relevance challenge of practice research.

Observations

The Company

The company is a company with global businesses, divided into several departments. One of them - the land based power plant department – has at any time during the year an average of 120 simultaneous projects, ranging from delivery to turnkey type projects to be delivered over periods from 3 - 24 months. An average project runs for 12 months. As a company that has been utilizing deliberate project management techniques in many projects for about 20 years, they are used to managing projects. The main weapon of competition is their concurrent engineering and ability to help the customer to benefit from their window of opportunity.

At the department there are about 30 project teams. One team consists of one project manager and a mechanical and electrical senior engineer. When the teams manage a turnkey project a senior civil engineer joins the team. When applicable, mainly for turnkey projects, a site team is associated with the project team. The site team manages the execution in the foreign country while the project team is responsible for planning, reporting, and customer contacts. This research has focused on the project team at the corporate office.

All in all, 15 deviations are presented in this article. They are summarized in exhibit 1 below. Two of the deviations are given extra scrutiny to aid the subsequent discussion.

NO	DEVIATION	DESCRIPTION	MANAGEMENT
1	Damaged equipment	Equipment found to be damaged, with a two month replacement process	Dummy pieces, rearrangement of tasks
2	Logistics delay	Equipment delayed by a month, including some critical equipment	Rearrangement of tasks, inform involved parties
3	Payment	The client did not pay according to an agreed schedule	Careful negotiations, involving support services and sales
4	Negotiations	Sensitive negotiations with a shipyard which were prolonged	Break down and discuss details
5	Demobilization of personnel	Personnel demobilized from construction site	Contact and force subcontractor to re-mobilize
6	Tank erection progress	Tank construction progress too slow	Contact and force subcontractor to increase speed
7	Pipe rack	Critical equipment within the client's scope was not finished in time	Careful reminders to customer, rearrangement of activities and demobilization
8	Switch gear delay	Critical equipment within the client's scope was not finished in time	Careful reminders to customer, rearrangement of activities and demobilization
9	Missing blueprint	Instructions for blueprint were missing when they should have been designed	Design parts around the specific blueprint and account for its features
10	Failing breaker	Breaker kept going of due to technical malfunction	Trial and Error problem solving, finally replaced failing equipment
11	Engine variance	Engine had a variance because the client had not finished some tasks of his	Contact expert, look into issue and explain to the customer
12	Language incompatibility	The entire contract including instructions in a foreign language. Demand for all documentation in that language	Appoint language support service to project and translate important parts of the contract. Try to renegotiate
13	Engine positioning	Engine positioning tools not present	Make use of alternative lifting arrangements
14	Change order	Various change orders from sub-contractors	Consider and decide
15	Heat recovery	Heat recovery did not function due to site conditions	Contact expert and explain the conditions

Exhibit 2. Deviations and their management.

Deviation 1 – Damaged Equipment

Damaged equipment was the cause of a deviation in a project with a twelve month schedule. The problem was discovered at a rather late stage of the project, partly because of the delay in the logistics. The damaged equipment consisted of a number of charge air silencers and switchgear cubicles. The deviation threatened to delay the entire project by three months, which would shred the budget to pieces.

The first thing the project team tried to understand was the extent of the damage through phone calls, photos, emails and by sending a junior engineer from the sister project to the project with the damaged equipment. Parallel to the these developments, the project manager started to try to contact the logistics company, which later made their own flawed investigation. In the end the insurance company had to be notified. Developments continued for ten weeks and during this time the project manager felt forced to order the material without full knowledge about the extent of the problem, although he made a probable calculation of it. He told one of his engineers in an email: “I understand that the two switchgear cubicles are so badly damaged that we need to supply new ones. Please act accordingly to not lose more time.” While waiting for the ordered material to arrive, the site team continued with the construction of the project plan using the damaged parts as dummy equipment which they could build around and replace later.

In the later stages of the developments the equipment deviation and the logistics deviation (see exhibit 1, deviation 2) became the subject of numerous reports, discussions and emails. The reason for the deviation emerged as a mishap during pre-loading which the logistics company was aware of – and had even photographed - but which was not reported to the project team. The logistics company eventually needed to meet the project manager to discuss reimbursement. It was clear to every one that there were no contracts stating the responsibilities in the case. Even though there had been organizational routines for reimbursement, everybody knew that they could not be implemented. Instead, any settlement had to be made by drawing on the goodwill between the companies. At the end, the company was reimbursed, and the reimbursement was acceptable to all parties even though it was about a tenth of the initial demands.

Deviation 2 – Payment

This deviation occurred in a quite novel project where the previous project had been done ten years earlier. The customer’s contract specified that the first and major payment was due two weeks after signing and if not then, after an additional two weeks. In the latter case, a daily extension would apply.

Problems with the payment were associated with the inability of the customer to make the initial payment on both the first and second opportunities. At the same time as the payment negotiations were proceeding, the project team had to negotiate with a number of shipyards, and choose one (see exhibit 1, deviation 4). The problem was that these negotiations were dependent upon the customer’s ability to pay. The dates that were negotiated with the shipyard were thus dependent on the dates of the payment. The payment negotiations were on the other hand dependent on the outcomes of the shipyard negotiations. Adding to the problem the missing payment threatened to delay the entire project which had very tight time and budget limitations.

The solution to the problem was to set a date on which the customer had to make the payment. If it was not made, the shipyard would receive a small payment which could be regained, albeit with deductions, with certain percentage intervals and dates. Provided that the payment was not prolonged too long, this solution would provide the company with some money from an earlier smaller payment. The negotiations and the payment finally worked out but it was not until the last possible date that the money arrived.

Exhibit 2 represents mechanisms of loose coupling during the decoupling process (when a deviation is becoming responsive and distinctive) and mechanisms of tight coupling during the recoupling process (when a deviation is losing its distinctiveness but remains responsive). Three loose coupling mechanisms were identified and eight tight coupling mechanisms. The process where these mechanisms were found was not sequential but parallel and intertwined.

#	DEVIATION	LOOSE COUPLING MECHANISMS DURING DECOUPLING	TIGHT COUPLING MECHANISMS DURING RECOUPLING
1	Damaged equipment	Alternative path (activities, equipment), Transfer (responsibility)	Reassume original path, Coordination, Negotiation, Acceptance (responsibility)
2	Payment	Alternative path (activities, time), Transfer (responsibility)	Create new path, Acceptance (responsibility)
3	Logistics delay	Alternative path (activities), Slack, Transfer (responsibility)	Reassume original path, Coordination, Negotiation, Acceptance (responsibility, decision)
4	Negotiations	Alternative path (activities), Embracing (responsibility)	Reassume original path
5	Demobilization of personnel	Transfer (responsibility)	Reassume original path
6	Tank erection progress	Alternative path (time, activities), Pushing (responsibility)	Reassume original path, Coordination, Acceptance (responsibility)
7	Pipe rack	Alternative path (time), Transfer (responsibility)	Create new path, Non-acceptance (responsibility)
8	Switch gear delay	Alternative path (time), Transfer (responsibility)	Create new path, Non-acceptance (responsibility)
9	Missing blueprint	Alternative path (equipment)	Reassume original path, Coordination
10	Failing breaker	Alternative path (time), Transfer (responsibility)	Acquire resources, Acceptance (responsibility)
11	Engine variance	Alternative path (time), Transfer (responsibility)	Coordination, Negotiation, Acceptance (responsibility)
12	Language incompatibility	Alternative path (activities), Embracing (responsibility)	Acquire resources, Create path
13	Engine positioning	Alternative path (equipment), Transfer (responsibility)	Reassume original path, Coordination, Negotiations, Acceptance (responsibility)
14	Change order	Embracing or Transfer (responsibility)	Acceptance (decision)
15	Heat recovery	Transfer (responsibility)	Acceptance (responsibility)

Exhibit 3. Deviations and mechanisms of coupling.

Loose Coupling Mechanisms

The findings from table 2 suggest that the decoupling process is achieved through three general mechanisms of loose coupling (number of occurrences within parenthesis);

1. Alternative path (15/15)
2. Responsibility (15/15)
3. Slack (1/15)

The first pattern refers to processes where alternative paths are created for activities, equipment and time. That is, the expected path is interrupted and work arounds are created to fit the situational need. The responsibility patterns refer to how the responsibility for a certain issue was transferred, embraced or argued for. *Transferring* the responsibility, (for example for the. logistics delay), implies that someone else will cover whatever the costs are or they will manage the situation. *Embracing* the responsibility, (for example for the negotiations) on the other hand, means that no one else is to blame and the problem belongs to the organization. *Arguing for* responsibility, (for example, tank construction) means that it is not clear cut who is to blame or carry responsibility and the situation has to be worked out and managed by mutual agreement. The creation and use of slack occurred, for example, when the logistics were delayed. The reason for using this buffer was basically that it was available and the issue was well-known. However, what made it distinctive was that the slack exceeded the intended buffer.

Tight Coupling Mechanisms

The findings from Table 2 suggest that the recoupling process is achieved through eight general mechanisms of tight coupling (number of occurrences within parenthesis);

1. Reassume original path (7/15)
2. Create path (4/15)
3. Coordination (6/15)
4. Negotiation (4/15)
5. Acceptance (10/15)
6. Non-acceptance (2/15)
7. Acquire resources (1/15)
8. Replacement (1/15)

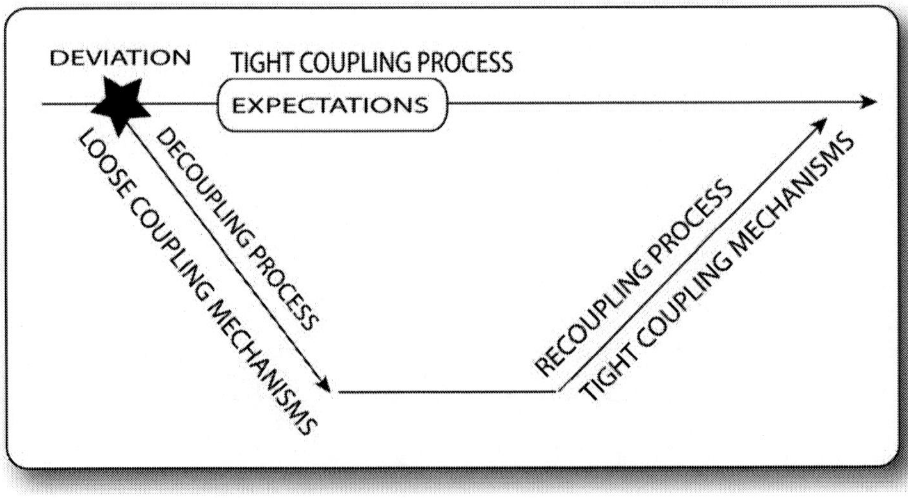

Exhibit 4. Processes and mechanisms of coupling.

Reassuming the original path meant that the project continued on the same path as before the deviation (e.g. the damaged equipment) in contrast to *creating an additional path* (e.g. payment). *Coordination* meant that activities were coordinated in such a way that all changes in the project were aligned (e.g. tank construction). *Negotiations* were less determined as the solution was dependent on a mutual understanding between several actors (e.g. the engine variance). *Acceptance* of responsibility or for a decision (e.g. engine positioning) and *non-*

acceptance of responsibility (e.g. pipe rack) determined whatever path the responsibility and subsequent activities would take. *Acquire resources* (language incompatibility) has to do with creating more resources for the resolution. It basically means capitulating and buying a way out of the deviation and thus is connected to returning to the original path.

The decoupling process is not mutually exhaustive meaning that one deviation can carry several loose coupling mechanisms whilst being tightly coupled to other activities within the project. Similarly, recoupling requires tight as well as loose coupling mechanisms. In addition, a deviation is not necessarily entirely decoupled before the recoupling process commences and this concurs with the basic arguments of Orton and Weicks (1990). The general patterns are found in exhibit 4 below.

Discussion

Traditionally projects are considered means for getting things done simultaneously striving for efficient and accurate methods – that is, doing more in less time. A consequence, which has been given extra scrutiny in this paper, is that doing more things in less time with a closer focus on cost, will inevitably lead to a more complex and tightly connected project execution system which is more sensitive to deviations and require more attention from the management.

Following that basic argument, since projects are a tool for managing change and unstable conditions (Lundin and Söderholm, 1995), they experience deviations more often than not as the context and task changes. (Dvir and Lechler, 2004). In the project it is vital that deviations are managed swiftly (Hällgren, 2007) That is why risk and change management together with the plan are essential to the execution of the project (compare Nicholas, 2001). It is apparent that even for companies such as the one in this paper used to managing projects over many years deviations are a problem. However, contemporary project management literature is quite limited in how it describes what is done during the implementation process (Kloppenborg and Opfer, 2002) on the "floor" , or - to paraphrase Geertz (1973) - how it describes the situated praxis of practitioners regarding to deviations. Therefore, the aim of this paper was to examine the logic of coupling processes in regard to deviations.

(Orton and Weick, 1990), establishing that something is loosely or tightly coupled is trivial but it becomes interesting when the detailed processes are investigated Therefore, within the boundaries of their argument the following is put forward:

Decoupling is achieved through the recognition of the deviation as a responsive and distinctive situation that needs to be managed. Once decoupled and governed by loosely coupled mechanisms, recoupling commences and tightly coupled mechanisms bring the deviation back on track so that they display a uni-dimensional behavior. From a management point of view, deviations are not mutually exhaustive and managing them needs to be sensitive to whatever changes are put in place as those changes will have an influence beyond the original deviation. The first logic that is displayed is thus one where old practices (norms, values, rules and routines (Whittington, 2006) are applied to bracket the deviation and decouple it through loose coupling mechanisms. The second logic is one where these old practices are applied to solve the problems generated by the deviation but also contribute to the stock of practices which can be utilized in future situations. From a management point of view this understanding means that in order to achieve a change in behavior or add new knowledge one should focus on what is coupled to what and in accordance with what logic.

Therefore, through the analysis I outline a practice associated with the effective management of deviations

Processes and Mechanisms of Coupling

It has been suggested that loosely coupled behavior is not loosely coupled for a long time (Roberts, 1990a; Snook, 2002). Upon decoupling, loose coupling follows. Loose coupling is achieved when something displays distinctiveness and responsiveness (Orton and Weick, 1990) which conforms with the idea that it is not a deviation until someone has said so in a social setting beyond the individuals' own reasoning (Weick, 1979; Engeström, 2000). The first step is thus that someone recognizes a deviation and announces it, or in other words, that someone `brackets and labels the event′ (Weick, 1979; Weick *et al.*, 2005). This process is referred to as the decoupling process of a deviation. It is necessary in order to protect the project as a common system. The deviation is separated from the organizing process as that allows for the project to continue relatively uninterrupted. Being identified as something that is different and requiring attention gives the issue responsiveness and contributes to a decoupling process making it loosely coupled.

Loose coupling seems to be achieved through responsibility being transferred (11/15), embraced (3/15) or argued for (1/15) in combination with creating alternative paths in activities (7/15), equipment (2/15) or time (6/15). Uni-dimensionality is achieved when the recoupling process and tightly coupling mechanisms are accepted as simultaneously present. That is, tight coupling mechanisms are present at the same time and coupled to other expectations of the project. These expectations are diverse but have to do with time, cost and quality and the goal of the project. Non-compliance may turn out to be expensive. Tight coupling seems not only to be more diverse but to follow one general pattern. The tendency for acceptance of responsibility is more common in combination with reassuming the original path. This seems to be likely as accepting responsibility creates less interference from other sources.

Coupling Processes and Deviation Solutions

Decoupling begins with bracketing and labeling the deviation (giving it distinctiveness) in a process of enactment and selection in response to "ecological change" (Weick, 1979). The logic that is displayed is one where the situated praxis is used to assess the situation and select a plausible explanation and a solution, followed by adding and preserving whatever knowledge that is created. This is the retention part of Weick's (1979) model. Thus, there are two processes, one of situated praxis (actions dependent on the situation) and one of practices preserved in a shared social system between several practitioners (Whittington, 2006) following the idea that the process by necessity is started.

The two processes have different features. Situated praxis can be mostly individual and/or mostly social dependent on whatever the case and need is (keeping in mind that it is framed against a social setting). Practices on the other hand are only shared between several practitioners and can be drawn upon later when the praxis is applied to another future deviation – or situation. The point is that the management of a deviation is both an individual and a social process that needs to be managed diligently in order to be successful.

Implications

Mechanisms of loose and tight coupling can probably be found in various combinations. The aim of this paper is not to delve into such issues but such an investigation may add to the understanding of the management process. A word of caution though, these investigations should be done by context-sensitive methods, such as observations (Weick, 1976; Orton and Weick, 1990).

Research has neglected the lion's share of Weick's (1979) work when it stops at saying that nouns are bad. Based on the findings in this paper it seems fruitful to take a closer look on the enactment, selection and retention process.

The consequences of not having mutually exhaustive deviations are that the management needs to be sensitive to whatever changes are done to the expectations surrounding the path of the project. As some parts will be tightly coupled they will be influenced by even minor changes, while other parts (and sometimes desired ones) may hardly be influenced at all.

Likewise, created knowledge is dependent on what is connected to what. To utilize whatever knowledge is created one should take a closer look at where in the process it is created, what it is connected to, and what logic it follows, whether that is an individual or social logic.

Conclusions

Whittington *et al* (2004) warned us about taking practice perspective too lightly. He argued that the practice perspective needs to be connected to aggregated levels of analysis. Through the analysis of 15 deviations this paper has showed that coupling is not a steady state but rather is dependent on the solution that is sought and its impact. Based on a limited number of deviations, one should of course be cautious about generalization but being a general organizing phenomenon some observations seem in order. Two major arguments were made. The management of deviations carries processes of decoupling and recoupling with loose and tight coupling mechanisms, displaying a uni-dimensional behavior. Secondly, a logic of individual and/or social praxis is followed by a social preservation of the knowledge that is created and used.

References

Alsakini, W. and Wikstrom, K. and Kiiras, J. (2004) Proactive schedule management of industrial turnkey projects in developing countries. *International Journal of Project Management,* **22** (1 SU -), 75-85.

Andersen, E. S. (2006) Toward A Project Management Theory For Renewal Projects. *Project Management Journal,* **37** (4), 15.

Barkley, B., T. (2004) *Project Risk Management,* New York, McGraw-Hill.

Blomquist, T. and Gällstedt, M. and Hällgren, M. and Nilsson, A. and Söderholm, A. (2006) Project as practice: Making project research matter. *IRNOP VII.* Xian, China.

Brusoni, S. and Prencipe, A. and Pavitt, K. (2001) Knowledge Specialization, Organizational Coupling, and the Boundaries of the Firm: Why Do Firms Know More Than They Make? *Administrative Science Quarterly,* **46** (4), 597-621.

Burns, T. and Stalker, G. M. (1961) *The management of innovation*, Tavistock.

Christensen, S. and Kreiner, C. (1991) *Projektledning, att leda och lära i en ofullkomlig värld,* Lund, Academia Adacta.

Cicmil, S. (2006) Understanding project management practice through interpretive and critical research perspectives. *Project Management Journal,* **37** (2), 27-37.

Clegg, S. and Kornberger, M. and Rhodes, C. (2005) Learning/Becoming/Organizing. *Organization,* **12** (2), 147-167.

Crawford, L. and Pollack, J. (2004) Hard and soft projects: a framework for analysis. *International Journal of Project Management,* **22** (8), 645-653.

Dubois, A. and Gadde, L.-E. (2002) The construction industry as a loosely coupled system: implications for productivity and innovation. *Construction Management and Economics,* **20** (7), 621-632.

Dvir, D. and Lechler, T. (2004) Plans are nothing, changing plans is everything: the impact of changes on project success. *Research Policy,* **33** (1), 1-15.

Ekstedt, E. and Lundin, R. A. and Söderholm, A. and Wirdenius, H. (1999) *Neo-Industrial Organising, renewal by action and knowledge formation in a project-intensive economy,* London, Routledge.

Engeström, Y. (2000) Activity Theory and the Social Construction of Knowledge: a Story of Four Umpires. *Organization,* **7** (2), 301-310.

Engwall, M. and Westling, G. (2004) Peripety in an RandD Drama: Capturing a Turnaround in Project Dynamics. *Organization Studies,* **25** (9), 1557-1578.

Geertz, C. (1973) *The Interpretation of Cultures,* New York, Basic Books.

Giddens, A. (1979) *Central problems in social theory: Action, structure, and contradiction in social analysis,* London, Macmillan.

Glassman, R. B. (1973) Persistence and loose coupling in living systems. *Behavioral Science,* **18** (2), 83–98.

Gray, C. F. and Larson, E. W. (2006) *Project Management: The managerial process,* New York, McGraw - Hill.

Hallett, T. and Ventresca, M. J. (2006) How Institutions Form: Loose Coupling as Mechanism in Gouldner's Patterns of Industrial Bureaucracy. *American Behavioral Scientist,* **49** (7), 908-924.

Hendry, J. and Seidl, D. (2003) The structure and significance of strategic episodes: Social systems theory and the routine practices of strategic change. *Journal of Management Studies,* **40** (1), 175-197.

Hällgren, M. (2007) Beyond the point of no return: On the management of deviations. *International Journal of Project Management,* **25** (8), 773-780.

Hällgren, M. and Maaninen - Olsson, E. (2005) Deviations, uncertainty and ambiguity in a project intensive organization. *EURAM,* Munich

Jarzabkowski, P. (2000) *Top management teams in action in three UK universities*, Doctoral Thesis, Warwick University, Warwick: UK

Jarzabkowski, P. (2005) *Strategy as Practice,* London, Sage Publishers.

Jarzabkowski, P. and Balogun, J. and Seidl, D. (2007) Strategizing: The challenges of a practice perspective. *Human Relations,* **60** (1), 5-27.

Johnson, G. and Melin, L. and Whittington, R. (2003) Micro Strategy and Strategizing: Towards an Activity-Based view. *Journal of Management Studies,* **40** (1), 3-22.

Kloppenborg, T. and Opfer, W., A. (2002) Forty years of Project Management Research: Trends, Interpretations, and predictions. IN Slevin, D., P. and Cleland, D., I. and Pinto, J. K. (Eds.) *The Frontiers of Project Management Research.* Newton Square, Project Management Institute.

Kreiner, C. (1995) In Search of Relevance: Project Management in Drifting Environments. *Scandinavian Journal of Management,* **11** (4), 335-346.

Lindahl, M. (2005) The little engine that could. On the "managing" qualities. On the managing qualities of technology. IN Czarniawska, B. and Hernes, T. (Eds.) *Actor-Network Theory and Organizing.* Malmö, Liber.

Lindkvist, L. and Söderlund, J. and Tell, F. (1998) Managing Product Development Projects: On the Significance of Fountains and Deadlines. *Organization Studies,* **19** (6), 931.

Love, P. and Gunasekaran, A. and Li, H. (1998) Concurrent engineering: a strategy for procuring construction projects. *International Journal of Project Management,* **16** (6), 375-383.

Luhmann, N. (1995) *Social systems,* Stanford, California, Stanford University Press.

Lundin, R. A. and Söderholm, A. (1995) A theory of the temporary organization. *Scandinavian journal of Management,* **11** (4), 437 - 455.

Nicholas, J. M. (2001) *Project Management for business and technology,* London, Prentice Hall.

Nilsson, A. (2005) The Change Masters: Project Managers in Short Duration Projects. *Project Perspectives,* **27** (1), 42-45.

Orton, D., J. and Weick, K. E. (1990) Loosely Coupled Systems: A Reconceptualization. *Academy of Management Review,* **15** (2), 203-223.

Perrow, C. (1984) *Normal Accidents: Living with High Risk Technology,* Princeton, Princeton University Press.

Pitsis, T., S. and Clegg, S. R. and Marossezeky, M. and Rura-Polley, T. (2003) Constructing the Olympic dream: A future perfect strategy of project management. *Organization Science,* **14** (5), 574 - 590.

Roberts, K., H. (1990a) Managing High Reliability Organizations. *California Management Review,* **32** (4), 101-113.

Roberts, K., H. (1990b) Some Characteristics of One Type of High Reliability Organization. *Organization Science,* **1** (2), 160-176.

Schatzki, T. R. and Knorr-Cetina, K. and Von Savigny, E. (2001) *The practice turn in contemporary theory,* New York, Routledge.

Scott, W. R. (2003) *Organizations,* New Jersey, Prentice Hall.

Shenhar, A. J. and Dvir, D. (1996) Toward a typological theory of project management. *Research Policy,* **25** (4), 607-632.

Simon, H. (1996) *The Sciences of the Artificial,* Cambridge, MA, The M.I.T Press.

Simon, L. (2006) Managing creative projects: An empirical synthesis of activities. *International Journal of Project Management,* **24** (2), 116-126.

Snook, S., A. (2002) *Friendly Fire: The Accidental Shootdown of U.S.Black Hawks Over Northern Iraq,* Princeton, Princeton University Press.

Söderlund, J. (2004) Building theories of project management: past research, questions for the future. *International Journal of Project Management,* **22** (3), 183-191.

Thompson, J. D. (1967) *Organizations in action,* New York, Mcgraw - Hill.

Tsoukas, H. and Chia, R. (2002) On Organizational Becoming: Rethinking Organizational Change. *Organization Science,* **13** (5), 567-582.

Turner, J. R. and Cochrane, R. A. (1993) Goals-and-methods matrix: coping with projects with ill defined goals and/or methods of achieving them. *International Journal of Project Management,* **11** (2), 93-102.

Weick, K. E. (1976) Educational Organizations as Loosely Coupled Systems. *Administrative Science Quarterly,* **21** (1).

Weick, K. E. (1979) *The Social Psychology of Organizing,* New York, McGraw-Hill.

Weick, K. E. (2004) Normal Accident Theory As Frame, Link, and Provocation. *Organization Environment,* **17** (1), 27-31.

Weick, K. E. and Roberts, K., H. (1993) Collective Mind in Organizations: Heedful Interrelating on Flight Decks. *Administrative Science Quarterly,* **38** (3), 357-381.

Weick, K. E. and Sutcliffe, K. M. and Obstfeld, D. (2005) Organizing and the Process of Sensemaking. *Organization Science,* **16** (4), 409-421.

Whittington, R. (2003) Practice perspectives on strategy: Unifying and developing a field. *Academy of Management Conference.* Denver, CO.

Whittington, R. (2006) Completing the Practice Turn in Strategy Research. *Organization Studies,* **27** (5), 613-634.

Whittington, R. (2007) Strategy Practice and Strategy Process: Family Differences and the Sociological Eye. *Organization Studies,* **28** (10), 1575-1586.

Whittington, R. and Johnson, G. and Melin, L. (2004) The emerging field of strategy practice: some links, a trap, a choice and a confusion. *EGOS,* Ljubljana, Slovenia

Yin, R., K. (1994) *Case study research: design and methods,* Thousand Oaks, Sage coperation.

In: Progress in Management Engineering
Editors: L.P. Gragg and J.M. Cassell, pp. 87-115

ISBN: 978-1-60741-310-3

Chapter 4

MONETIZING PROCESS CAPABILITY

Fred Spiring[1,*] and Bartholomew Leung[2•]
[1] Department of Statistics, University of Manitoba,
Winnipeg, Manitoba, Canada
[2] Department of Applied Mathematics, The Hong Kong Polytechnic University,
Kowloon, Hong Kong

Abstract

A major concern among managers and administrators has been the lack of cost assessment/financial implications associated with process improvement and process capability. The impact of process control frequently gets treated more as good will than actual cost savings. In this manuscript we provide methods for quantifying cost savings through use of the metrics used to assess and improve process performance and capability. Initially we develop the general relationship between process capability indices and financial costs using the process capability index *Cpw* and various loss functions. The relationship between the unified approach for some common process capability indices (PCIs) through the use of a non-stochastic weight function and the expected weighted squared error loss provides an intuitive interpretation of *Cpw*. Using different values of the non-stochastic weights, w, the distributions of the estimated loss associated with the measures of process capability indices can be determined. Upper confidence limits for the expected loss associated with *Cpw* as well as its generalization *Cpw**, and special cases such as *Cp*, *Cp**, *Cpm*, *Cpm**, *Cpk* and *Cpk** are discussed. Quality practitioners and manufacturers need only specify the target, maximum loss, the estimated process mean and standard deviation, in order to determine an estimate of the expected loss associated with the process. Examples are demonstrated.

Introduction

The use of loss functions in quality assurance settings has grown with the introduction of Taguchi's philosophy. Decision theoretic statisticians and economists have for many years used the squared error loss function in making decisions or evaluating decision rules. With

---

* E-mail addresses: spiring@cc.umanitoba.ca and fspiring@mts.net.
• E-mail address: maleung@polyu.edu.hk.

the increasing importance of clustering around the target, rather than conforming to specification limits, and the understanding of loss functions there appears to be an alternative to PCIs. Rather than numbers or percentage non-conforming, economic/production costs or losses may provide improved opportunities to assess, monitor and compare process capability.

A general approach to process capability indices will be integrated with a general class of loss functions in order to provide economical/financial assessments of process performance and process improvements. An overview of the most commonly used PCIs and an introduction to a related class of Loss and Risk functions is followed by a section combining the two concepts. Several examples are used to illustrate the results of integrating PCIs and Loss functions.

Process Capability

Historically process capability was synonymous with process variation and expressed as either 6σ (σ being the population standard deviation) or the population range. Neither of these measures considers customer requirements nor permit general comparisons among processes as both measures are unit dependent. Juran (1979) suggested that Japanese companies initiated the use of process capability indices by relating process variation to customer requirements in the form of the ratio

$$Cp = \frac{\text{USL} - \text{LSL}}{6\sigma}$$

where the difference between the upper specification limit (USL) and the lower specification limit (LSL) provides a measure of allowable process spread (i.e., customer requirements) and 6σ, a measure of actual process spread (i.e., process performance). Bissell (1990) suggests that the British Standards Institution proposed an analogous measure referred to as the relative precision index, which was withdrawn and replaced by *Cp* in 1942.

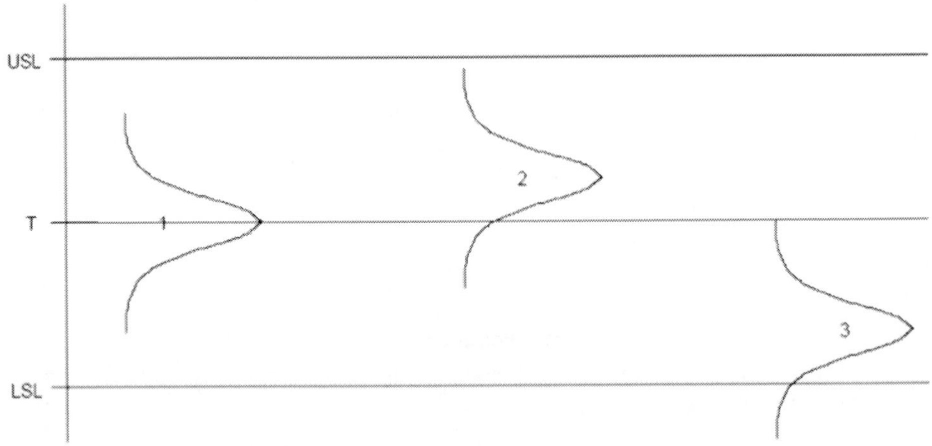

Figure 1. Three processes with Identical Values of *Cp*.

Incorporating customer specification limits into the assessment of process capability results in a more meaningful measure that fosters comparisons across all types of processes. However *Cp* uses only the customer's upper and lower specification limits in its assessment and fails to consider a target (or nominal) value. The three processes depicted in Figure 1 have identical process spreads and therefore identical values of *Cp*. However processes 2 and 3 deviate from the target (T) and additional costs due to these departures will be incurred. As a result processes 2 and 3 are considered less capable of meeting customer requirements than process 1.

Processes with small variability, but poor proximity to the target, sparked the derivation of several indices that incorporate targets into their assessment. The most common of these measures assume the target (T) to be the midpoint of the specification limits and include

$$Cpu = \frac{\mathrm{USL} - \mu}{3\sigma},$$

$$Cpl = \frac{\mu - \mathrm{LSL}}{3\sigma},$$

$$Cpk = \min(Cpl,\ Cpu),$$

$$Cpm = \frac{\mathrm{USL} - \mathrm{LSL}}{6\sqrt{\sigma^2 + (\mu - \mathrm{T})^2}}$$

and

$$Cpk = (1 - \mathrm{k})\ Cp$$

where μ is the process mean, $\mathrm{k} = \frac{2|\mathrm{T} - \mu|}{\mathrm{USL} - \mathrm{LSL}}$, $0 \le \mathrm{k} \le 1$ and $\mathrm{LSL} < \mu < \mathrm{USL}$. The two definitions of *Cpk* are presented interchangeably and are equivalent when $0 \le \mathrm{k} \le 1$.

The generalized analogues of these measures do not assume T to be the midpoint of the specifications and are

$$Cpu^* = \frac{\mathrm{USL} - \mathrm{T}}{3\sigma}\left(1 - \frac{|\mathrm{T} - \mu|}{\mathrm{USL} - \mathrm{T}}\right),$$

$$Cpl^* = \frac{\mathrm{T} - \mathrm{LSL}}{3\sigma}\left(1 - \frac{|\mathrm{T} - \mu|}{\mathrm{T} - \mathrm{LSL}}\right),$$

$$Cpk^* = \min(Cpl^*,\ Cpu^*)$$

and

$$Cpm^* = \frac{\min[\mathrm{USL} - \mathrm{T}, \mathrm{T} - \mathrm{LSL}]}{3\sqrt{\sigma^2 + (\mu - \mathrm{T})^2}}.$$

A hybrid of these measures, *Cpmk*, is defined to be

$$Cpmk = \frac{\min[\mathrm{USL} - \mu, \mu - \mathrm{LSL}]}{3\sqrt{\sigma^2 + (\mu - \mathrm{T})^2}}.$$

Individually *Cpu, Cpu*, Cpl* and *Cpl** consider only unilateral tolerances (i.e., USL or LSL respectively) in assessing process capability. Each uses 3σ as a measure of actual process spread, while the distance from the process center (μ) to the USL (for *Cpu*) or LSL (for *Cpl*) is used as a measure of allowable process spread. Both *Cpu* and *Cpl* compare the length of one tail of the normal distribution (3σ) with the distance between the process mean and the respective specification limit. In the case of bilateral tolerances *Cpu* and *Cpl* have an inverse relationship and individually do not provide a complete assessment of process capability. However conservatively taking the minimum of *Cpu* and *Cpl* results in the bilateral tolerance measure defined as *Cpk*. Similarly *Cpu** and *Cpl** use adjusted distances between their respective specification limits and μ, that incorporate the target in assessing allowable spread.

Cpm incorporates a measure of proximity to the target by replacing the process variance in the definition of *Cp*, with the process mean square error around the target. *Cp* and *Cpm* are identical when the process is centered at the target (i.e., μ=T). But since *Cp* is not a valid measure of process capability when the process is not centered at the target, *Cpm* dominates *Cp* as a measure of process capability. *Cpl*, Cpu*, Cpk*, Cpm** and *Cpmk* extend the class of allowable/customer process requirements versus actual process spread/performance by allowing the target to be other than the midpoint of the specification limits.

By definition, each of *Cp, Cpl, Cpu, Cpk, Cpm*, their generalized analogues and *Cpmk* are unit less, thereby fostering comparisons among and within processes regardless of the underlying mechanics of the product or service being monitored. In all cases, as process performance improves, either through reductions in variation and/or moving closer to the target, these indices increase in magnitude. As process performance improves relative to customer requirements, customer satisfaction increases as the process has greater ability to be near the target. In all cases larger index values indicate a more capable process.

Spiring (1997) proposed an unified index, *Cpw*, of the form

$$Cpw = \frac{USL - LSL}{6\sqrt{\sigma^2 + w(\mu - T)^2}}$$

which, by varying *w*, can be used to represent a wide spectrum of capability indices. Allowing *w* to take on various values permits *Cpw* to assume the format of a variety of indices including *Cp, Cpm, Cpk* and *Cpmk*. For example, setting *w*=0, results in

$$Cpw = \frac{USL - LSL}{6\sqrt{\sigma^2}} = Cp,$$

while for *w*=1, $Cpw = \dfrac{USL - LSL}{6\sqrt{\sigma^2 + (\mu - T)^2}} = Cpm$.

Defining $d = \frac{USL-LSL}{2}$, $a = \mu - \frac{USL+LSL}{2}$ and $p = \frac{|\mu - T|}{\sigma}$ then

$$w = \begin{cases} \left(\frac{d^2}{(d - |a|)^2} - 1\right)\frac{1}{p^2} & 0<p \\ 0 & \text{elsewhere} \end{cases}$$

allows *Cpw* = *Cpk*. Similarly

$$w = \begin{cases} \frac{k(2-k)}{(1-k)^2 p^2} & 0<k<1 \\ 0 & \text{elsewhere} \end{cases}$$

where $k = \frac{2|T-\mu|}{USL - LSL}$, allows *Cpw* = *Cpk**,

while $$w = \begin{cases} \left(\frac{d}{d-|a|}\right)^2\left(\frac{1}{p^2}+1\right)-\frac{1}{p^2}, & 0 < p \\ 0 & \text{elsewhere} \end{cases}$$

results in *Cpw* = *Cpmk*. Other weights associated with other common process capability indices include:

w	*Resulting Index*
$\begin{cases} \frac{6C_p - p}{(3C_p - p)^2 p}, & 0 < \frac{p}{3} < C_p \\ 0, & elsewhere \end{cases}$	$Cpk^* = (1-k)\frac{USL - LSL}{6\sigma}$
$\begin{cases} \left(\frac{d}{d-\|a\|}\right)^2\left(\frac{1}{p^2}+1\right)-\frac{1}{p^2}, & 0 < p \\ 0, & elsewhere \end{cases}$	$Cpmk = \frac{\min(USL-\mu, \mu-LSL)}{3\sqrt{\sigma^2+(\mu-T)^2}} = \frac{d-\|\mu-M\|}{3\sqrt{\sigma^2+(\mu-T)^2}}$
$\begin{cases} \left(\frac{d}{d-u\|a\|}\right)^2\left(\frac{1}{p^2}+v\right)-\frac{1}{p^2}, & 0 < p \\ 0, & elsewhere \end{cases}$	$Cp(u,v) = \frac{d-u\|\mu-M\|}{3\sqrt{\sigma^2+v(\mu-T)^2}}$

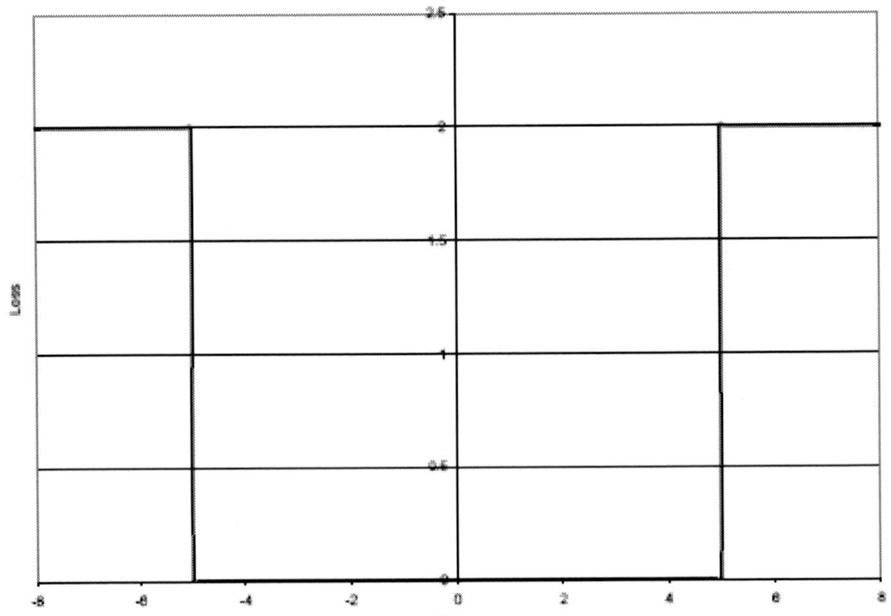

Figure 2. Square-Well Loss Function.

It is important to note that the universal assumptions associated with the unified index *Cpw* include the underlying process is stable and the process measurements are normally distributed. As well, none of these process assessments on their own provide information regarding the financial/costs performance associated with the process.

Process capability measures have traditionally been used to provide insights into the number (or proportion) of non-conforming product (i.e., yield). Practitioners cite a *Cp* value of one as representing 2700 parts per million (ppm) non-conforming, while 1.33 represents 63 ppm; 1.66 corresponds to .6 ppm; and 2 indicates < .1 ppm. *Cpk* has similar connotations, with a *Cpk* of 1.33 representing a maximum of 63 ppm non-conforming. Practitioners use the value of the process capability index and its associated number non-conforming to identify capable processes. A process with a *Cp* greater than or equal to one has traditionally been deemed capable. While a *Cp* of less than one indicates that the process is producing more than 2700 ppm non-conforming and used as an indication that the process is not capable of meeting customer requirements. In the case of *Cpk*, the auto industry frequently uses 1.33 as a benchmark in assessing the capability of a process.

In practice the magnitudes of *Cp, Cpl, Cpu, Cpk,* their generalized analogues and *Cpmk* are interpreted as a measure of non-conforming or process yield. Any change in the magnitude of these indices (holding the customer requirements constant) is due to changes in the distance between the specification limits and the process mean. By design *Cp, Cpl, Cpu, Cpk, Cpl*, Cpu*, Cpk** and *Cpmk* are used to identify changes in the amount of product beyond the specification limits, not proximity to the target. Inherent in any discussion of yield as a measure of process capability, is the assumption that product produced just inside the specification limit is of equal quality to that produced at the target and equivalent to assuming a square-well loss function for the quality variable (see Figure 2).

Process Capability: Yield (Television Process)

A crude assessment of process costs/losses is frequently fashioned by first assessing the process capability, converting the derived value into a process yield (i.e., ppm non-conforming) and then calculating the costs/losses by multiplying the yield by the cost associated with a complete loss. Consider an example from Taguchi, Elsayed and Hsiang (1989) where the maximum in-factory cost of repairing a failed television in the North American plant was known to be \$2 per unit. The maximum repair costs were incurred once the measurement of interest (color concentration) exceeded the tolerance of $\pm$ 5 units from its target value of 0. Assuming the process variable under investigation follows a N(μ_R =0, σ_R^2 = 4), *Cp* for the North American factory was

$$Cp = \frac{USL - LSL}{6\sqrt{\sigma^2 + (\mu - T)^2}} = \frac{5-(-5)}{6\sqrt{2^2 + (0-0)^2}} = 0.833.$$

A *Cp* of 0.8333 translates into 12,419 ppm beyond specifications. The repair costs associated with product beyond specifications was known to be \$2 per unit, suggesting a repair cost/loss of \$24,838 per million units or \$0.024838 per unit. This type of assessment assumes a) parts are either good or bad, b) the cost associated with good product is zero and a maximum (\$2) for bad and c) ppm are exact (i.e., not an estimate or an upper limit). Note also that the calculation of *Cp* and the resulting translation to ppm assumes μ and σ are known and not estimates. In practice these values are estimated and their stochastic behaviour needs to be reflected in any inference drawn from the estimates.

Loss Functions

In decision theory, loss functions are used to describe the deviation of an estimator from a parameter value. Loss functions traditionally take forms such as square-well loss (or 0-1 loss, see Figure 2), quadratic loss (i.e., $L(X) = w(X-T)^2$, see Figure 3), absolute error loss and weighted loss. Each of these forms tacitly assumes that the larger the error made in estimating parameter values the larger the loss incurred. Different levels of penalties are inherent to each form the loss function takes. Keeping these in mind, statisticians and practitioners make use of this concept to develop new applications in quality and reliability settings. This idea helps to stress the importance of being on target for both customers and suppliers. The use of loss functions has increased steadily in industrial applications.

Returning to the example from Taguchi, Elsayed and Hsiang (1989) where the maximum in-factory cost of repairing a failed television in the North American plant was known to be \$2 per unit and the maximum repair costs were incurred once the measurement of interest (color concentration) exceeded the tolerance of $\pm$ 5 units from its target value of 0. Again assuming the process variable under investigation follows a N(μ_R =0, σ_R^2 = 4), Figure 4 illustrates the implied square-well loss function associated with tolerances as well as the assumed distribution of the process variable.

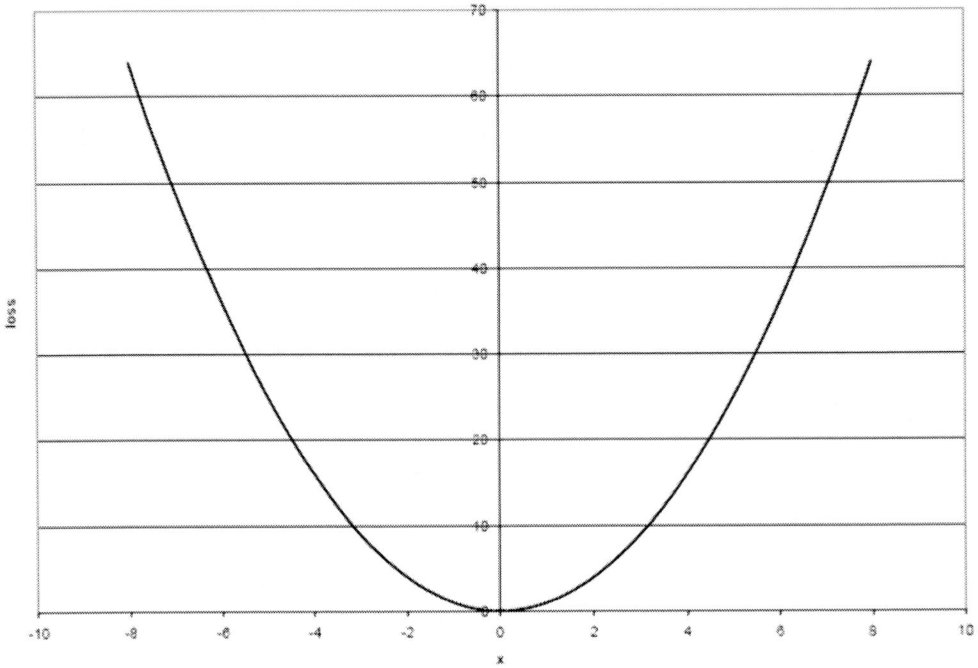

Figure 3. Quadratic Loss Function.

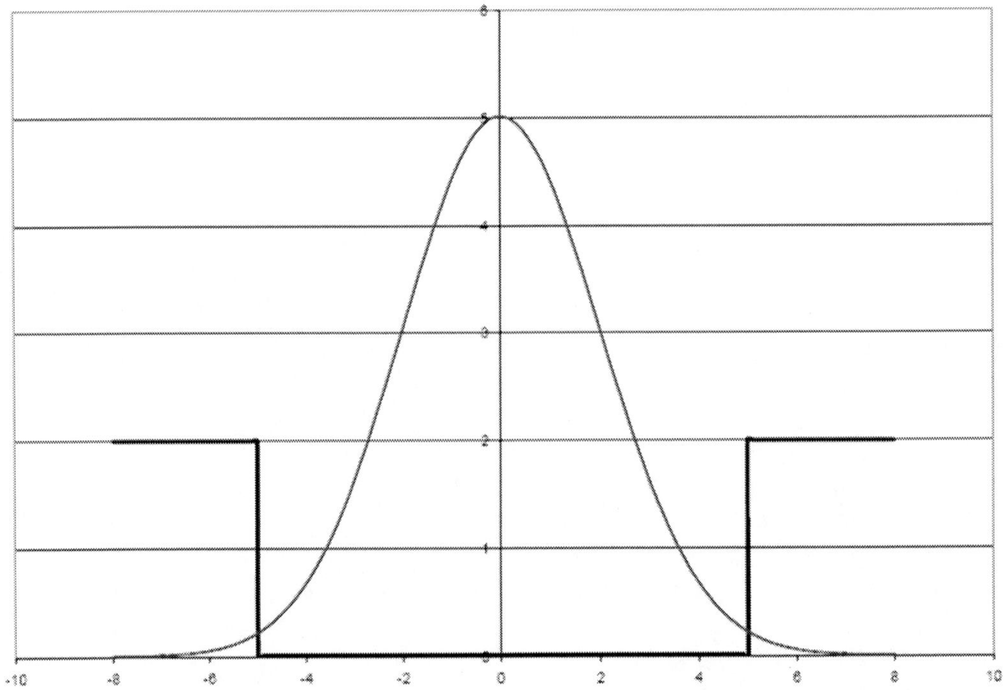

Figure 4. Loss Function and assumed distribution for example.

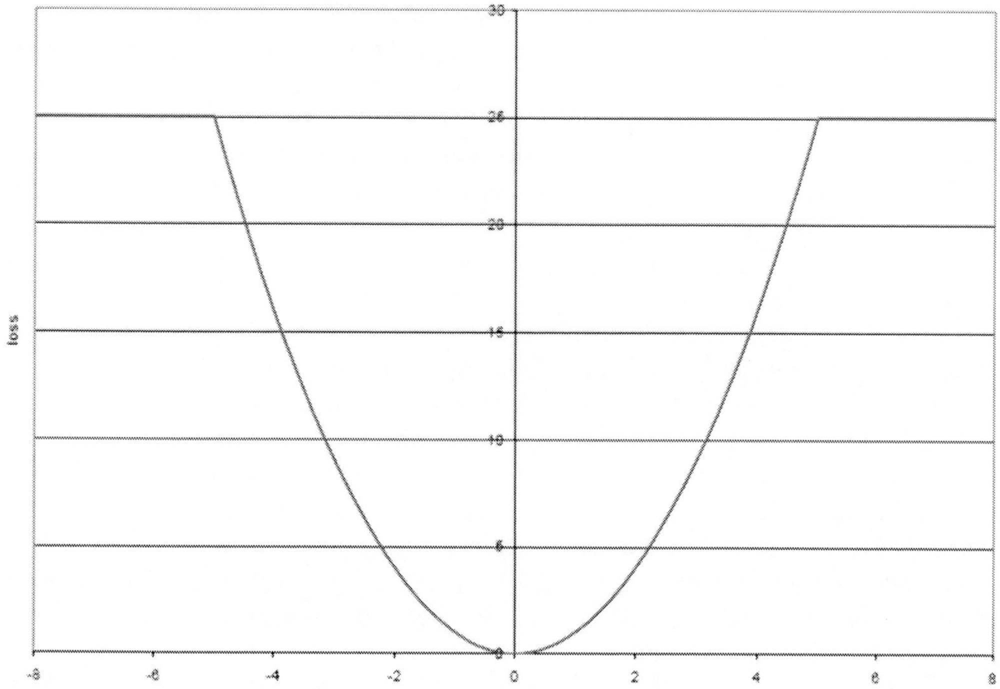

Figure 5. Modified Quadratic Loss Function.

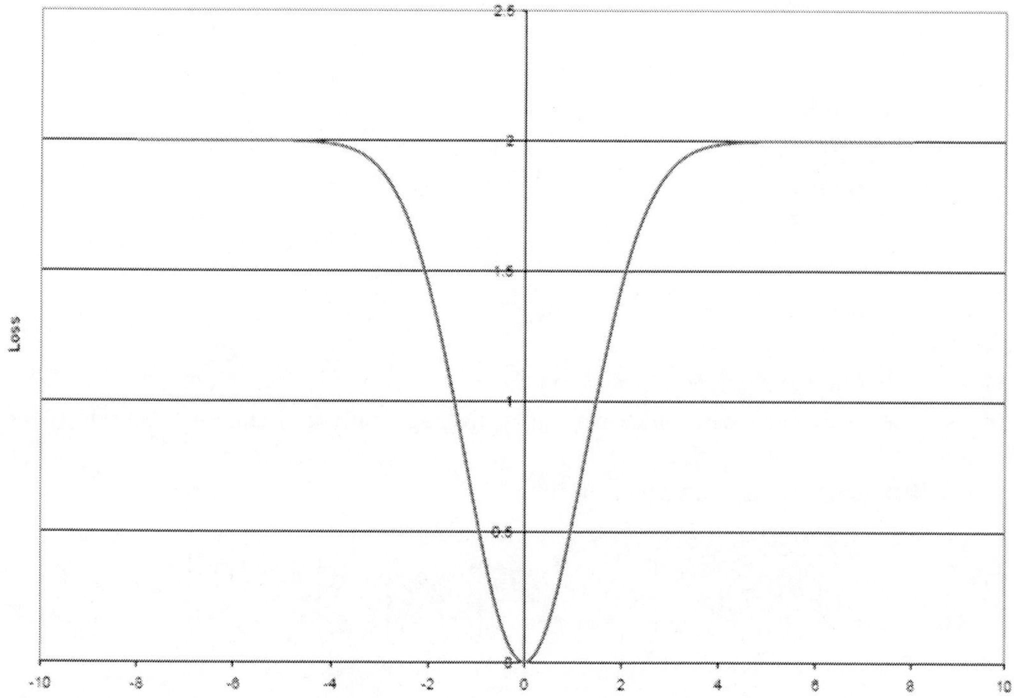

Figure 6. Inverted Normal Loss Function.

Loss functions have been studied for several decades and have been widely used for various purposes such as business decision making, quality assurance and reliability settings. Taguchi (1986) modified the quadratic loss function (see Figure 5) to illustrate the need to consider the target while assessing quality. Stressing that a loss in quality occurs as the process drifts away from the target and loss increases as the distance from target increases. He motivates the use of loss functions by suggesting that a product imparts no loss only if the product characteristic meets its target. Taguchi maintained that even small deviations from the target result in a loss of quality, and as the deviations from the target increase there are larger and larger losses in quality.

Spiring (1993) further modified this loss function approach by inverting the normal probability density function (INLF) (see Figure 6) bearing a unique maximum at the target value. It satisfies the usual loss function requirements and has varieties on the shape of the loss function that assesses loss accurately over a specified region, and extended to the case of asymmetric loss around the target. Subsequently, Sun, Laramee and Ramberg (1996) refined this loss function using least squares estimation. Spiring and Yeung (1998) have further extended this methodology to include an entire class of loss functions (IPLF) including gamma and Tukey's symmetric lambda distributions. Leung and Spiring (2002) developed the inverted beta loss function (IBLF) and associated properties, while Leung and Spiring (2004) developed properties for the general class of IPLF.

The general form of the IPLF is based on the inversion of common probability density functions. This family of loss functions satisfies the criteria that the loss must be non-negative, is zero worth at the target value, is monotonically increasing as the process drifts from either side of target and attains a quantifiable maximum near the lower and/or upper specification limits of the process. Let f(x) be a probability density function possessing a unique maximum at x=T (where T is the target value) and m be such that $m = \sup_{x \in X} f(x) = f(T)$. Defining $\pi(x,T) = f(x)$, the IPLF then takes the form

$$L(X,T) = K\left[1 - \frac{\pi(x,T)}{m}\right] \tag{1}$$

where K is the maximum loss incurred. K, $\pi(x,T)$ and T can then be chosen to represent various losses associated with processes under investigation (see Leung and Spiring (2004) for details).

The Risk function associated with an IPLF is

$$E[L(X,T)] = E\left\{K\left[1 - \frac{\pi(X,T)}{m}\right]\right\} = K\left\{1 - E\left[\frac{\pi(X,T)}{m}\right]\right\}, \tag{2}$$

while the variance associated with an IPLF is

$$V[L(X,T)] = K^2 \left[E\left\{ \left[\frac{\pi(X,T)}{m} \right]^2 \right\} - \left\{ E\left[\frac{\pi(X,T)}{m} \right] \right\}^2 \right]. \tag{3}$$

Different choices of IPLFs can reveal different levels of costs/penalties for similar deviations from a target. Similarly, different process characteristics (conjugate distributions) with suitable choice of IPLF can succinctly reflect the correct loss incurred by practitioners and hence to society. We introduce a monetary evaluation of the quality of products, assuming that the tolerances are correct and the process measurements are in-control. The IPLF class of loss function was developed to provide practitioners with a set of tools that could depict and represent actual process losses. The variance of the IPLF class of loss functions provides assessment of permissible error associated with the actual process losses. Potential applications include situations where a practitioner may wish to determine the optimal operating conditions for the process, to determine the average loss per unit produced, or to monitor the loss associated with the process all allowing for stochastic behaviour. In the subsequent examples, selected loss functions with statistical distributions associated with the process measurements are studied and compared in order to provide practitioners' with suitable choice of IPLF.

Loss Functions: Symmetric Loss (Television Process)

Again consider the example from Taguchi, Elsayed and Hsiang (1989) where the maximum in-factory cost of repairing a failed television in the North American plant was known to be \$2 per unit and the maximum repair costs were incurred once the measurement of interest (color concentration) exceeded the tolerance of $\pm$ 5 units from its target value of 0. Now assume only those televisions produced at the target have zero additional costs and that costs/losses follow an IPLF pattern with K = 2, T = 0 and σ = 1.25. The IPLF takes the functional form

$$L(X,T) = 2\left[1 - \exp\left(-\frac{x^2}{3.125}\right)\right]$$

in an attempt to reflect the actual costs associated with not being on target. Figure 7 compares the shapes of the squared-well loss function with the IPLF.

The IPLF provides a more conservative and potentially more realistic approach to representing the actual costs/losses associated with most process characteristic. Again assuming the process variable under investigation in the North American plant follows a N(μ_R =0, σ_R^2 = 4), the expected loss per unit (i.e., E[L(x, T)]) for the IPLF is \$0.94 and the standard error \$0.7024. This assessment a) allows repair costs to vary relative to the distance from the target value and b) does not assume μ and σ are known. In practice these values are estimated and their stochastic behaviour needs to be reflected in any inference drawn from the estimates.

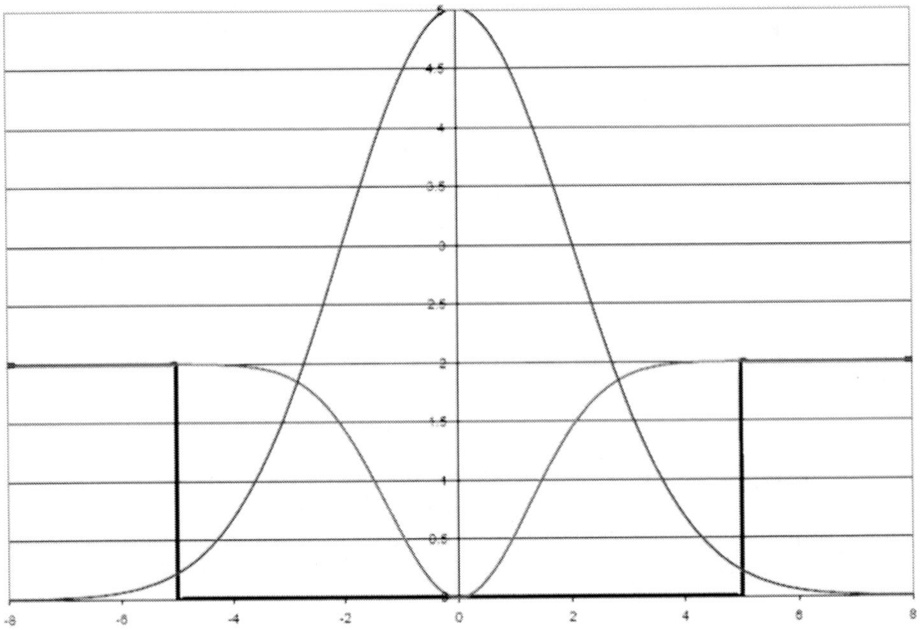

Figure 7. Inverted Normal & Square-well Loss Function with assumed distribution

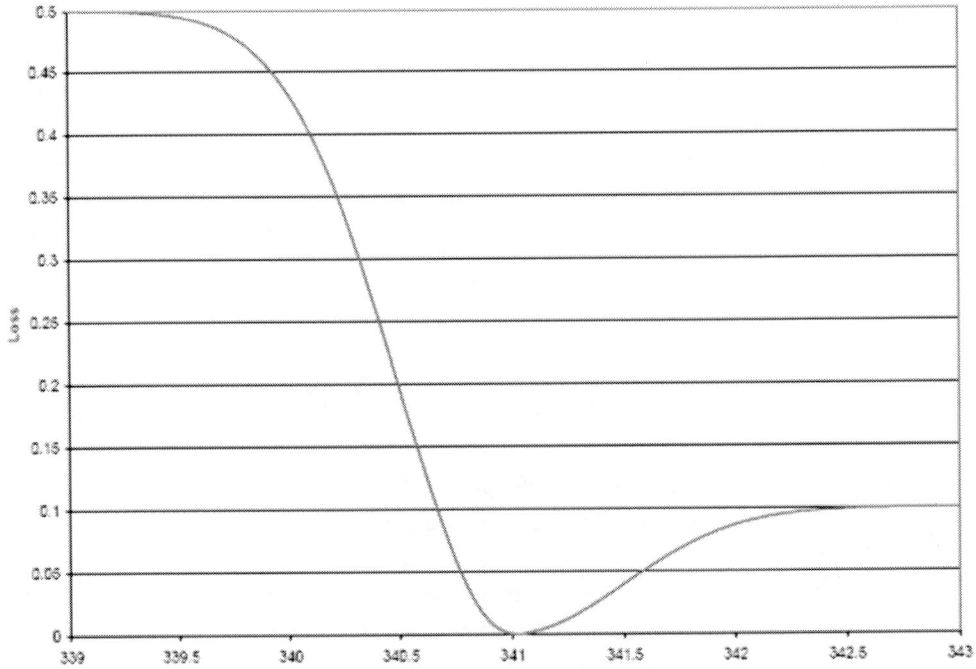

Figure 8. Asymetric Loss Function for Filling Process.

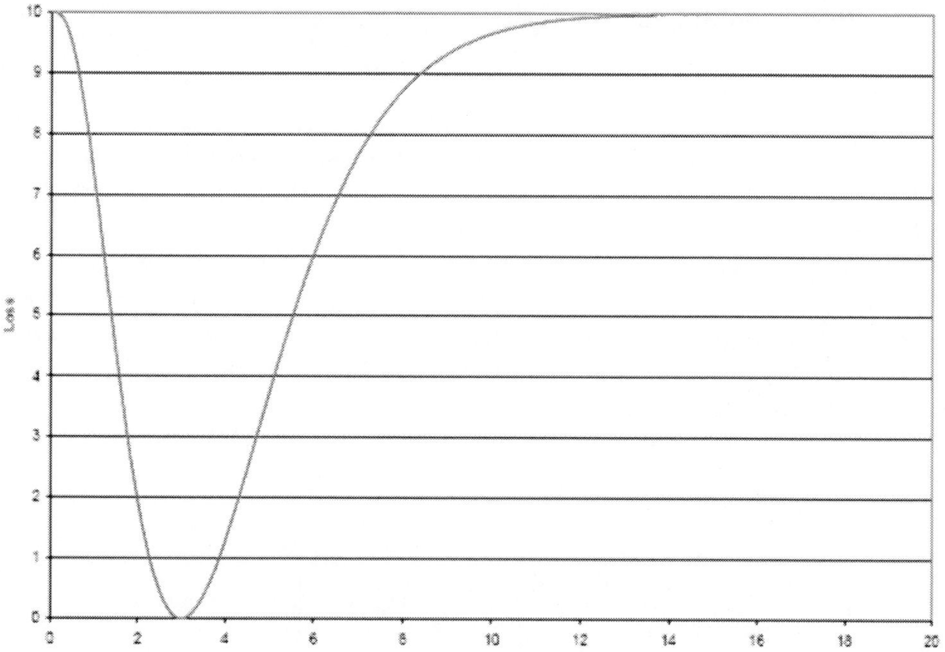

Figure 9. Loss Function for Registration Process.

Loss Functions: Asymmetric Loss (Filling Process)

Consider the process of filling bottles with a liquid, where each bottle has a target capacity of 341 ml. A bottle must be topped up if it is under the lower fill limit of 339 ml (set by governmental agency), while a bottle which has an amount equal or greater than the lower limit can be shipped directly to the market place. Since under fill is more serious than overfill in terms of loss to the producer, the economic loss around the target is asymmetric. Under regular operating conditions, data collected from the process suggest that fill volume, the characteristic of interest, follows a normal distribution with σ_R =0.5 ml with an adjustable mean selected to minimize the average loss (which turns out to be μ_R = 342 ml). With T = 341 ml and an IPLF of the form (see Figure 8)

$$L(x,T)=\begin{cases} 0.5\left\{1-\exp\left[-2(x-341)^2\right]\right\}, & 0<x<341; \\ 0.1\left\{1-\exp\left[-2(x-341)^2\right]\right\}, & 341<x<\infty. \end{cases}$$

The expected loss is \$0.09 and associated standard errors are \$0.0149 on the left-hand-side and \$0.0277 on the right-hand-side of the target.

Loss Functions: Monitoring Loss (Printing Process)

The characteristic of interest was the registration of two images. One image was laid down at station one, while the other at station twelve of the process. Registration was defined to be the distance between the images. Subgroups of four sheets were sampled regularly from the process and the distance between the images measured (in tenths of an inch) for each sheet. Fifteen subgroups of size four were used in the analysis.

If the images touch or overlap, the registration was set to zero. All sheets with a registration value of zero are discarded as scrap (i.e., maximum loss). The target registration is 3, with registrations greater than 10 also treated as scrap. Price reductions were negotiated based on the appearance of the printed sheets, with registration being the critical characteristic. After lengthy discussion, a loss function based on f(x, T) from the gamma family was agreed upon. The loss function used in this case, with K=10, T = 3 and α = 4, was (see Figure 9)

$$L(x,T) = 10\left\{1 - \left[\frac{x \exp\left(1 - \frac{x}{3}\right)}{3}\right]^3\right\}$$

where the maximum loss was \$0.10 per sheet.

On fitting the measurements from the printing process, a gamma distribution with α_R = 4.6557, β_R = 1.0732 is verified (Kolmogorov-Smirnov test, K = 0.1044, p-value = 0.5076; and Anderson-Darling test, A^2 = 0.8738, p-value = 0.4749). From equations [2] and [3], the expected value of the loss function is \$0.0393 with standard errors of \$0.011134 on the left-hand-side of the target and \$0.033633 on the right-hand-side of target.

Integrating Capability with Loss

Similar to the Johnson's (1992) development for *Cpm, Cpw* can be expressed as a function of deviations from the target. Consider a general loss function where the loss is zero when the process is on target and depicted as a weighted (w) squared deviation from target

$$L(X) = w(X-T)^2.$$

The Risk (i.e., average loss) function associated with this loss function is

$$\begin{aligned} E[L(X)] &= E\left[\ w(X-T)^2\ \right] \\ &= E\left[\ w(X-\mu+\mu-T)^2\ \right] \\ &= w\sigma^2 + w(\mu-T)^2 \\ &= (w-1)\sigma^2 + \sigma^2 + w(\mu-T)^2 \end{aligned}$$

where w is considered non-stochastic. Further, defining $E[L_w(X)]$ to be

$$E[L_W(X)] = \sigma^2 + w(\mu - T)^2,$$

and rewriting in the form

$$E[L_w(X)] = E[L(X)] + (1-w)\sigma^2,$$

$E[L_w(X)]$ is the equivalent of $E[\hat{L}_W(X)]$, where $\hat{L}_W(X) = \frac{n-w}{n-1}\hat{\sigma}^2 + w(\overline{X}-T)^2$ and $\hat{\sigma}^2 = \frac{\sum_{i=1}^{n}(X_i - \overline{X})^2}{n}$.

$\hat{L}_W(X)$ is a linear combination in w that incorporates the variability of the process measurements and average off-targetness in its determination. $\hat{L}_W(X)$ is an unbiased estimator of the loss function parameters for any w (i.e., $E[L_W(X)]$, for finite mean μ and variance σ^2). Further, if $X \sim N(\mu, \sigma^2)$, $\hat{L}_W(X)$ is a function of jointly complete sufficient statistics, hence it is a uniformly minimum variance unbiased estimator (UMVUE) for $E[L_W(X)]$.

Rewriting *Cpw* in terms of $E[L_W(X)]$

$$Cpw = \frac{USL - LSL}{6\sqrt{E[L_W(X)]}}.$$

It follows that $\hat{Cpw}$ can be written as

$$\hat{Cpw} = \frac{USL - LSL}{6\sqrt{\hat{L}_W(X)}}.$$

If $X_1, X_2, \ldots, X_n$ is a random sample from $N(\mu, \sigma^2)$, then

$$(\overline{X}-T)^2 \sim \frac{\sigma^2}{n}\chi^2_{1,\lambda}$$

$$w\left(\overline{X}-T\right)^2 \sim \frac{w\sigma^2}{n}\chi^2_{1,\lambda},$$

$$\hat{\sigma}^2 \sim \frac{\sigma^2}{n}\chi^2_{n-1},$$

$$\frac{n-w}{n-1}\hat{\sigma}^2 \sim \frac{n-w}{n-1}\frac{\sigma^2}{n}\chi^2_{n-1}$$

and $\overline{X}$, $\hat{\sigma}^2$ are independent, where $\lambda = \dfrac{n(\mu-T)^2}{\sigma^2}$ is the non-centrality parameter.

Analogous to the development in Spiring (1997), $Q^2_{n,\lambda}$, where

$$Q^2_{n,\lambda} = \frac{n}{\sigma^2}\left[\frac{n-w}{n-1}\hat{\sigma}^2 + w\left(\overline{X}-T\right)^2\right],$$

is a linear combination of two independent chi-square distributions, that is

$$Q^2_{n,\lambda} \sim \frac{n-w}{n-1}\chi^2_{n-1} + w\chi^2_{1,\lambda}.$$

The distribution of $Q^2_{n,\lambda}(x)$ is defined if both $\frac{n-w}{n-1}$ and w are greater than 1. Hence it is reasonable to define

$$w = \begin{cases} 1+w^*, & 0 < w \le 1 \\ w & 1 < w < n-1 \\ 1+w^* & n-1 \le w \end{cases}$$

where $w^* = \dfrac{n}{1+(n-1)(1+2\lambda)}$ is chosen such that the variance of $\hat{L}_W(X)$ would be minimized. If w = 0, then the distribution of $\hat{L}_W(X)$ is χ^2_{n-1} while when w = 1 the distribution is $\chi^2_{n,\lambda}$. Weights can be selected or determined using existing process information. In those cases where w > n, the distribution cannot be determined and hence of limited use. Moreover it can be approximated by taking w = 1 + w*, where w* is the optimal value that minimizing the variance of $\hat{L}_W(X)$.

Denoting $Q^2_{n,\lambda}(x)$ as the cumulative distribution function (cdf) associated with $Q^2_{n,\lambda}$, Press (1966) showed that the $Q^2_{n,\lambda}$ can be expressed as a mixture of central chi-square distribution with general form

$$Q^2_{n,\lambda}(x) = \sum_{j=0}^{\infty} d_j \chi^2_{n+2j}(x)$$

with the d_j's being the weights such that $\sum_{j=0}^{\infty} d_j = 1$, where the d_j's are the functions of the degrees of freedom (i.e., n - 1 and 1), the non-centrality parameter λ, and the weight function w. The functional form of the d_j's are given in Press (1966), which for the general $Q^2_{n,\lambda}$, are as follows:

$$d_0 = \left(\frac{w(n-1)}{n-w}\right)^{-\frac{1}{2}} \exp\left(-\frac{\lambda}{2}\right)$$

$$d_i = \sum_{j=0}^{i} \sum_{k=0}^{j} \frac{\exp\left(-\frac{\lambda}{2}\right)\left(\frac{\lambda}{2}\right)^{j-k}}{(j-k)!} \left(\frac{w(n-1)}{n-w}\right)^{-\frac{1}{2}} \left(1-\frac{n-w}{w(n-1)}\right)^{i-j}$$
$$\times \frac{\Gamma\left(i-j+\frac{1}{2}\right)}{\Gamma(i-j+1)\Gamma\left(\frac{1}{2}\right)} \left(\frac{w(n-1)}{n-w}\right)^{k-j} \left(1-\frac{n-w}{w(n-1)}\right)^{k} \binom{j-1}{k}$$
$$= \sum_{j=0}^{i} \sum_{k=0}^{j} \frac{\exp\left(-\frac{\lambda}{2}\right)\left(\frac{\lambda}{2}\right)^{j-k}\binom{j-1}{k}}{(j-k)!} \frac{\Gamma\left(i-j+\frac{1}{2}\right)}{\Gamma(i-j+1)\Gamma\left(\frac{1}{2}\right)} \left(\frac{w(n-1)}{n-w}\right)^{k-j-\frac{1}{2}} \left(\frac{n(w-1)}{w(n-1)}\right)^{i-j+k}$$

$i = 1, 2, 3, \ldots, \infty$

where λ denotes the value of the non-centrality parameter and w the value of the weight function. See APPENDIX A for *Mathematica* (Wolfram (1999)) programs used to assist in determining the d_i's ($i \geq 1$) and to approximate the value of $Q^2_{n,\lambda}(x)$.

A (1 - α) 100% confidence interval for *Cpw* can be constructed as follows:

$$P\left[Q^2_{n,\lambda;1-\frac{\alpha}{2}} \leq \frac{n}{\sigma^2}\left[\frac{n-w}{n-1}\hat{\sigma}^2 + w(\overline{X}-T)^2\right] \leq Q^2_{n,\lambda;\frac{\alpha}{2}}\right] = 1-\alpha$$

$$P\left[\sqrt{Q^2_{n,\lambda;\,1-\frac{\alpha}{2}}} \le \sqrt{\frac{n}{\sigma^2}\left[\frac{n-w}{n-1}\hat{\sigma}^2 + w\left(\overline{X}-T\right)^2\right]} \le \sqrt{Q^2_{n,\lambda;\,\frac{\alpha}{2}}}\right] = 1-\alpha$$

$$P\left[\frac{USL-LSL}{6\sigma\sqrt{Q^2_{n,\lambda;\,\frac{\alpha}{2}}}} \le \frac{USL-LSL}{6\sigma\sqrt{\frac{n}{\sigma^2}\left[\frac{n-w}{n-1}\hat{\sigma}^2 + w\left(\overline{X}-T\right)^2\right]}} \le \frac{USL-LSL}{6\sigma\sqrt{Q^2_{n,\lambda;\,1-\frac{\alpha}{2}}}}\right] = 1-\alpha .$$

For $Cpw = \frac{USL-LSL}{6\sigma\sqrt{1+\frac{\lambda w}{n}}}$ it follows that $\sqrt{1+\frac{\lambda w}{n}}\,Cpw = \frac{USL-LSL}{6\sigma}$,

then

$$P\left[\sqrt{\frac{n\left(1+\frac{\lambda w}{n}\right)}{Q^2_{n,\lambda;\,\frac{\alpha}{2}}}}\,Cpw \le \hat{C}pw \le \sqrt{\frac{n\left(1+\frac{\lambda w}{n}\right)}{Q^2_{n,\lambda;\,1-\frac{\alpha}{2}}}}\,Cpw\right] = 1-\alpha$$

$$P\left[\sqrt{\frac{Q^2_{n,\lambda;1-\frac{\alpha}{2}}}{n+\lambda w}}\,\hat{C}pw \le Cpw \le \sqrt{\frac{Q^2_{n,\lambda;\frac{\alpha}{2}}}{n+\lambda w}}\,\hat{C}pw\right] = 1-\alpha$$

where $P\left[\;Q^2_{n,\lambda} > Q^2_{n,\lambda;\alpha}\right] = \alpha$. A similar development for *Cpw** results in a (1 - α) 100% confidence interval of the form

$$P\left[\sqrt{\frac{Q^2_{n,\lambda;1-\frac{\alpha}{2}}}{n+\lambda w}}\,\hat{C}pw^* \le Cpw^* \le \sqrt{\frac{Q^2_{n,\lambda;\frac{\alpha}{2}}}{n+\lambda w}}\,\hat{C}pw^*\right] = 1-\alpha .$$

As a decision maker who may be interested in an upper limit on the loss from the process rather than just a point estimate of the loss, $\hat{L}_W(X)$, the upper (1 - α) 100% confidence limit for the loss function parameter $E[L_W(X)]$ is the same for the loss associated with $\hat{C}pw$ and

$\hat{C}pw^*$. An upper confidence limit for the loss function parameter can be found by considering the ratio

$$\frac{\hat{L}_W(X)}{E[L_W(X)]}=\frac{\left[\frac{n-w}{n-1}\hat{\sigma}^2+w(\overline{X}-T)^2\right]}{\sigma^2+w(\mu-T)^2}\cdot\frac{\frac{n}{\sigma^2}}{\frac{n}{\sigma^2}}=\frac{Q^2_{n,\lambda}}{n+\lambda w}.$$

The distribution of $\frac{n+\lambda w}{E[L_W(X)]}\hat{L}_W(X)\sim Q^2_{n,\lambda}$ and

$$P\left[\ Q^2_{n,\lambda}\geq Q^2_{n,\lambda;1-\alpha}\ \right]=1-\alpha$$

$$P\left[\ \frac{n+\lambda w}{E[L_W(X)]}\hat{L}_W(X)\geq Q^2_{n,\lambda;1-\alpha}\ \right]=1-\alpha$$

$$P\left[\ E[L_W(X)]\ \leq\ \frac{n+\lambda w}{Q^2_{n,\lambda;1-\alpha}}\hat{L}_W(X)\right]=1-\alpha.$$

Therefore, $\left[0,\frac{n+\lambda w}{Q^2_{n,\lambda;1-\alpha}}\hat{L}_W(X)\right]$ is an upper (1 - α) 100% confidence limit for the loss function parameter, $E[L_W(X)]$. This is an exact upper confidence limit of $E[L_W(X)]$ provided tables of $Q^2_{n,\lambda}$ are available. Otherwise, approximation of $Q^2_{n,\lambda}$ by means of a scaled chi-square is an alternative to the solution. Applying the classical Patnaik (1949) approximation by matching the first two moments of a scaled chi-square of the form $c\chi^2_v$, where the constants c and v are determined by equating the means and variances of the two distributions, i.e., to solve the equations

$$\frac{n-w}{n-1}(n-1)+w(1+\lambda)=cv\quad 2\left(\frac{n-w}{n-1}\right)^2(n-1)+2w^2(1+2\lambda)=2c^2v$$

$$c=\frac{\frac{(n-w)^2}{n-1}+w^2(1+2\lambda)}{n+\lambda w}=\frac{(n-w)^2+w^2(n-1)(1+2\lambda)}{(n-1)(n+\lambda w)},$$

and $$\nu = \frac{(n+\lambda w)^2}{\frac{(n-w)^2}{n-1} + w^2(1+2\lambda)} = \frac{(n-1)(n+\lambda w)^2}{(n-w)^2 + w^2(n-1)(1+2\lambda)}.$$

So that $$Q^2_{n,\lambda} = \frac{n-w}{n-1}\chi^2_{n-1} + w\chi^2_{1,\lambda} \approx c\chi^2_\nu$$

$$\frac{n+\lambda w}{Q^2_{n,\lambda}} \approx \frac{n+\lambda w}{c\chi^2_\nu} = \frac{\nu}{\chi^2_\nu}$$

and results in an approximate upper (1 - α) 100% confidence limit for the loss function parameters,

$$P\left[E[L_W(X)] \le \frac{\hat{\nu}}{\chi^2_{\hat{\nu};1-\alpha}} \hat{L}_W(X) \right] = 1-\alpha.$$

Now consider the special cases when w is assigned with some specified values. For w = 0

$$Cpw = \frac{USL-LSL}{6\sigma} = Cp$$

with (1 - α) 100% confidence interval

$$P\left[\sqrt{\frac{\chi^2_{n-1;1-\frac{\alpha}{2}}}{n-1}}\,\hat{C}p \le Cp \le \sqrt{\frac{\chi^2_{n-1;\frac{\alpha}{2}}}{n-1}}\,\hat{C}p \right] = 1-\alpha.$$

Then the loss function ratio becomes

$$\frac{\hat{L}_W(X)}{E[L_W(X)]} = \frac{\hat{L}_0(X)}{E[L_0(X)]} = \frac{\left[\frac{n-0}{n-1}\hat{\sigma}^2 + 0\cdot(\overline{X}-T)^2\right]}{\sigma^2 + 0\cdot(\overline{X}-T)^2} = \frac{1}{n-1}\chi^2_{n-1},$$

$$\frac{(n-1)\hat{L}_0(X)}{E[L_0(X)]} \sim \chi^2_{n-1}.$$

The upper (1 - α) 100% confidence limit for the loss function parameter is

$$P\left[\frac{(n-1)\hat{L}_0(X)}{E[L_0(X)]} \geq \chi^2_{n-1;1-\alpha}\right] = 1-\alpha$$

$$P\left[E[L_0(X)] \leq \frac{(n-1)}{\chi^2_{n-1;1-\alpha}}\hat{L}_0(X)\right] = 1-\alpha .$$

Therefore $\left[0, \frac{(n-1)}{\chi^2_{n-1;1-\alpha}}\hat{L}_0(X)\right]$ is an upper (1 - α) 100% confidence limit for the loss function parameter $E[L_0(X)]$.

If we define $Cp^* = \frac{\min[USL-T, T-LSL]}{3\sigma}$, then $Cpw^* = Cp^*$ and the (1 - α) 100% confidence interval is $P\left[\sqrt{\frac{\chi^2_{n-1;1-\frac{\alpha}{2}}}{n-1}}\hat{C}p^* \leq Cp^* \leq \sqrt{\frac{\chi^2_{n-1;\frac{\alpha}{2}}}{n-1}}\hat{C}p^*\right] = 1-\alpha$. The upper (1 - α) 100% confidence limit of the loss associated with *Cp** is given by

$$P\left[E[L_0(X)] \leq \frac{(n-1)}{\chi^2_{n-1;1-\alpha}}\hat{L}_0(X)\right] = 1-\alpha .$$

For w = 1, *Cpw* = *Cpm and Cpw** = *Cpm**, the confidence intervals are respectively

$$P\left[\sqrt{\frac{\chi^2_{n,\lambda;1-\frac{\alpha}{2}}}{n+\lambda}}\hat{C}pm \leq Cpm \leq \sqrt{\frac{\chi^2_{n,\lambda;\frac{\alpha}{2}}}{n+\lambda}}\hat{C}pm\right] = 1-\alpha ,$$

$$P\left[\sqrt{\frac{\chi^2_{n,\lambda;1-\frac{\alpha}{2}}}{n+\lambda}}\hat{C}pm^* \leq Cpm^* \leq \sqrt{\frac{\chi^2_{n,\lambda;\frac{\alpha}{2}}}{n+\lambda}}\hat{C}pm^*\right] = 1-\alpha .$$

The ratio

$$\frac{\hat{L}_1(X)}{E[L_1(X)]} = \frac{\left[\hat{\sigma}^2 + (\overline{X}-T)^2\right]\frac{n}{\sigma^2}}{\left[\sigma^2 + (\mu-T)^2\right]\frac{n}{\sigma^2}} = \frac{\frac{n}{\sigma^2}\left[\hat{\sigma}^2 + (\overline{X}-T)^2\right]}{n+\lambda}$$

so that $$\frac{n+\lambda}{E[L_1(X)]}\hat{L}_1(X) \sim \chi^2_{n,\lambda}.$$

Then an upper (1 - α) 100% confidence limit for the loss function parameters, $E[L_1(X)]$ is

$$P\left(\frac{n+\lambda}{E[L_1(X)]}\hat{L}_1(X) \geq \chi^2_{n,\lambda;1-\alpha}\right) = 1-\alpha$$

or $$P\left(E[L_1(X)] \leq \frac{(n+\lambda)}{\chi^2_{n,\lambda;1-\alpha}}\hat{L}_1(X)\right) = 1-\alpha.$$

For w = 1, the values of c and ν of the scaled chi-square $c\chi^2_\nu$ become

$$c=\frac{n+2\lambda}{n+\lambda} \quad \nu=\frac{(n+\lambda)^2}{n+2\lambda}$$

hence $\chi^2_{n,\lambda} \approx c\chi^2_\nu$ implies $\frac{n+\lambda}{\chi^2_{n,\lambda}} \approx \frac{n+\lambda}{c\chi^2_\nu} = \frac{\nu}{\chi^2_\nu}$ and results in an approximate upper (1 - α) 100% confidence limit for the loss function parameter $E[L_1(X)]$ of the form

$$P\left(E[L_1(X)] \leq \frac{\hat{\nu}}{\chi^2_{\hat{\nu};1-\alpha}}\hat{L}_1(X)\right) = 1-\alpha.$$

The approximate (1 - α) 100% confidence intervals of *Cpm and Cpm** are respectively

$$P\left[\sqrt{\frac{\chi^2_{\hat{\nu};1-\frac{\alpha}{2}}}{\hat{\nu}}}\,\hat{Cpm} \leq Cpm \leq \sqrt{\frac{\chi^2_{\hat{\nu};\frac{\alpha}{2}}}{\hat{\nu}}}\,\hat{Cpm}\right] = 1-\alpha,$$

$$P\left[\sqrt{\frac{\chi^2_{\hat{\nu};1-\frac{\alpha}{2}}}{\hat{\nu}}}\,\hat{Cpm}^* \leq Cpm^* \leq \sqrt{\frac{\chi^2_{\hat{\nu};\frac{\alpha}{2}}}{\hat{\nu}}}\,\hat{Cpm}^*\right] = 1-\alpha.$$

Defining $d = \dfrac{USL - LSL}{2}$, $a = \mu - \dfrac{USL + LSL}{2}$ and $p = \dfrac{|\mu - T|}{\sigma}$ while setting $w = \left[\left(\dfrac{d}{d-|a|}\right)^2 - 1\right]\dfrac{1}{p^2}$, then $1 + \dfrac{\lambda w}{n} = \left(\dfrac{d}{d-|a|}\right)^2$ allows *Cpw* = *Cpk*. The (1 - α) 100% confidence interval becomes

$$P\left[\sqrt{\frac{Q^2_{n,\lambda;1-\frac{\alpha}{2}}}{n\left[\frac{d}{d-|a|}\right]^2}}\,\hat{Cpk} \le Cpk \le \sqrt{\frac{Q^2_{n,\lambda;\frac{\alpha}{2}}}{n\left[\frac{d}{d-|a|}\right]^2}}\,\hat{Cpk}\right] = 1 - \alpha .$$

An approximate (1 - α) 100% confidence interval of *Cpk* is given by

$$P\left[\sqrt{\frac{\chi^2_{\hat{v};1-\frac{\alpha}{2}}}{\hat{v}}}\,\hat{Cpk} \le Cpk \le \sqrt{\frac{\chi^2_{\hat{v};\frac{\alpha}{2}}}{\hat{v}}}\,\hat{Cpk}\right] = 1 - \alpha .$$

An exact upper (1 - α) 100% confidence limit for the loss function parameters $E[L_W(X)]$ is

$$P\left[E[L_W(X)] \le \frac{n\left[\frac{d}{d-|a|}\right]^2}{Q^2_{n,\lambda;1-\alpha}}\hat{L}_W(X)\right] = 1 - \alpha .$$

With w defined above, an approximate upper (1 - α) 100% confidence limit for the loss function parameters can be constructed as

$$P\left[E[L_W(X)] \le \frac{\hat{v}}{\chi^2_{\hat{v};1-\alpha}}\hat{L}_W(X)\right] = 1 - \alpha ,$$

where $\hat{v}$ is $\dfrac{(n-1)(n+\lambda w)^2}{(n-w)^2 + w^2(n-1)(1+2\lambda)}$.

Proceeding similarly, allow w = $\frac{k(2-k)}{(1-k)^2 p^2}$, 0 < k < 1, (or w = $\frac{6Cp-p}{(3Cp-p)^2 p}$, with k = $\frac{p}{3Cp}$) with 1 + $\frac{\lambda w}{n}$ = 1 + wp^2, then *Cpw* = *Cpk**. A (1 - α) 100% exact and approximate confidence interval and for *Cpk** are respectively

$$P\left[\sqrt{\frac{Q^2_{n,\lambda;1-\frac{\alpha}{2}}}{n(1+wp^2)}}\,\hat{C}pk^* \le Cpk^* \le \sqrt{\frac{Q^2_{n,\lambda;\frac{\alpha}{2}}}{n(1+wp^2)}}\,\hat{C}pk^*\right] = 1-\alpha,$$

$$P\left[\sqrt{\frac{\chi^2_{\hat{v};1-\frac{\alpha}{2}}}{\hat{v}}}\,\hat{C}pk^* \le Cpk^* \le \sqrt{\frac{\chi^2_{\hat{v};\frac{\alpha}{2}}}{\hat{v}}}\,\hat{C}pk^*\right] = 1-\alpha.$$

An exact upper (1 - α) 100% confidence limit for the loss function parameters is

$$P\left[\;E[L_W(X)] \le \frac{n(1+wp^2)}{Q^2_{n,\lambda;1-\alpha}}\hat{L}_W(X)\right] = 1-\alpha.$$

For w = $\frac{k(2-k)}{(1-k)^2 p^2}$, an approximate upper (1 - α) 100% confidence limit for the loss function parameters can be constructed as

$$P\left[\;E[L_W(X)] \le \frac{\hat{v}}{\chi^2_{\hat{v};1-\alpha}}\hat{L}_W(X)\right] = 1-\alpha.$$

Now, define the expected weighed Taguchi loss function as follows

$$E[L_{WT}(X)] = K\,E[L_W(X)] = K\left[\;\sigma^2 + w(\mu-T)^2\right]$$

where K is the monetary loss for the process and w is the additional penalty of off-targetness.
An unbiased estimator of $E[L_{WT}(X)]$ is

$$\hat{L}_{WT}(X) = K\hat{L}_W(X) = K\left[\frac{n-w}{n-1}\hat{\sigma}^2 + w(\overline{X}-T)^2\right].$$

If $X \sim N(\mu, \sigma^2)$, then the distribution of $\hat{L}_{WT}(X)$ is proportional to (a multiple of K, the maximum monetary loss per unit). Therefore the (1 - α) 100% confidence limits for $E[L_{WT}(X)]$ associated with each PCI used is also a multiple by K.

Integrating Capability with Loss: Capability Analysis (Blow Moulding Process)

Consider the process capability studies from Tarver (1986). The process under scrutiny is a blow-molding procedure for plastic bottles where the quality characteristic of interest is the outside lip diameter measured in inches. A sample of 100 observations obtained from the process with the following process statistics summarized in Table 1.

Further assume that there is a 50 cents loss including cost, scrap, rework, etc. for the blow-molding process, i.e., K = \$0.50. The 95% confidence interval estimates for the PCIs and the associated weighed losses are summarized in Tables 3 and 4. The critical values used in the calculations are listed in Table 2.

Table 1. Process Statistics

n	100	$\hat{k}$	0.2875
$\overline{X}$	0.8254	$\hat{d}$	0.016
$\hat{\sigma}$	0.005	$\lvert\hat{a}\rvert$	0.0046
X_{max}	0.838	$\hat{p}$	0.920
X_{min}	0.814	$\hat{\lambda}$	84.64≅85
USL	0.846	$\hat{C}p=\hat{C}p^*$	1.067
LSL	0.814	$\hat{C}pm=\hat{C}pm^*$	0.785
T	0.830	$\hat{C}pk=\hat{C}pk^*$	0.760

Table 2. Critical values of $P(X>c_\alpha)=\alpha$

α	χ^2_{99}	χ^2_{121}	χ^2_{127}	$\chi^2_{100,85}$	$Q^2_{100,85}$
0.975	73.361	92.446	97.698	142.004	160.302
0.950	77.046	96.598	101.971	148.329	165.805
0.025	128.422	153.338	160.086	232.978	238.032

Table 3. 95% Confidence Intervals of Process Capability Indices

w	$\hat{C}pw$	Exact	Approximate
0	$\hat{C}p=\hat{C}p^{*}=1.067$	(0.9185, 1.2153)	N.A.
1	$\hat{C}pm=\hat{C}pm^{*}=0.785$	(0.6878, 0.8809)	(0.6885, 0.8813)
#	$\hat{C}pk=\hat{C}pk^{*}=0.760$	(0.6856, 0.8354)	(0.6643, 0.8556)

$$\# \ \hat{w} = \frac{\hat{k}(2-\hat{k})}{(1-\hat{k})^2\,\hat{p}^2} = \frac{6\hat{C}p-\hat{p}}{(3\hat{C}p-\hat{p})^2\,\hat{p}} = \left[\left(\frac{\hat{d}}{\hat{d}-|\hat{a}|}\right)^2 - 1\right]\frac{1}{\hat{p}^2} = 1.1458$$

Table 4. 95% Upper Confidence Limit of $E[L_{WT}(X)]$

w	$\hat{L}_{WT}(X)$	Exact	Approximate
0	1.2626×10^{-5}	$(0, 1.6225\times10^{-5})$	N.A.
1	2.3080×10^{-5}	$(0, 2.8786\times10^{-5})$	$(0, 2.8745\times10^{-5})$
#	2.4605×10^{-5}	$(0, 2.9231\times10^{-5})$	$(0, 3.0820\times10^{-5})$

Table 5. Changes in Process Statistics when T = 0.825

T	0.825
$\hat{k}$	0.025
$\hat{p}$	0.080
$\hat{\lambda}$	$0.64 \cong 1$
$\hat{C}p^{*}$	0.733
$\hat{C}pm^{*}$	0.731
$\hat{C}pk^{*}$	1.040

Table 6. Critical values of $P(X > c_\alpha) = \alpha$

X α	χ^2_{41}	χ^2_{100}	$\chi^2_{100,1}$	$Q^2_{100,1}$
0.975	25.215	74.222	74.965	85.844
0.950	27.326	77.929	78.710	89.826
0.025	60.561	129.561	130.855	144.608

Table 7a. 95% Confidence Intervals of *Cpw*

w	$\hat{Cpw}$	Exact	Approximate
0	$\hat{Cp} = 1.067$	(0.9185, 1.2153)	N.A.
1	$\hat{Cpm} = 0.785$	(0.6763, 0.8919)	(0.6763, 0.8935)
#	$\hat{Cpk} = 0.760$	N.A.	(0.6542, 0.8656)

$\hat{w} = \left[\left(\frac{\hat{d}}{\hat{d} - |\hat{a}|}\right)^2 - 1\right]\frac{1}{\hat{p}^2} = 151.537$. This value of $\hat{w}$ is sufficiently large to make the exact distribution of $\hat{Cpk}*$ indeterminable, however the approximate distribution of $\hat{Cpk}*$ can be used.

Table 7b. 95% confidence Intervals of *Cpw**

w	$\hat{Cpw}*$	Exact	Approximate
0	$\hat{Cp}* = 0.733$	(0.6313, 0.8352)	N.A.
1	$\hat{Cpm}* = 0.731$	(0.6298, 0.8306)	(0.6298, 0.8321)
##	$\hat{Cpk}* = 1.04$	(0.9395, 1.2194)	(0.8156, 1.2640)

$$\#\# \ \hat{w} = \frac{\hat{k}(2-\hat{k})}{(1-\hat{k})^2 \hat{p}^2} = \frac{6\hat{Cp}-\hat{p}}{(3\hat{Cp}-\hat{p})^2 \hat{p}} = 8.11555$$

Table 8. 95% Upper Confidence Limit of $E[L_{WT}(X)]$

w	$\hat{L}_{WT}(X)$	Exact	Approximate
0	1.2626×10^{-5}	(0, 1.6225×10^{-5})	N.A.
1	1.2580×10^{-5}	(0, 1.6200×10^{-5})	(0, 1.6143×10^{-5})
#	1.2560×10^{-5}	(0, 3.1259×10^{-5})	(0, 1.6138×10^{-5})
##	1.2251×10^{-5}	(0, 1.4347×10^{-5})	(0, 1.8381×10^{-5})

Now assume the target value has changed to T=0.8250. Changes in the process statistics are tabulated in Table 5 with the critical values used in the calculations of the confidence intervals are included in Tables 7a, 7b. The upper confidence limits associated with the developed loss (see Table 8) are listed in Table 6.

Conclusion

A technique to integrate process capability indices and loss functions has been developed. The technique is robust in that by using different weights, various process capability indices, their associated confidence intervals and average loss can be determined. The additional penalty of w carries a special meaning for process measurements that are not on target, allowing quality practitioners greater flexibility in assessing a process's actual costs.

The general PCI relationship with expected loss and the expanding research effort in the area of more applicable loss functions offers both practical and research opportunities for developing improved assessment, monitoring and comparisons methods in the area of monetizing process capability. Research efforts relating PCIs and loss would appear to offer opportunities that could address practitioners, managers and researchers concerns and differences in the area of process capability.

Appendix A

Mathematica (Wolfram (1999)) can be used to a) determine and print the d_i's ($i \geq 1$) for the number of specified **i**'s using the requested values of λ and ω (*In[1]*) and b) approximate the value of $Q^2_{n,\lambda}(x)$ by replacing the infinite sum with the finite sum of i+1 terms using the requested values of n, α (proportion), λ and ω (*In[2]).*

In[1]:

```
λ= ;ω= ;
Do[Print[Sum[Sum[Exp[-(λ)/2](((λ)/2)^(b-k))(((b-k)!)^-1)*
(ω^(-.5-b+k))((1-ω^(-1))^(k+g-b))Gamma[(.5+g-b)]*
Binomial[b-1,k]/(Gamma[(g-b+1)]Gamma[.5]),
{k,0,b}],{b,0,g}]],{g,1,i}]
```

In[2]:

```
<<Statistics`ContinuousDistributions`
λ= ;ω= ;n= ;α= ;
Sum[Quantile[ChiSquareDistribution[n+2g],α]*
Sum[Sum[Exp[−(λ)/2](((λ)/2)^(b-k))(((b-k)!)^-1)*
(ω^(-.5-b+k))((1-ω^(-1))^(k+g-b))Gamma[(.5+g-b)]*
Binomial[b-1,k]/(Gamma[(g-b+1)]Gamma[.5]),
{k,0,b}],{b,0,g}],{g,1,i}]+
(Exp[-λ/2](ω^(-0.5))*Quantile[ChiSquareDistribution[n],α])
```

References

Bissell, A. F. (1990). How Reliable is Your Capability Index?, *Applied Statistics*, Vol. 39, pp 331-340.

Johnson, T. (1992). The Relationship of *Cpm* to Squared Error Loss. *Journal of Quality Technology*, Vol. 24(4), pp211-215.

Juran, J.M. (1979). *Quality Control Handbook*, McGraw-Hill, New York, New York.

Leung, B.P.K. and Spiring, F.A. (2002). The Inverted Beta Loss Function: Properties and Applications, *IIE Transactions*, Vol. 34, pp 1101-1109.

Leung, B.P.K. and Spiring, F.A. (2004). Some Properties of the Family of Inverted Probability Loss Functions, *International Journal of Quality Technology and Quantitative Management*, Vol. 1(1), pp 125-147.

Patnaik, P.B. (1949). The Non-Central χ^2 and F Distribution and their Applications, *Biometrika*, Vol. 36, pp 202-232.

Press, S.J. (1966). Linear Combinations of Non-Central Chi-Square Variables, *Annals of Mathematical Statistics*, Vol. 37, pp 480-487.

Spiring, F.A. (1993). The Reflected Normal Loss Function. *The Canadian Journal of Statistics*, Vol. 21(3), pp 321-330.

Spiring, F.A. (1997). A Unifying Approach to Process Capability Indices. *Journal of Quality Technology*, Vol. 29(1), pp 49-58.

Spiring, F.A., and Yeung, A. (1998). A General Class of Loss Functions with Industrial Applications. *Journal of Quality Technology*, Vol. 30(2), pp 107-187.

Sun, F., Lamaree, J., and Ramberg, J. (1996). On Spiring's Inverted Normal Loss Function. *The Canadian Journal of Statistics*, Vol. 24, pp 241-249.

Taguchi, G. (1986). *Introduction to Quality Engineering: Designing Quality into Products and Processes*. Kraus, White Plains, New York.

Taguchi, G, Elsayed, E.A. and Hsiang, T. (1989). *Quality Engineering In Production Systems*, McGraw-Hill, New York, New York.

Tarver, M. (1986). Process Capability Studies in *Quality Management Handbook* edited by L.M. Walsh, R. Wurster and R.J. Kimber. Marcel Dekker, New York, New York, pp 175-196.

Wolfram, S. (1999). *Mathematica: A System for Doing Mathematics by Computers*, 4th edition, Addison Wesley.

In: Progress in Management Engineering
Editors: L.P. Gragg and J.M. Cassell, pp. 117-134
ISBN: 978-1-60741-310-3

Chapter 5

PROJECT SCHEDULING

Jorge J. Magalhães Mendes[1]
Civil Engineering Department, School of Engineering,
Polytechnic of Porto, Porto, Portugal

Abstract

Nowadays, construction projects grow in complexity and size. So, finding feasible schedules which efficiently use scarce resources is a challenging task within project management.

Project scheduling consists of determining the starting and finishing times of the activities in a project. These activities are linked by precedence relations and their processing requires one or more resources. The resources are renewable, that is, the availability of each resource is renewed at each period of the planning horizon.

The objective of the well-known resource constrained project scheduling problem is minimizing the makespan. While the exact methods are available for providing optimal solution for small problems, its computation time is not feasible for large-scale problems [20].

This paper presents two approaches for the project scheduling problem. The first approach combines a new implementation of a genetic algorithm with a discrete system simulation. This approach generates non-delay schedules. This study also proposes applying a local search procedure trying to yield a better solution (GA-RKV-ND).

The second approach combines a new implementation of a genetic algorithm with a discrete system simulation. This approach generates active schedules. This study also proposes applying a local search procedure trying to yield a better solution (GA-RKV-AS).

The chromosome representation of the problem is based on random keys. The dynamic behaviour of the system simulation is studied by tracing various system states as a function of time and then collecting and analysing the system statistics. The events that change the system state are generated at different points in time, and the passage of time is represented by an internal clock which is incremented and maintained by the simulation program. The simulation strategy is the event oriented simulation [27].

The good computational results on benchmark instances enlighten the interest of the best approach (GA-RKV-AS).

Keywords: Construction management, project management, evolutionary algorithms, simulation, scheduling, genetic algorithms, random keys, RCPSP.

[1] E-mail address: jjm@isep.ipp.pt. Rua Dr. António Bernardino de Almeida, 431, 4200-072 Porto, PORTUGAL.

1. Introduction

As the complexity of projects increases, the requirement of an organized planning and scheduling process is enhanced.

The need for organized planning and scheduling of a construction project is influenced by a variety of factors (e.g., project size and number of project activities).

To plan and schedule a construction project, activities must be defined sufficiently so that adequate communication is provided to all those who will use the information. The level of detail determines the number of activities contained within the project plan and schedule. As the number of project activities increases and thus the complexity of their sequential ordering, the need for organized planning and scheduling also increases. This need increases even further when a large number of project activities are considered relative to the uniqueness of each construction project in terms of the dynamic plant and nonstandardized nature of the work.

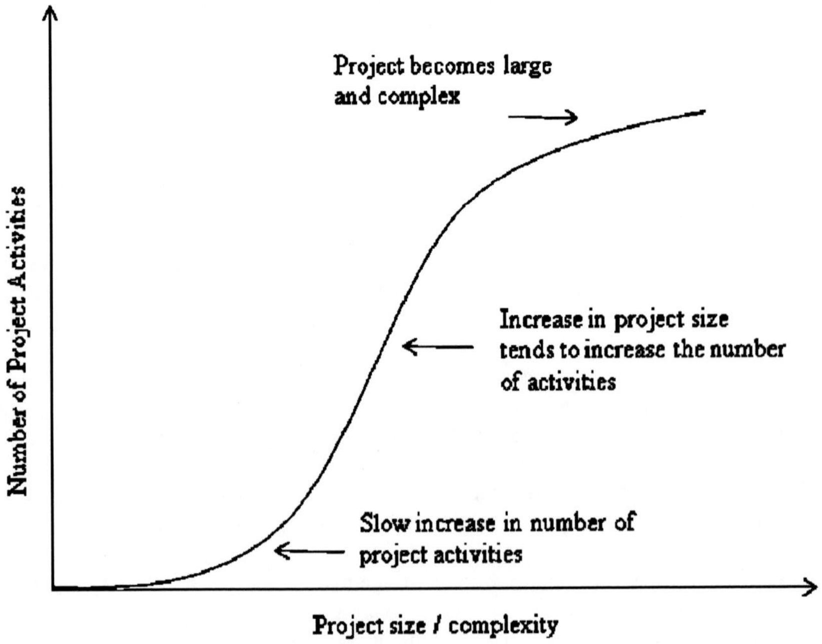

Figure 1. Effect of project size/complexity on number of project activities, Patrick [25].

A relationship exists between project size and number of activities, as represented in Figure 1. Small projects require a relatively small number of activities. As project size increases, so does the number of required activities, but slowly. The lower one-third portion of the curve in Figure 1 represents this. As project size continues to increase, project complexity also increases (i.e., sequential ordering, activity relationships) and thus the number of activities representing the project increases more rapidly – the middle portion of the curve. At some point, the number of activities begins to become unmanageable for the project planner, slowing the rate of growth – the upper portion of the curve, Patrick [25].

A project management problem consists typically of planning and scheduling decisions. The analysis of resources, particularly time, materials, labor and equipment is the key to good project management.

Project scheduling allows to determine the project duration and involves the allocation of the limited resources to projects to determine the start and completion times of the detailed activities.

The use of microcomputers and project scheduling computer software is commonplace in the construction industry. This is mainly true for scheduling project activities and managing resources.

The problem of scheduling activities under the restrictions of resources and precedence relationships with the objective of minimizing the total project duration is referred in the literature as the resource constrained project scheduling problem (*RCPSP*).

Project scheduling involves the allocation of the given resources to projects to determine the start and finish of detailed activities. There are sometimes multiple activities (e.g., excavate, set foundation, erect steel) contending for limited resources (e.g., human resources, crane vehicles), which will make the solution method complex. The allocation of scarce resources becomes a major objective of the problem.

The *RCPSP* problem can be stated as follows. A project consists of $n+2$ activities where each activity has to be processed in order to complete the project. Let $J = \{0, 1, \ldots, n, n+1\}$ denote the set of activities to be scheduled and $K = \{1, \ldots, k\}$ the set of resources. The activities 0 and $n+1$ are dummy, have no duration and represent the initial and final activities. The activities are interrelated by two kinds of constraints:

1. The *precedence constraints*, which force each activity j to be scheduled after all predecessor activities, P_j, are completed.
2. Performing the activities requires resources with *limited capacities*.

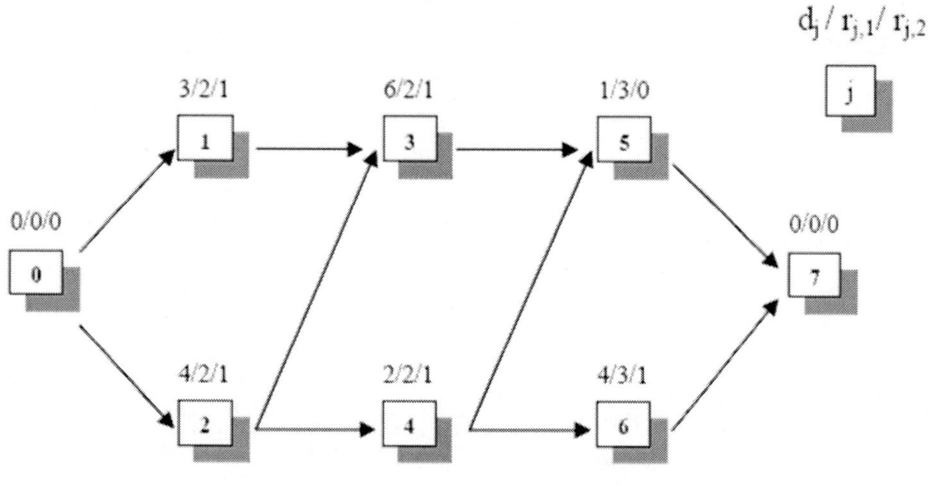

Figure 2. Project network example.

While being processed, activity j requires $r_{j,k}$ units of resource type $k \in K$ during every time instant of its non-preemptable duration d_j. Resource type k has a limited capacity of R_k at any point in time. The parameters d_j, $r_{j,k}$ and R_k are assumed to be non-negative and deterministic. For the project start and end activities we have $d_0 = d_{n+1} = 0$ and $r_{0,k} = r_{n+1,k} = 0$ for all $k \in K$.

The problem depends on finding a schedule of the activities, taking into account the resources and the precedence constraints, which minimize the *makespan* (C_{max}).

Let F_j represent the finish time of activity *j*. A schedule can be represented by a vector of finish times $(F_1, \ldots, F_m, \ldots, F_{n+1})$. The *makespan* of the solution is given by the maximum of all predecessors activities of activity *n+1*, i.e. $F_{n+1} = Max_{l \in P_{n+1}} \{F_l\}$.

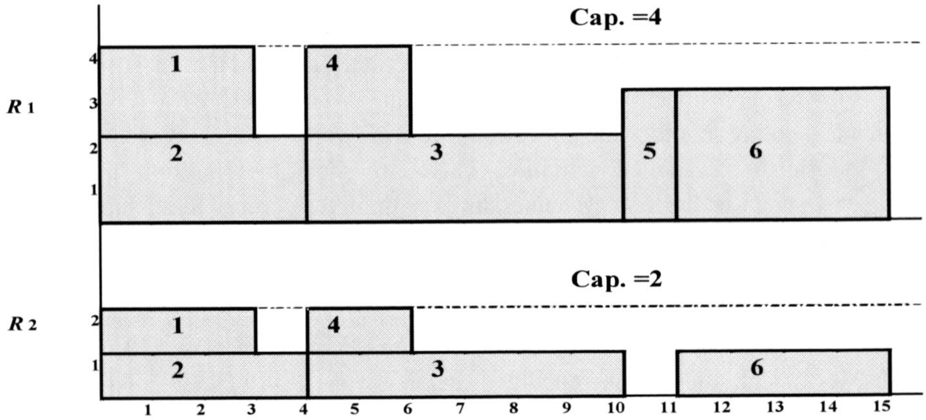

Figure 3. Feasible schedule.

Figure 2 shows an example of a project comprising n=6 activities which have to be scheduled, subject to two renewable resource types with a capacity of four and two units. A feasible schedule with an optimal value of 15 time-periods is represented in Figure 3, see Mendes [20].

The conceptual model of the RCPSP was described by Christofides et al. [23] in the following way:

$$\textbf{Min } F_{n+1} \tag{1}$$

subject to:

$$F_l \le F_j - d_j \qquad j = 1, \ldots, N+1 \; ; \; l \in P_j \tag{2}$$

$$\sum_{j \in A(t)} r_{j,k} \le R_k \qquad k \in K \; ; \; t \ge 0 \tag{3}$$

$$F_j \ge 0 \qquad j = 1, \ldots, N+1 \tag{4}$$

The objective function (1) minimizes the finish time of activity n+1, and therefore minimizes the makespan. Constraints (2) impose the precedence relations between activities and constraints (3) limit the resource demand imposed by the activities being processed at time t to the capacity available. Finally (4) forces the finish times to be non-negative.

The *RCPSP* problem belongs to the class of NP-hard optimization problems, therefore justifying the indispensable use of heuristic solution procedures when solving large problem instances.

Recent classification and survey can be found in Brucker et al. [22] and Kolisch and Hartmann [16]. The survey provided by Kolisch and Hartmann [16] presents more than eighty models and algorithms for complex scheduling problems and discusses the RCPSP.

More recent work is due to Debels et al. [3], Debels and Vanhoucke [4], Mendes et al. [6], Fleszar and Hindi [8], Palpant et al. [9], Kochetov and Stolyar [11], Valls et al. [12], Valls et al. [13], Ranjbar [14], Mendes and Gonçalves [15], Seda [17], Kljajc et al. [18] and Yeh and Pan [19].

2. Types of Schedules

Classifying schedules is the basic work to be done before attacking scheduling problems [21]. Schedules can be classified into one of the following three types of schedules:

i. *Feasible* schedules. A schedule is said to be feasible if it is non-preemptive and if the precedence and resource constraints are satisfied.
ii. *Semi-active schedules*. These are feasible schedules obtained by sequencing activities as early as possible. In a semi-active schedule the start time of a particular activity is constrained by the processing of a different activity on the same resource or by the processing of the directly preceding activity on a different resource.
iii. *Active schedules*. These are feasible schedules in which no activity could be started earlier without delaying some other activity or breaking a precedence constraint. Active schedules are also semi-active schedules. An optimal schedule is always active.
iv. *Non-delay schedules*. These are feasible schedules in which no resource is kept idle at a time when it could begin processing some activity. Non-delay schedules are active and hence are also semi-active.

In this work are generated active and non-delay schedules.

3. New Approach

The new approach combines a genetic algorithm, a discrete system simulation that generates active or non-delay schedules and a local search procedure. The genetic algorithm is responsible for evolving the chromosomes which represent the priorities of the activities. For each chromosome the following three phases are applied:

1. Schedule parameters - this phase is responsible for transforming the chromosome supplied by the genetic algorithm into the priorities of the activities and delay times;
2. Schedule generation - this phase makes use of the priorities and the delay times and constructs active or non-delay schedules;

3. Schedule improvement - this phase makes use of a local search procedure to improve the solution obtained in the schedule generation phase.

After a schedule is obtained, the quality is feedback to the genetic algorithm. Figure 4 illustrates the sequence of steps applied to each chromosome. Details about each of these phases will be presented in the next sections.

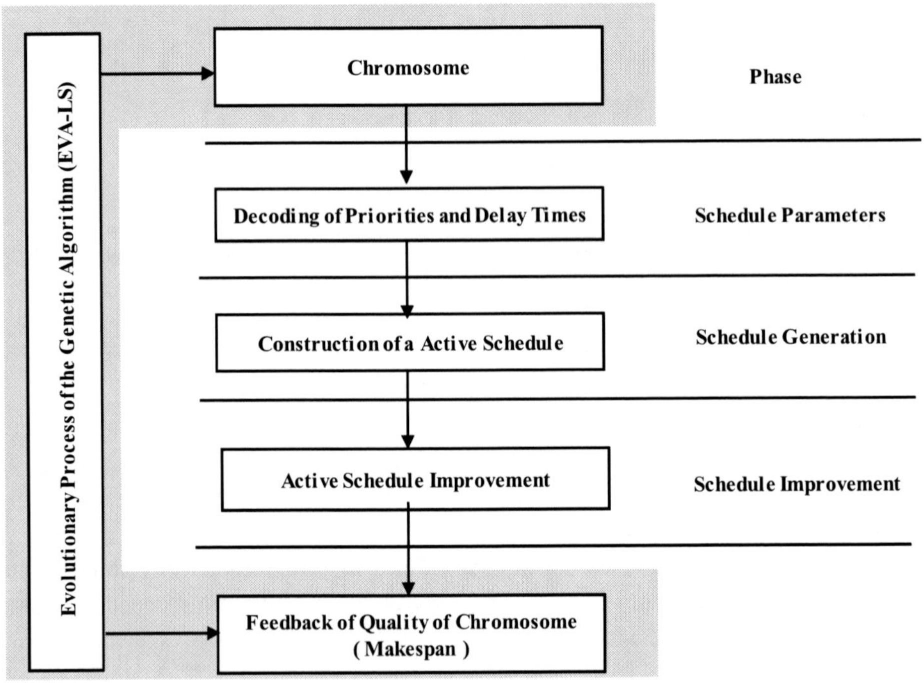

Figure 4. Architecture of the new approach.

3.1. Genetic Algorithm

The genetic algorithm uses a random key alphabet U (0, 1). A chromosome represents a solution to the problem and it is encoded as a vector of random keys (random numbers). Each solution chromosome is made of $2n$ genes where n is the number of activities:

$$Chromosome = (gene_1, gene_2, ..., gene_n, gene_{n+1}, ..., gene_{2n})$$

3.2. Decoding

In this section, we describe how the chromosomes supplied by the genetic algorithm are decoded (transformed) into activity priorities and delays. In this approach, we consider the following two solution alternatives:

GA-RKV-ND –a decoding procedure where activities priorities are evolved by the genetic algorithm and the schedules are non-delay;

GA-RKV-AS –a decoding procedure where activities priorities and delay times are evolved by the genetic algorithm and the schedules are active;

The next sub-section presents the decoding procedures for the activity priorities and delay times for each of the above solution alternatives.

3.2.1. Decoding the Priorities of the Activities

The priority decoding expression uses the following expression:

$$PRIORITY_j = \frac{LLP_j}{LCP} \times \left[\frac{1+gene_j}{2} \right] \qquad j=1,\ldots,n$$

where LLP_j is the longest length path from the beginning of the activity j to the end of the project and LCP is length along the critical path of the project, see Mendes [20] and Mendes et al. [6].

3.2.2. Decoding the Delay Times

The genes between $n+1$ and $2n$ are used to determine the delay times, $Delay_g$, used by each scheduling iteration g. Below we present the decoding procedures for the delay times according to each of the above proposed solution alternatives.

GA-RKV-AS

For these solutions alternatives, the delay schedules generated are given by the following decoding expression:

$$Delay_g = gene_g \times 1.5 \times MaxDur$$

where $MaxDur$ is the maximum duration amongst all activity durations. The factor 1.5 was obtained after experimenting with values between 1.0 and 2.0 in increments of 0.1.

GA-RKV-ND

For this solution alternative, the schedules generated are non-delay. Therefore, all delays are zero, i.e.

$$Delay_g = 0.$$

3.3. Evolutionary Strategy

To breed good solutions, the random key vector population is operated upon by a genetic algorithm.

There are many variations of genetic algorithms obtained by altering the reproduction, crossover, and mutation operators. Reproduction is a process in which individual (chromosome) is copied according to their fitness values (*makespan*). Reproduction is accomplished by first copying some of the best individuals from one generation to the next, in what is called an elitist strategy.

Table 1. Selection probability and fitness value

Number of chromosome	Fitness value	Selection probability
1	14	0,20
2	12	0,17
3	10	0,14
4	9	0,13
5	8	0,11
6	7	0,10
7	4	0,06
8	3	0,04
9	2	0,03
10	1	0,01

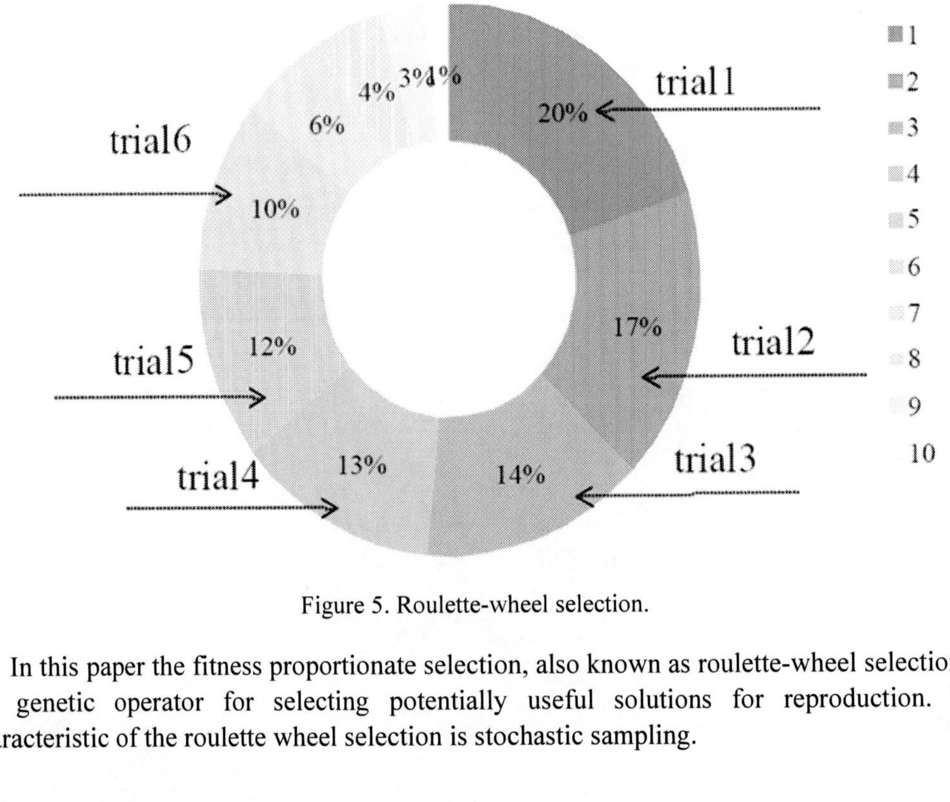

Figure 5. Roulette-wheel selection.

In this paper the fitness proportionate selection, also known as roulette-wheel selection, is the genetic operator for selecting potentially useful solutions for reproduction. The characteristic of the roulette wheel selection is stochastic sampling.

The fitness value is used to associate a probability of selection with each individual chromosome. If f_i is the fitness of individual i in the population, its probability of being selected is,

$$p_i = \frac{f_i}{\sum_{i=1}^{N} f_i} \quad , \qquad i = 1, \ldots, n$$

An example is presented in Table 1.

The mutation operator preserves diversification in the search. This operator is applied to each offspring in the population with a predetermined probability. We assume that the probability of the mutation in this paper is 0.001. With 60 genes positions we should expect 60 x 0.001 = 0.06 genes to undergo mutation for this probability value.

The general schema of genetic algorithms (GAs) may be illustrated as follows (Figure 7).

Random position *k* = 3 Chromosome length *l* = 8

Chromosome 1	0.32	0.22	0.34	0.89	0.23	0.76	0.78	0.45
Chromosome 2	0.12	0.65	0.38	0.47	0.31	0.56	0.88	0.95

swapping all the genes between *4* and *8*

Offspring 1	0.32	0.22	0.34	0.47	0.31	0.56	0.88	0.95
Offspring 2	0.12	0.65	0.38	0.89	0.23	0.76	0.78	0.45

Figure 6. Crossover operator example.

```
procedure GENETIC-ALGORITHM

Generate initial population P0;
  Evaluate population P0;
  Initialize generation counter g ←0;

  While stopping criteria not satisfied repeat
        Select some elements from Pg to copy into Pg+1;
        Crossover some elements of Pg and put into Pg+1;
        Mutate some elements of Pg and put into Pg+1;
        Evaluate some elements of Pg and put into Pg+1;
        Increment generation counter: g ← g+1;
  End while
End GENETIC-ALGORITHM;
```

Figure 7. Pseudo-code of a genetic algorithm, Mendes [20].

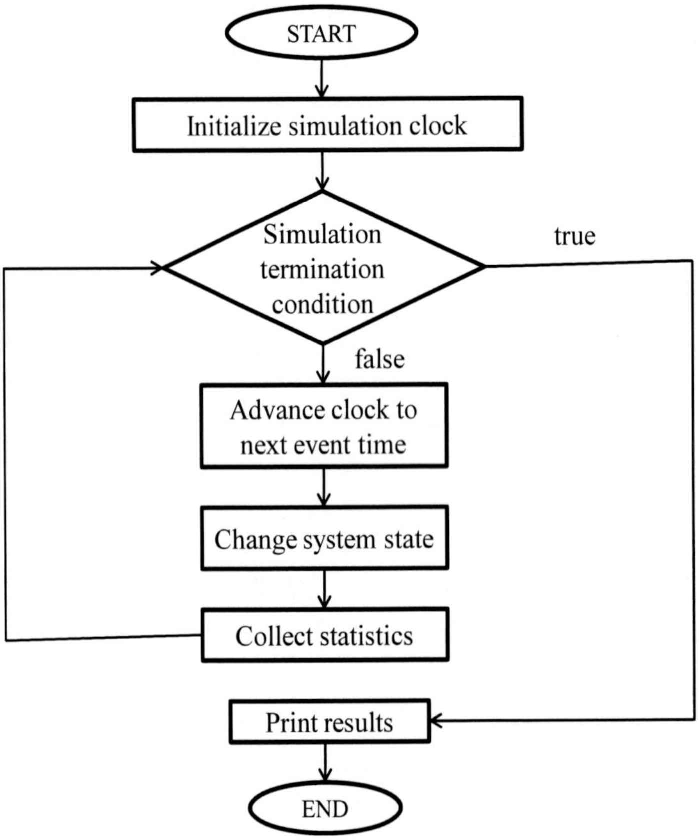

Figure 8. General model for discrete simulation system.

3.4. Discrete System Simulation

The idea of this new approach is combining a new genetic algorithm with a discrete system simulation. The feasible schedules are constructed using the discrete event simulation in which the priorities of the activities are defined by genetic algorithm.

3.4.1. Discrete Event Model

In the discrete approach to system simulation, states changes in the physical system are represented by a series of discrete changes or events at specific instants of time and such models are known as *discrete event models*. In this case the time and state are the two important coordinates used in describing simulation models. Between events, the states of the entities remain constant. The change in state is brought about by events which form the driving force behind every discrete simulation model, see Neelamkavil [27].

3.4.2. Representation of Time

The dynamic behaviour of the system simulation is studied by tracing various system states as a function of time and then collecting and analysing the system statistics. The events that change the system state are generated at different points in time, and the passage of time is represented by an internal clock which is incremented and maintained by the simulation program. The simulation strategy is the event oriented simulation where the clock is incremented from time *t* to the next event time *t′*, see Figure 8.

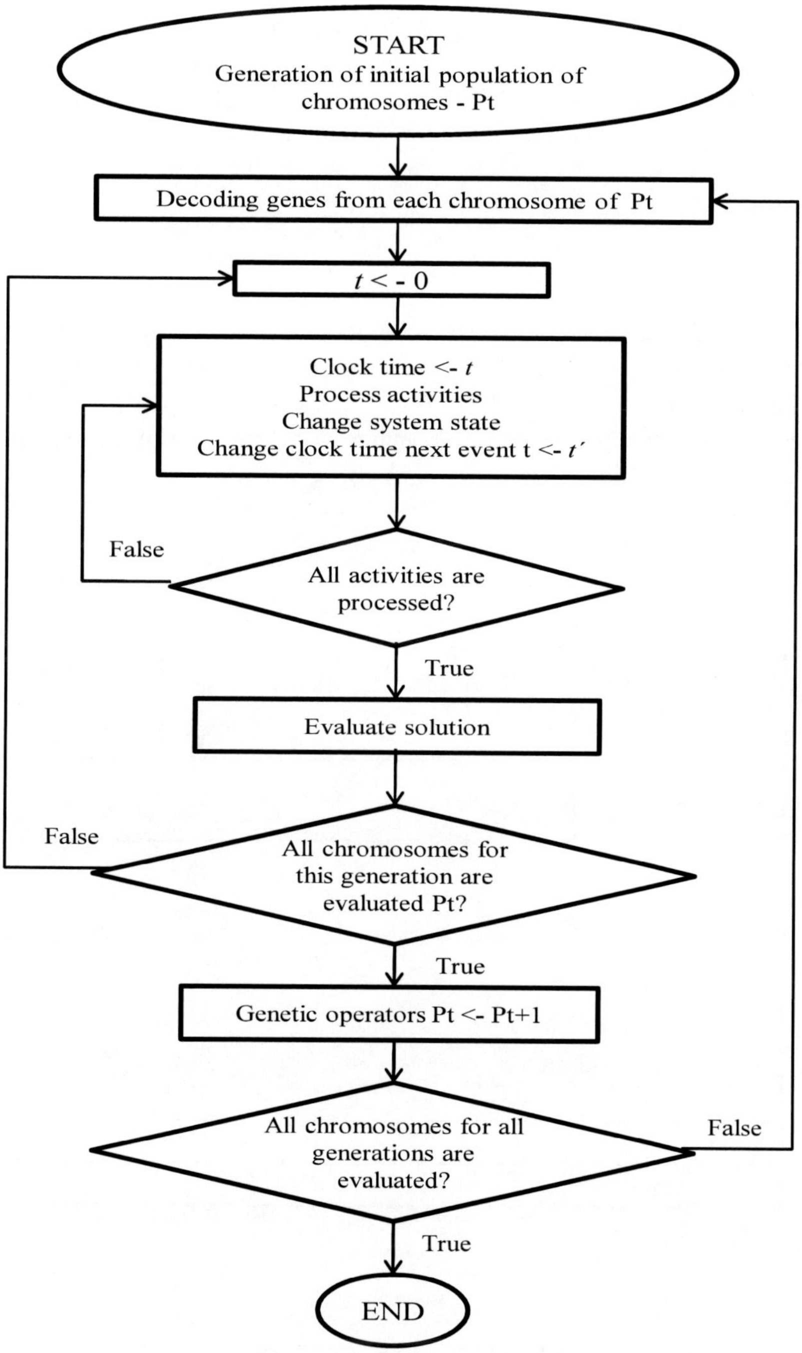

Figure 9. Application of GA using discrete simulation.

3.4.3. Simulation

A discrete simulation program must provide efficient mechanisms for the generation of the time and the type of the next event.

Figure 9 is an example of simulation by event scheduling (resources begin processing, resources end processing, begin changing between resources, end changing between resources). The time is advanced to the time of the occurrence of the next event and simulation is accomplished by the execution of ordered (by time) event sequences. This method involves sorting of event activation times and maintaining *current* and *future event* lists.

3.5. Local Search

Local search algorithms move from solution to solution in the space of candidate solutions (the search space) until an optimal solution or a stopping criterion is found. In this paper we apply *backward and forward improvement* based on Klein [10].

Initially is constructed a schedule by planning in a forward direction starting from the project's beginning, see Figure 10. After is applying backward and forward improvement based on Klein [10].

The backward planning consists in reversing the project network and applying the scheduling generator scheme.

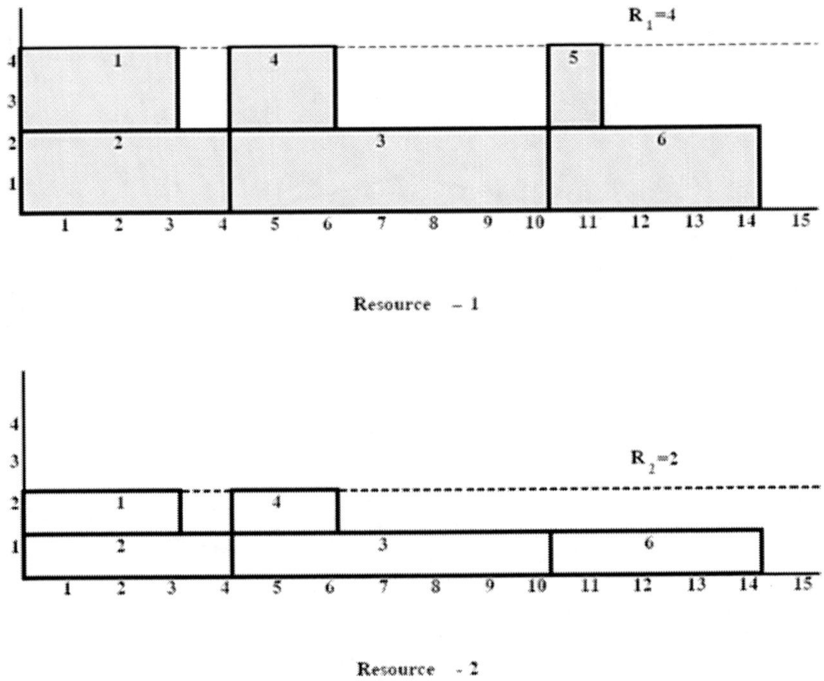

Figure 10. Feasible schedule with a makespan of 14.

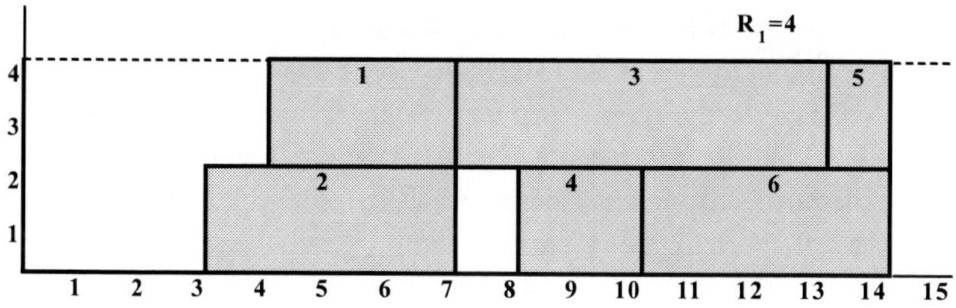

Resource - 1

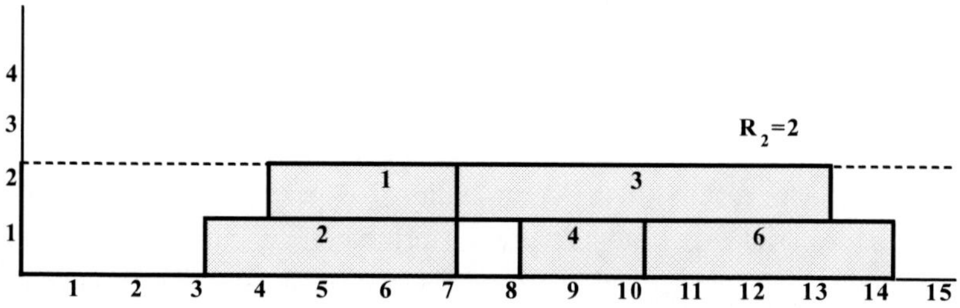

Resource - 2

Figure 11. Improved schedule with a makespan of 11.

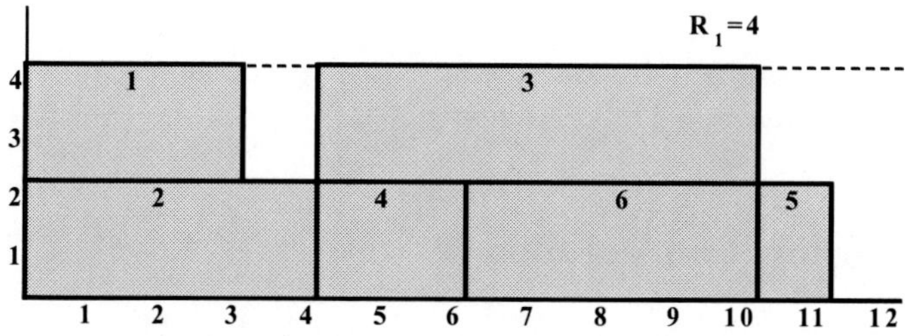

Resource - 1

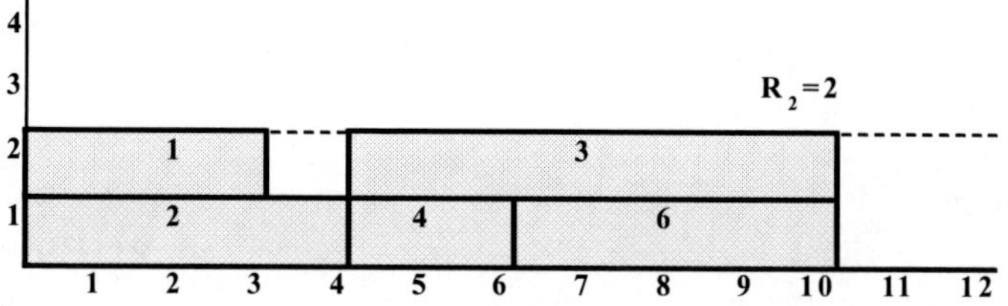

Resource - 2

Figure 12. Final schedule with a makespan of 11.

Figure 11 presents the solution obtained by backward planning. Having scheduled the dummy end activity 7, activities 5 and 6 which are backward eligible at $t = 14$ can be executed in parallel. Due to the precedence constraint, activities 3 and 4 are scheduled that they finishes at $t = 13$ and $t = 10$, respectively. Finally activities 1 and 2 are scheduled. By reducing all the scheduled starting and finisinh times by 3, a schedule with a makespan of 11 is obtained, i.e., the initial schedule (Figure 10) is improved.

Figure 12 presents the solution obtained by forward planning. Having scheduled the dummy initial activity 0, activities 1 and 2 which are eligible at $t = 0$ can be executed in parallel. Due to the precedence constraint, activities 3 and 4 are scheduled that they finishes at $t = 10$ and $t = 6$, respectively. Finally activities 5 and 6 are scheduled. A schedule with a makespan of 11 is obtained without improving the schedule in Figure 11.

4. Computational Experiments

This section presents results of the computational experiments done with the algorithm proposed in this paper. The experiments were performed on an Intel Core 2 Duo CPU T7250 @2.00 GHz. The algorithm was coded in Visual Basic 6.0. The GA-RKV (Genetic Algorithm - Random Key Variant) was tested on the instance sets:

- J30 (480 instances each with 30 activities)
- J60 (480 instances each with 60 activities)
- J120 (600 instances each with 120 activities)

available in PSPLIB. All problem instances require four resource types.

Instances details are described in Kolisch et al. [24].

4.1. Genetic Algorithm Configuration

Though there is no straightforward way to configure the parameters of a genetic algorithm, we obtained good results with values: population size of 5 × number of activities in the problem; mutation probability of 0.001; top (best) 1% from the previous population chromosomes are copied to the next generation; stopping criterion of 250 generations.

4.2. Experimental Results

Table 2 summarizes the experimental results with the criteria's average deviation percentage from the optimal *makespan* (D_{OPT}) for the instance set J30 and the average deviation percentage from the well-known critical path-based lower bound (D_{LB}) for the instance set J60 and J120, respectively. The lower bound values (D_{LB}) are reported by Stinson et al. [7].

Table 2. Computational results for J30, J60 and J120 instances

Algorithm	Types of schedules	J30 D_{OPT}	J60 D_{LB}	J120 D_{LB}
GA-RKV-AS	Actives	0.01	10.81	32.47
GA-RKV-ND	Non-delays	0.03	11.06	32.55

Table 3. Top-ten computational results for J30 instances

Algorithm	Reference	J30
MAPS	Mendes and Gonçalves [15]	0.00
GA-TS path relinking	Kochetov and Stolyar [11]	0.00
Decomp. and local opt.	Palpant et al. [9]	0.00
FandF(5)	Ranjbar [14]	0.00
GA – RKV-AS	This paper	0.01
GAPS	Mendes et al. [6]	0.01
Scatter Search - FBI	Debels et al. [3]	0.01
VNS-activity list	Fleszar and Hindi [8]	0.01
GA - DBH	Debels and Vanhoucke [4]	0.02
GA – hybrid, FBI	Valls et al. [12]	0.02
GA - FBI	Valls et al. [13]	0.02

The algorithm GA-RKV-AS is the best relatively to the GA-RKV-ND, for all the instance sets. For the instance set J30, GA-RKV-AS obtained D_{OPT} = 0.01, for J60 obtained D_{LB} = 10.81 and for J120 obtained D_{LB} = 32.47. We can conclude that actives schedules produce better solutions than non-delay schedules.

The comparative results with the best approaches are given in Tables 3, 4 and 5.

Table 3, column 3, summarizes the average deviation percentage from the optimal *makespan* (D_{OPT}), for the instance set J30. GA-RKV-AS obtained D_{OPT} = 0.01. The number of instances for which the algorithm obtains the optimal solution is 476. For the set J30, GA-RKV-AS ranks five.

Table 4. Top-ten computational results for J60 instances

Algorithm	Reference	J60 D_{LB}
FandF(5)	Ranjbar [14]	10.56
MAPS	Mendes and Gonçalves [15]	10.64
GAPS	Mendes et al. [6]	10.67
GA - DBH	Debels and Vanhoucke [4]	10.68
Scatter Search - FBI	Debels et al. [3]	10.71
GA – hybrid, FBI	Valls et al. [12]	10.73
GA, TS – path relinking	Kochetov and Stolyar [11]	10.74
GA - FBI	Valls et al. [13]	10.74
GA – RKV-AS	This paper	10.81
Decomp. and local opt.	Palpant et al. [9]	10.81
VNS-activity list	Fleszar and Hindi [8]	10.94

Table 5. Top-ten computational results for J120 instances

Algorithm	Reference	J120
		D_{LB}
GA-DBH	Debels and Vanhoucke [4]	30.82
MAPS	Mendes and Gonçalves [15]	31.19
GAPS	Mendes et al. [6]	31.20
GA – hybrid, FBI	Valls et al. [12]	31.24
FandF(5)	Ranjbar [14]	31.42
Scatter Search - FBI	Debels et al. [3]	31.57
GA - FBI	Valls et al. [12]	31.58
GA, TS – path relinking	Kochetov and Stolyar [11]	32.06
Decomp. and local opt.	Palpant et al. [9]	32.41
GA – RKV-AS	This paper	32.47
GA - Self adapting	Kolisch and Hartmann [16]	33.21

Tables 4 and 5, columns 3, summarize the average deviation percentage from the well-known critical path-based lower bound (D_{LB}) for the instance set J60 and J120, respectively. For the instances set J60 GA-RKV-AS ranks nine and for J120 GA-RKV-AS ranks ten. The lower bound values (D_{LB}) are reported by Stinson et al. [7].

The maximum computational time dispended is 120 seconds for each instance of J60 and 300 seconds for each instance of J120.

5. Conclusions and Further Research

This paper presents a new genetic algorithm (a variant of the genetic algorithm proposed by Goldberg [1]) for the resource constrained project scheduling problem. The chromosome representation of the problem is based on random keys. Reproduction, crossover and mutation are applied to successive chromosome populations to create new chromosome populations. These operators are simplicity itself, involving random number generation, chromosome copying and partial chromosome exchanging.

The schedules are constructed using a discrete simulation system with a priority rule in which the priorities are defined by the genetic algorithm. The actives schedules produced better solutions than the non-delay schedules.

The discrete simulation system for constructing feasible schedules is extended by the flexible use of different planning directions including the backward and forward planning. For some instances, a combination of the discrete simulation system and the genetic algorithm may yield a good result, but in some cases the local search can improve the schedule.

The approach was tested on a set of 1560 standard instances taken from the literature and compared with the best state-of-the-art approaches. The algorithm GA-RKV-AS produced good results when compared with other approaches therefore validating the effectiveness of the proposed algorithm.

Further work could be conducted to explore the possibility of using activities with multi-mode usage of limited resources.

References

[1] D. E. Goldberg. *Genetic Algorithms in Search, Optimization and Machine Learning,* Addison-Wesley, 1989.

[2] D. Beasley, D.R. Bull and R.R. Martin, *An Overview of Genetic Algorithms: Part 1, Fundamentals,* University Computing, Department of Computing Mathematics, University of Cardiff, UK, Vol. 15(2), 1993, pp. 58-69.

[3] D. Debels, B. De Reyck, R.Leus and M.Vanhoucke, A Hybrid Scatter Search/Electromagnetism Meta-Heuristic for Project Sheduling, *European Journal of Operational Research,* Vol. 169, 2006, pp. 638-653.

[4] D. Debels and M. Vanhoucke. A Decomposition-Based Heuristic for the Resource-Constrained Project Scheduling Problem. *Working Paper* 2005/293, Faculty of Economics and Business Administration, University of Ghent, Ghent, Belgium, 2005.

[5] J.F. Gonçalves, J.M. Mendes and M.C.G. Resende. A hybrid genetic algorithm for the job shop scheduling problem. *European Journal of Operational Research,* Vol. 167, 2005, pp. 77-95.

[6] J.J.M. Mendes, J.F. Gonçalves and M.G.C. Resende, A random key based genetic algorithm for the resource constrained project scheduling problem, *Computers and Operations Research,* Vol. 36, 2009, pp. 92-109.

[7] J.P. Stinson, E.W. Davis and B.M. Khumawala, Multiple Resource-Constrained Scheduling Using Branch and Bound, *AIIE Transactions,* Vol. 10, 1978, pp. 252-259.

[8] K. Fleszar and K.S. Hindi, Solving the resource-constrained project scheduling problem by a variable neighbourhood search, *European Journal of Operational Research,* Vol. 155, 2004, pp. 402-413.

[9] M. Palpant, C. Artigues and P. Michelon, LSSPER: Solving the resource–constrained project scheduling problem with large neighbourhood search, *Annals of Operations Research,* Vol.131, 2004, pp. 237-257.

[10] R. Klein, Bidirectional planning : improving priority rule-based heuristics for scheduling resource-constrained projects, *European Journal of Operational Research,* Vol. 127, 2000, pp. 619-638.

[11] Y. Kochetov and A. Stolyar. Evolutionary local search with variable neighborhood for the resource constrained project scheduling problem. In *Proceedings of the 3rd International Workshop of Computer Science and Information Technologies,* Russia, 2003.

[12] V. Valls, F. Ballestin and M.S. Quintanilla. *A hybrid genetic algorithm for the RCPSP.* Technical report, Department of Statistics and Operations Research, University of Valencia, 2003.

[13] V. Valls, F. Ballestin and M.S. Quintanilla, Justification and RCPSP: A technique that pays, *European Journal of Operational Research,* Vol.165, 2005, pp. 375-386.

[14] M. Ranjbar, Solving the resource-constrained project scheduling problem using filter-and-fan approach, *Applied Mathematics and Computation,* Vol. 201, 2008, pp. 313–318.

[15] J.J.M. Mendes and J.F. Gonçalves, A Memetic Algorithm-Based Heuristics for the Resource Constrained Project Scheduling Problem, *Proceedings of II International Conference on Computational Methods for Coupled Problems in Science and Engineering,* Spain, 2007, pp. 644-648.

[16] R. Kolisch and S. Hartmann, Experimental investigation of heuristics for resource-constrained project scheduling: an update, *European Journal of Operational Research*, Vol.174 (1), 2006, pp. 23-37.

[17] M. Seda, Solving Resource-Constrained Project Scheduling Problem as a Sequence of Multi-Knapsack Problems, *WSEAS Transactions on Information Science and Applications,* Issue 10, Vol. 3, 2006, pp.1785-1791.

[18] M. Kljajc, U. Breskvar and B. Rodic, Computer aided scheduling with use of genetic algorithms and a visual discrete event simulation model, *WSEAS Transactions on Systems*, Issue 3, Vol. 3, 2004, pp. 1021-1026.

[19] C.H. Yeh and H. Pan, System Development for Fuzzy Project Scheduling, *WSEAS Transactions on Business and Economics, World Scientific and Engineering Academy and Society, USA*, Vol. 1(4), 2005, pp. 311-317.

[20] J.J.M. Mendes, "Sistema de Apoio à Decisão para Planeamento de Sistemas de Produção do Tipo Projecto", Ph.D. Thesis, Departamento de Engenharia Mecânica e Gestão Industrial, Faculdade de Engenharia da Universidade do Porto, Portugal, 2003. (In portuguese)

[21] R. Kolisch, *Project Scheduling under Resource Constraints*, Physica-Verlag, Germany, 1995.

[22] P. Brucker, A Drexl, R. Mohring, K. Neumann, E. Pesch, Resource-constrained project scheduling: Notation, classification, models and methods, *European Journal of Operational Research,* Vol.112 (1), 1999, pp. 3-41.

[23] N. Christofides, R.Alvarez-Valdés and J. Tamarit, Problem scheduling with resource constraints: A branch and bound approach, *European Journal of Operational Research*, Vol. 29, 1987, pp. 262-273.

[24] R. Kolisch, Schwindt, A.Sprecher, Benchmark instances for scheduling problems. In J.Weglarz, (ed.) *Handbook on recent advances in project scheduling*, Kluwer, Amsterdam, 1998, pp. 197-212.

[25] C. Patrick, *Construction Project Planning and Scheduling*, PEARSON Prentice Hall, Columbus, Ohio, 2004.

[26] R. Kolisch and S. Hartmann, *Heuristic Algorithms for Solving the Resource-Constrained Project Scheduling Problem: Classification and Computational Analysis*, J. Weglarz (editor), Kluwer, Amsterdam, the Netherlands, 1999, pp. 147–178.

[27] F. Neelamkavil, "Computer Simulation and Modelling", John Wiley and Sons Ltd, 1990.

In: Progress in Management Engineering
Editors: L.P. Gragg and J.M. Cassell, pp. 135-172
ISBN: 978-1-60741-310-3

Chapter 6

COMPUTERIZED BLOOD BANK INFORMATION MANAGEMENT AND DECISION MAKING SUPPORT

***Bing Nan Li*[1,2,*], *Ming Chui Dong*[2,3] and *Mang I. Vai*[2]**

[1] NUS Graduate School for Integrative Science and Engineering,
National University of Singapore, Singapore
[2] Department of Electrical and Electronics Engineering,
University of Macau, Macau
[3] Institute of Systems and Computer Engineering of Macau, Taipa 1356, Macau

Abstract

Blood donation and transfusion service is an indispensable part of contemporary medicine and healthcare. It involves collecting, processing, storing and providing human blood intended for transfusion, performing pre-transfusion testing, cross-matching, and finally infusing into the patients. In view of the life-threatening nature of blood and blood components, it entails the rigorous controlling, monitoring and the complete documentation of the whole procedure from blood collection to blood infusion. The introduction of information and computer technology facilitates the overall procedure of blood donation and transfusion service, and improves its efficiency as well.

In general, a computerized blood bank information system refers to acquiring, validating, storing, and circulating various data and information electronically in blood donation and transfusion service. With regard to its unique service objects, the blood bank information system should pay enough attention on the following characteristics of blood bank data and information: information credibility, information integrity, information coordination, and information security. This chapter firstly surveys the development of computerized blood bank information systems, elucidates their rationale and infrastructures, and then exemplifies a real-world blood bank information system. The relevant engineering implementation will be discussed too.

Other than consistency and security, another challenge in computerized blood bank information management is, in face of explosive data and information, how to make good use of them for decision making support. In this chapter, we will further address the underlying mechanisms of decision making support in blood bank information systems. The unique

* E-mail address: bingoon@ieee.org. Corresponding address: Bing Nan LI, INESC-Macau, Block 3, 1/F, University of Macau, Taipa 1356, Macau.

properties of blood bank data and decisions are firstly examined. Then, with special concerns on blood donation and transfusion service, we shift to the development of computerized decision making support. Finally, a case study will be presented to evidence our understanding on computerized decision making support in blood bank information systems.

1. Introduction

Blood donation and transfusion service involves collecting, processing, storing and providing human blood intended for transfusion, performing pre-transfusion testing, cross-matching, and finally infusing into a patient (AABB 2002). So an official blood bank infrastructure generally consists of the independent blood centers, where human blood is collected, stored and distributed, as well as the hospital blood banks in charging of transfusion-related services. Few health affairs are as complex as blood donation and transfusion service. Even the first part, namely blood donation service, involves various interdependent operations at a blood center: donor registration, donation evaluation, blood collection, blood screening, component production, inventory management, and blood dissemination, etc.

With regards to its life-threatening nature, blood donation and transfusion service entails the rigorous, complete documentation of the whole procedure from blood collection to blood infusion. In face of the tremendous amount of data and information, there are various errors and risks in the mentioned procedure. It has been reported that the errors at the time of administration of blood or blood components are the most frequently documented errors accumulating in the transfusion of the wrong blood (Sazama 1990). In addition, the dominant errors during blood sampling, laboratory testing and especially inventory management of blood components were found to be an important factor in many of such accidents (Linden *et al* 1992; Linden and Kaplan 1994).

Information and computer technology (ICT) has been widely deployed in medicine. And it exhibits the great potential to improve working efficiency as well as service quality. In terms of blood donation and transfusion service, combined with various automation apparatus, information and computer technology can facilitate and secure most procedures of donor screening, blood collection, laboratory testing and cross-matching. For instance, the complete documentation for possible backward inspection has challenged blood donation and transfusion service for years. Fortunately, it has been evidenced that information and computer technology is able to relieve the workload of blood banks and reduce the incidence of “wrong blood” episodes (Pietersz 1995; Kern and Bennett 1996; CLBTS 2001; Butch 2002).

Hence, the researchers have been keen on the pilot projects early in 1960s–1980s (Singman *et al* 1965; Kempf 1967; Oba *et al* 1971; Peyretti 1971; Moore 1973; Chambers *et al* 1975; Brodheim 1978). Then, with the systemization of computers in blood banks (Page 1980; Brodheim 1983; Sapountzis 1984; Myhre and Ritland 1986), US Food and Drug Administration (FDA) officially recommended the implementation of blood bank computerization in 1988 (CBER 1988). It pointed out “*Automated or electronic data systems used in blood and plasma establishments should have the capacity to trace the history of every donation forward through final disposition of each component and from each transfusion, infusion or sale backward to the original donor*”. And FDA published the detailed requirements for blood bank computerization in 1989 (CBER 1989). Further, in accordance with the development of information and computer technology, FDA has to keep updating the instructive specifications, including the subsequent publications in 1994, 1997 and 2005 (CBER 1994; 1997; 2005). Meanwhile, other developed countries and regions, including the United Kingdom and France, also developed corresponding blood bank computing standards (BCSH 1995; 1999; 2000; Moncharmont *et al* 1999; ANZSBTI 2004).

Nowadays, it is well established about the necessity and feasibility of blood bank computing. As a consequence, many institutes and vendors are active in this field. For instance, merely in the United States, FDA has licensed a series of computer software for blood bank computing (CBER 2008). However, different from hospital information systems, it is noteworthy that most of the released computer software are designed and optimized for the specific blood banks, although there are a few industrial consensuses in the specific aspects of blood donation and transfusion service (ICCBBA 2004). In other words, lacking of widely accepted standards definitely impairs the development of blood bank information systems.

In this chapter, what we are concerned with covers the underlying role, mission and infrastructure of blood bank information systems, and explore the feasible standards and consensus. We will pay special attention on the following characteristics of blood bank data and information: information credibility, information integrity, information coordination, and information security. Both the infrastructure and its engineering implementation of a computerized blood bank information system, including barcode technology, electronic donor cards and decision making support, will be elucidated in this chapter.

2. Blood Bank Information Management

2.1. Role of Blood Centers in Transfusion Service

According to its formal definition, the whole lifecycle of blood donation and transfusion service involves collection, processing, storage, transportation, pre-transfusion testing, and final infusion. Although those operations may take place in a single hospital blood bank, as a matter of fact, they are often performed in two separate places. For example, the blood for transfusion is usually collected in the independent blood centers (or blood establishments), where blood components are then processed and disseminated to hospital transfusion units. Thus, unless with special notification, the term "blood bank" hereinafter is defined as a blood center responsible for maintaining an adequate supply of needed blood components, and releasing the blood for transfusion service. And a hospital blood bank refers to the transfusion unit in that hospital.

However, from the viewpoint of blood circulation, the role of independent blood banks seems questionable. Generally speaking, it is appropriate to sketch the integral transfusion service as Figure 1. Due to the separate operations in blood centers and hospital blood banks, many factors, such as secondary infection and misplacement, unavoidably increase the risk of blood transfusion service. As a matter of fact, hospital blood banks were firstly introduced into blood donation and transfusion service. In addition, the autologous blood donations are often settled in hospital blood banks. Then, why the independent blood centers? Benefited from the advances of blood preservation and circulation, nowadays the mode of independent blood centers and hospital blood banks is widely accepted for voluntary, directed and aphaeresis blood donation service in most countries and regions (Bloodbooks 2008).

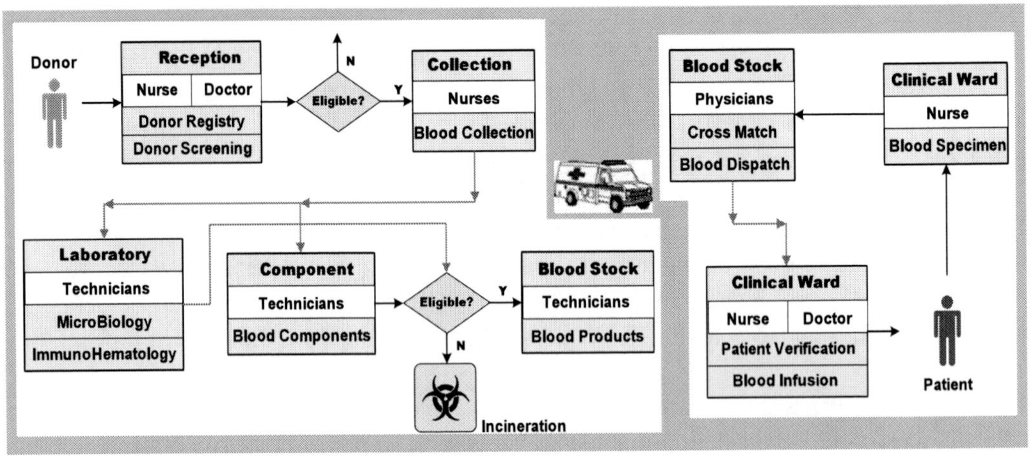

Figure 1. The operational model of blood donation and transfusion service.

Obviously, an environment-friendly blood center, compared with hospital blood banks, is more contributive to the mood of blood donors, which is of vital importance for donor recruitment and retention. On the other hand, with the discovery of more and more transfusion-related diseases, the secure blood donation and transfusion service requires a series of sophisticated blood testing and analyzing apparatus. The centralized blood banks thus contribute to decreasing national medical expenditure for those apparatus and related professional training. The reason is justifiable from the viewpoint of blood processing and production too. Finally, the independent blood banks can effectively guarantee the impartial blood dissemination and supply. Such model of blood donation and transfusion service has been proved effective to optimize the national blood utilization.

At the same time, hospital blood banks also play a key role in blood donation and transfusion service. The first of all, to guarantee safe blood transfusion, the indispensable operations before formal blood infusion include cross-matching and cross-validation. In the second place, although most blood centers operate in the 24/7 mode, any hospital has to preserve an appropriate portion of blood in its blood bank so as to guarantee the quick response to medical emergencies. Finally, it is a common strategy that the autologous blood donation and transfusion should be conducted within a hospital blood bank according to the patient's residence. Therefore, although blood analyzing and processing can be submitted to a tertiary blood center, the hospital blood bank is in essence a self-contained system.

In a whole, the integral network of blood donation and transfusion service is comprised of blood centers, hospital blood banks and small or ambulatory blood stations. If we understand their individual advantages and disadvantages, the success of modern blood donation and transfusion service should be attributed to information and computer technology. In the first place, the donor screening procedure looks for the computerized and networked donor information systems.

Then blood banks could preclude the illegible donors in time. Secondly, if all blood inventory information systems can run transparently and subject to the surveillance of independent committees, it is possible to revise and optimize the traditional network of blood donation and transfusion service. For instance, the blood center can be affiliated as a complex hospital blood bank. But such kind of revisions must depend on the advances of computerized and networked blood bank information systems.

2.2. Data and Information in a Blood Bank

Few healthcare affairs are as complex as managing blood and blood components, especially taking the decentralized affairs into account. However, independent blood centers and hospital blood banks are necessary to meet the needs of contemporary blood donation and transfusion service. Therefore, a competent blood bank information system often adapts itself to the specific blood bank. But even a single blood bank has to cope with daily hundreds and thousands pieces of data and information, including those from blood collection to blood dissemination. To develop a robust blood bank information system, the first-of-all task is how to archive those related heterogeneous data and information well. In this section, it will be addressed in accordance with the lifecycle of blood components from original donor to final patient, as shown in Figure 2.

Recruiting healthy donors is the first whilst crucial step to guarantee the safe blood donation and transfusion service. So it is necessary for any blood bank information system to provide the effective solutions for donor screening and tracking. Traditionally, donor screening is based on the self-conscious questionnaire covering general information of personal data, medical records and donation history. The additional information includes simple physical examination, post-donation reactions and subsequent blood testing results (CBER 2003). However, in contemporary blood bank information systems, more and more objective data and information, such as accurate donation parameters and confident medical information from hospitals, are included to improve the safety of blood donation and transfusion service (CBER 2004).

Similar as other medical facilities involved in life-threatening affairs, any related apparatus, materials and staffs in blood donation and transfusion service should be rigorously labeled and recorded for possible backward inspection. This part of information consists of the unique identifications and their connection with every unit of blood and blood components, for example, blood containers, staff's decisions, and operations, etc.

Another appreciated role of blood bank information systems is able to optimize blood distribution and utilization. It is built on the integral information of blood inventories in blood banks and hospital blood banks. The peripheral information includes lists of transfusion facilities and blood distribution records. Again, the computerized and networked management substantially promotes the quick response of blood banks to various medical emergencies.

Final part of information is comprised of the feedback data and messages from outside transfusion facilities with the help of Electronic Data Interchange (EDI). More than the substituting role of text documents, a contemporary blood bank information system contributes to blood donation and transfusion service through the complete chain of information interchange and share. As a consequence, the feedback information, including blood transfusion records and patients' transfusion reaction, enriches the blood information substantially.

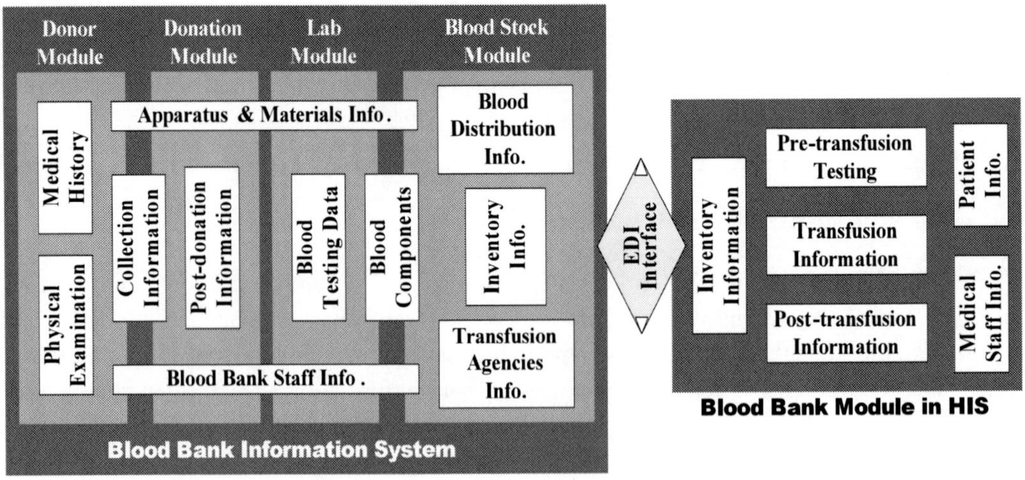

Figure 2. Data and information in the blood donation and transfusion service.

The essential divisions in a qualified blood bank and their concomitant data have been enumerated in Table I. The daily operations in these divisions inevitably lead to considerable amounts of data and information, most of which should be recorded clearly for possible audit and backward inspection. In light of its role and responsibilities, each division usually complies with the distinct operational strategy and results in different datasets. Most of reported blood bank information systems adopt the client/server infrastructure in order to cope with the decentralized affairs within a blood bank. In other words, various datasheets in a database are generally organized in accordance with the individual division and related operations. In spite of different structures and volumes, those datasheets exhibit a few common properties:

- *Heterogeneity*: The diversified blood bank operations naturally lead to the complicated data structure in blood bank information systems. At least three types of languages are necessary to describe the incurred data and results in blood banks: quantitative for amounts and results; pseudo-quantitative for encoded data and information; and qualitative for information.
- *Interrelationship*: To assure the safety of blood transfusion service, it is mandatory to track and control each unit of blood products. In other words, the complicated datasheets could be organized by some unique keys and indices as a whole. And only with such associations, it is possible to discover the underlying information and knowledge with the help of modern computers.
- *Consistency*: It is mandatory to record all operations in blood donation and transfusion service for possible backward inspections. Hence, those records related to the specific object, despite their diversified representations, should be consistent in a blood bank information system. Data consistency is an important factor in the implementation of computer analysis and decision making support.
- *Dynamics*: It is also necessary to take the dynamic nature of blood bank data and information into account. In the first side, the database system keeps updating with the increment of data and results. In the other side, with the process of blood

collecting, testing, processing, storing, and disseminating, the records related to every blood component keep updating too.

2.2.1. Essences of Blood Bank Information

As a whole, modern blood bank information systems should pay more attention on the following characteristics of information in blood donation and transfusion service:

- *Information credibility*: In the first place, besides donor's subjective response, more objective items, including electronic patient/health records, are introduced to secure information credibility so that the eligibility of blood donation could be improved. In the second place, donation and transfusion information is checked and verified throughout the whole procedure with the help of barcode technology. In the third place, coming to blood information, not only is it guaranteed by barcode technology but also it benefits from various automatic testing and processing apparatus in blood donation and transfusion service.
- *Information integrity*: It is mentioned that blood bank information systems appreciate the close information flow in order to improve information integrity. Thanks to information share, candidate donors can be evaluated in all sides because of donation reaction, blood testing results, subsequent blood utilization, and electronic patient/health records. Moreover, the integral blood information can be built from initial blood donor, involved staffs and materials to the final blood recipient.
- *Information coordination*: A salient contribution of blood bank information systems lies in promoting blood bank working efficiency. However, the concurrent operations of blood analysis and processing impose various threats to the safety of blood donation and transfusion service. So information coordination plays a crucial role in this field. Here barcode technology is recommended and has been practically proved effective for such synergic workflow.
- *Information security*: Same as other medical facilities, information security is always a top issue with regard to life-threatening products in blood banks. The related topics cover protecting donor privacy, preventing unauthorized data operation, information system disaster recovery, and so on. A robust blood bank information system should guarantee not only the consistent data recovery but also quick disaster response due to its uninterrupted 24/7 operation mode.

Table I. The divisions and related data of a blood bank

Division	Responsibilities	Subjects and datasets
Administration	Supervision	Staffs
		Compartments
		Logistics
	Quality control	Donation suspension and elimination
		Product movement
		Quarantines and wastes

Table I. Continued

Division	Responsibilities	Subjects and datasets
Reception	Donor registration	Donors identification
		Barcode labels
		Electronic donor cards
	Donor screening	Historical records
Medical areas	Donor screening	Nurses' results
		Physicians' results
		Eligibility
	Blood collection	Donation records
		Donation monitoring
Mobile stations	Blood collection	Donation records
		Service sites
Laboratories	Blood screening	Immunohematological testing
		Microbiological testing
		Blood validation rules
	Blood grouping	
Manufacturers	Blood production	Blood components
	Waste processing	Quarantine
		Incineration
Stocks	Blood storage	Inventory records (Move in, transfer, and move out, etc)
		Expiry monitoring
	Blood dissemination	Recipient institutes
		Dissemination records (Send out, reception, return, etc)
		Information exchanging records
Property	Raw material management	Blood bags and labels
		Syringes and other disposable materials
	Apparatus management	Blood screening apparatus
		Blood manufacturing apparatus
		Information and computer apparatus
		Logistic equipments

3. Computerized Blood Bank Information Systems

In a blood bank, the automation benefits from a broad range of systems and apparatus, including automated manufacturing equipments, control systems, automated laboratory systems, computers, and laboratory or manufacturing database system, etc. All of them are organized in a hierarchy of hardware, software and network components (ISBT 2003). In essence, those systems firstly attempt to automate various processing, testing and producing activities in order to avoid human-introduced contamination and errors. At the same time, those automation systems are designed to streamline diverse operations taking place in a blood bank, such as donor screening, data analyzing, blood management and dissemination, etc. (BCSH 1995; CBER 2001) Nowadays all of above activities have been deeply intertwined with the computerized and networked information system in a blood bank.

The first of all, the blood bank information system enables the genuine autonomous blood processing without human intervention. Those processing results can be automatically imported into the blood bank information system as electronic data in database via various hard/soft interfaces. In the second place, thanks to the integral information system, the heterogeneous operations in the blood donation and transfusion service can be streamlined due to the consistent data and information. Similarly, the close network of blood donation and transfusion service can be built to cope with the challenging issues such as global optimization of blood dissemination, tracking adverse transfusion reaction, and the possible backward inspection.

Coming back to blood bank information systems, there are two alternative paradigms widely accepted in blood donation and transfusion service. The first one is to build the blood bank information system as a subsystem of hospital information system, exemplified by the blood bank module in Sushrut© (http://www.cdac.in/html/his/ bbank.asp). Such paradigm is recommended for hospital blood banks because it close integrates with other modules in hospital information system, such as in-patient management, billing, and out-patient management, etc. On the contrary, most independent blood banks prefer to the standalone blood bank information systems, which exchange information with outside systems via the special interfaces.

To understand the intrinsic difference between two aforementioned paradigms, it is necessary to capture their respective missions of hospital information systems and blood bank information systems. A hospital information system is generally developed with the objective of streamlining the treatment of a patient in the hospital, which allows physicians and other medical personnel to work in an optimized and efficient manner. Hence, the hospital information system is in essence a patient-oriented system with the underlying objectives to improve hospital efficiency. As to blood bank information systems, one of their objectives is no doubt to optimize the streamline of blood donation and transfusion service. However, the most important is to track every unit of blood and blood components from donation to infusion so that the safety could be fundamentally guaranteed. Obviously, such difference rooting in their missions brings the distinct models and frameworks in nature.

3.1. Autonomous Information System

A blood bank information system is in essence an autonomous system. And it exhibits its autonomy in two major aspects: the interfaces for autonomous data acquisition and the modules of intelligent decision-making support. The autonomous data acquiring interfaces, a salient feature of modern blood bank information system, can effectively reduce human-introduced errors. For instance, with the help of barcode printers and readers, most materials in that blood bank can be labeled and verified through their unique barcodes (FDA 2004). At the same time, owing to the autonomous interfaces to various blood testing and processing apparatus, most of raw data and detailed information can be archived without human intervention, which not only guarantees the consistency of blood bank information but also improves blood bank working efficiency (CBER 2001).

As to intelligent decision-making support, two intrinsic challenges call for the development of autonomous information analyzing technology in blood banks. The first one stems from the fact that most blood bank staff lack of professional computer skills. And the other one comes from the challenge of tremendous and heterogeneous data and information. Without the effective decision-making support tools, any blood bank information system will not be fully embraced by blood bank staffs, which no doubt impairs its deserved esteem. Consequently, two paradigms including data-driven and knowledge-based decision support modules have been advanced to answer the above challenging issues (Kros and Pang 2004; Li *et al* 2008).

3.2. Streamlined Information System

Barcode technology is an essential tool for the implementation of automated blood bank information systems. The full participation of barcode technology promotes blood bank working efficiency too. For instance, the procedures such as laboratory testing and component producing can be undertaken simultaneously in that all blood materials have been uniformly barcoded. Of course, besides barcode technology, the streamlined workflow should be attributed to the application of information and computer technology in blood banks. Thanks to network and information technology, it is possible to synergize various activities in blood bank via information share. As a consequence, one feature of modern blood bank information system is able to streamline overall blood bank procedure in order to improve working efficiency whilst keeping the high-level safety of blood donation and transfusion service (Brataas *et al* 1998).

3.3. Close Information System

The term “close” implies two aspects of blood bank management: In the first aspect, a blood bank information system should accomplish the complete information flow from blood donation to blood infusion; In the other aspect, the blood bank information system should keep as a black box to unauthorized users. Workflow coordination and information synchronization are particularly appreciated in modern blood bank information systems. It means that any data or information in blood donation and transfusion service will play a global role. On the other hand, in face of the network environment permeated with Trojan and virus programs, the blood banks should pay more attention on data security. It is necessary to build blood bank information systems in a hierarchical framework. Any operation on database should be rigorously verified and monitored. Similarly, the privilege of every group of users should be clearly configured so that even the authorized users are able to undertake the permitted operations only.

3.4. Open Information System

Paradoxically, at the same time, blood bank information systems should comply with the open platform for information share. As discussed in above, to form a complete information flow, blood bank information systems have to exchange various data and messages with outside subjects, for example, the hospital blood banks at least. However, due to the diversified information systems in blood donation and transfusion service, it is no doubt a challenging issue to implement the effective open interface for electronic data interchange. In this field, the related advances include "United Nations/Electronic Data Interchange for Administration, Commerce and Transport" (UN/EDIFACT), "American Society of Testing and Materials" (ASTM), and "Health Level Seven" (HL7) (Hirsch and Brodheim 1981; Weisshaar 1991; Larson 1999).

3.5. SIBAS @ Macau

SIBAS, the acronym of "Sistema Integrado de BAncos de Sangue", is a blood bank information system specially designed for Macau Blood Transfusion Center (Li *et al* 2007). It is equipped with many advanced technologies, including electronic donor cards (Li and Dong, 2006) and the ISBT 128 barcode technology (Li *et al* 2006). In this section, the overall blood bank information system will be firstly examined in detail. Then, we will further elucidate the solutions of barcode technology and electronic donor card system in that blood bank information system.

3.5.1. Infrastructure of SIBAS

Any process or procedure within a blood bank must comply with the safety of blood donation and transfusion service. Once a blood donor arrives at the blood bank, a unique donor number and donation number will be assigned to that donor; before his/her donation, the nurses will conduct a series of simple physical examination; then, based on the donor's background information and the nurses' examination, the physician will make the final decision whether the donor is suitable for blood donation without influence on his/her health

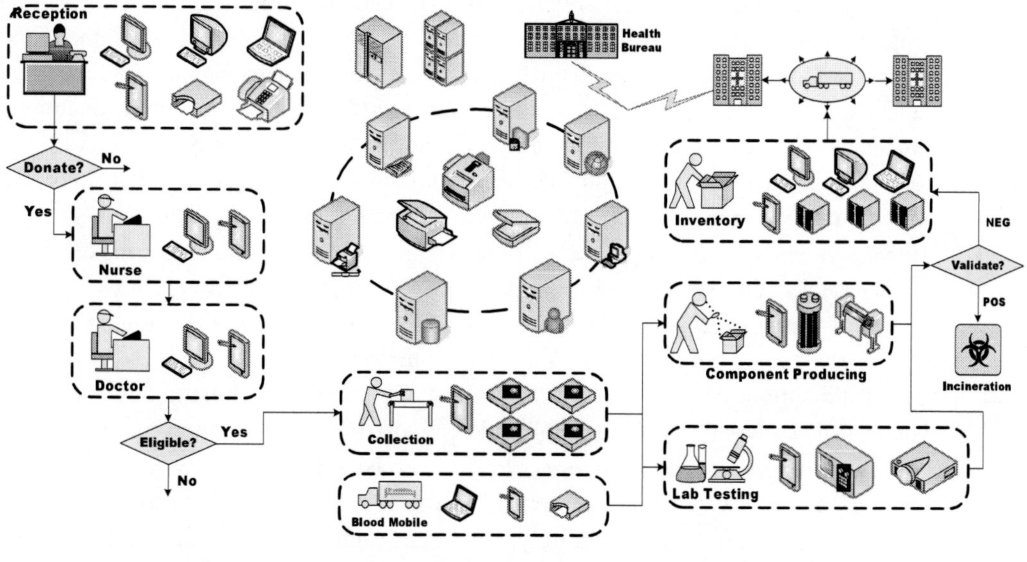

Figure 3. Workflow and infrastructure of SIBAS.

condition. If donor screening is successful, that blood donor is eligible to enter the area of blood collection. After that, the blood donor will be requested to have a rest and the nurse will record his/her post-donation reactions.

The blood specimen is then transferred to the professional laboratories for blood ABO/Rh grouping and examinations. It pays special attention on those blood-transmitted diseases, for example, hepatitis (a liver infection), HIV (the virus that causes AIDS), HLTV I/II (the virus associated with a rare form of leukemia), and syphilis, etc. Meanwhile the whole blood is submitted to be separated into red blood cells, platelets, plasma, and other human blood clotting agents. If the validating results from blood testing laboratories are negative, the produced blood components will be conveyed to blood stock for inventory management. Otherwise, those blood components and blood specimen should be incinerated right away.

As introduced in previous, automation systems have permeated in blood banks, and are improving their operating efficiency substantially. Coming to Macau Blood Transfusion Center, it owns the full-function intranet/internet network, all-round autonomous blood testing and producing apparatus, the barcode-based blood tracking and managing system. Its computerized and networked blood bank information system can be illustrated as in Figure 3. It is hereby appropriate to categorize the involved apparatus and systems as following:

- *Personal computers*: This kind of apparatus includes desktop computers, mobile notebooks, and tablet computers. They run Windows-compliant environment, including Windows 98©, Windows 2000© and Windows XP©. Those computers scatter in every division of the blood bank, including reception, collection area, laboratories, stocks, and even mobile blood stations. Most of them connect to blood bank databases through the intranet network in order to share and exchange blood bank information.
- *Servers*: The major components support information share and workflow coordination. In the blood bank information system, two measures are adopted to guarantee data safety and network robustness. The first one is one-to-one service strategy, namely, a serve (maybe as virtual machine) is specially designed and optimized for a specific service, for example, data management (independent database for donor, donation, laboratory, stock, staff, and materials, etc.), document management (printing, scanning, and faxing, etc.), Virtual Private Network (VPN) to hospitals and health bureau, and Internet information accessing. Secondly, there is a server for synchronous backup and a magnetic tape server for daily backup.
- *Automated blood testing apparatus*: Macau Blood Transfusion Center makes good use of various automated blood testing and producing apparatus in order to secure the quality of testing results and final blood products. For instance, there are a series of sophisticated instruments for blood tests, including Microdom®'s "Mitis2" for blood typing and antibody screening test, Vitros®'s "ECi" for blood immunodiagnostic test, Roche®'s "MagNA Pure LC System" and "Cobas Amplicor" for microbiologic test.
- *Peripheral apparatus and instruments*: Macau Blood Transfusion Center adopts Terumo®'s "T-RAC System" to control donation procedure and record the blood donor's health condition accurately, including duration, volume, and average/ minimal/maximal flow rate, etc. Other major components include barcode reader/printer and electronic donor card reader/writer, which will be further examined in the following sections.

Figure 4. Comparison of barcodes in a component label.

3.5.2. Barcode Technology

According to the barcode labeling guideline in blood banks, the objective of uniform blood component label is to reduce the danger of incompatible transfusions caused through human errors via presenting important information in a clear, logical and easily recognizable format (CBER 2006). Nowadays the barcode labeling technology has participated in most divisions of a blood bank from the areas of reception and collection to laboratories, stocks, and even subsequent blood dissemination. In summary, the role of barcode technology in blood donation and transfusion service can be concluded as follows:

- *Controlling workflow*: Given the biohazardous nature of blood products with potential virus, the procedure of blood donation and transfusion service should be controlled rigorously in every step. Barcode labeling technology has been proven effective to coordinate and control the mentioned processes. For instance, once a blood donor arrives at Macau Blood Transfusion Center, he/she will be assigned a unique donor number and donation number. Then, benefited from the printed barcode labels, the subsequent processing in parallel, such as testing and screening in laboratories, manufacturing in components and validating in stocks, can contribute to working efficiency fairly.
- *Managing blood components*: The barcode technology comprising barcode and labels provides an effective manner to manage the confusable blood components in that it is impractical to distinguish them by naked eyes. Firstly, the machine-readable barcode labels enable information automation so that the human errors could be excluded. Similarly, it enhances information security to acquire and validate information without human intervention. Finally, the barcode label including readable texts improves the sorting and settling capability in terms of mass storage and management of blood components.

- *Tracking blood donation and transfusion service*: Besides its applications in blood management, barcode labeling technology plays a critical role throughout the procedure of blood donation and transfusion service, which effectively prevents the potentially incompatible blood transfusion. In blood banks, the critical information about blood components, such as donation number, blood group, product code and expiration date, has been printed out as the barcode label attached to that blood bag. Then, from the initial blood bank to the final transfusion service, the circulation procedure can be organized and coordinated in a whole with the help of machine-readable barcode labels. The information about blood usage in transfusion service can also be sent back to the initial blood bank with those unique donation identification numbers. Finally, in blood transfusion services, barcode labeling technology should be deployed as in blood banks. Not only for information feedback, but also it is necessary for the security of blood transfusion practice (Turner 2003).

The barcode paradigm at Macau Blood Transfusion Center has been transformed from Codabar to ISBT 128 (Li *et al* 2006). Not only information security but also data content is hereby enhanced. In the first place, the component label of ISBT 128, comparing with the one of Codabar, is in a thoroughly different style. It is necessary to rearrange the format of barcode label including font typeface and content layout as shown in Figure 4. In the second place, to comply with ISBT 128, more critical information, such as donation type and blood bag, is encoded in the label of blood component. Finally, due to different barcode symbologies and standards, the major transformation is related to the barcode itself, including barcode format as well as barcode content as shown in Figure 4 and Table II.

- *Donation identification number*: The number, identifying a blood donation uniquely, takes two different roles in blood banks. The first one is to organize and coordinate all procedures related to that donation. Another role is to identify a blood component in order to facilitate the globally uniform management.
- *Blood group*: Generally speaking, there are totally 8 types of ABO/Rh combinations as the blood groups. Hence, 1-bit digit is reserved for blood-group codification in original Codabar paradigm. Besides 8 blood groups, the code "0" is designated for any uncertain or invalid blood group. Nevertheless, in ISBT 128, the blood group of a component will be coded with the type or status of blood donation. In other words, there are more than 128 different codes for encoding and interpretation of blood groups.
- *Product code*: The original Codabar paradigm merely provides 1-bit digit for product encoding and interpretation. Obviously, it is too limited for the expanded classes of blood components, let alone the combination of various modifier(s) and/or attribute(s). Instead, ISBT 128 uses a scheme of Class, Modifier(s), Attribute(s) and Core Conditions to identify the specific component. In the meantime, the new coding scheme can optionally encode the information about donation type and whether the product has been further processed.
- *Expiration and collection date*: The expiration date, and even collection date, is of vital importance for any blood component. They should be pinpointed in any barcode labeling paradigm.

Finally, it is noteworthy that barcode technology should be integrated in the blood bank information system as an essential component. Take SIBAS© for example. It equips the professional "Zebra®" printers for barcode label production and the barcode readers for information acquisition. As shown in Figure 3, the barcode printers and readers, in spite of distribution in different divisions, are actively linked to the central database. Therefore, all information can be rigorously checked and verified in the procedure of barcode label production. Meanwhile, to secure the high fidelity, all barcode labels are dynamically generated and produced by the proprietary programming language "ZPL II©".

Table II. Comparison of barcode contents in a component label

	Format		Bits		Content	
	Macau Codabar	ISBT 128	Macau Codabar	ISBT 128	Macau Codabar	ISBT 128
Donation No	yynnnnn	appppyynnnnnnff	7	15	year + *SN*	*FIN* + year + *SN* + *FF*
Blood Group	g	ggre	1	4	ABO+Rh	ABO + Rh + *DT*
Product Code	o	aoooottds	1	8	*PDC*	*PDC* + *DT*
Expiration	dd-mm-yyyy	cyyjjj	10	6	day + month + year	century + year + Julian day
Donor No	vvvvvvv	appppvvvvvvvvvvvvvvvv	7	21	*SN*	*FIN* + *SN*
Staff No	uuu	appppuuuuu	3	10	*SN*	*FIN* + *SN*

FIN: Facility Identification Number; SN: Serial Number; FC: Flag Characters;
DT: Donation Type; PDC: Product Description Code.

3.5.3. Electronic Donor Card System

It is known that blood donors should subject to the rigorous screening procedure with the concerns on blood transfusion safety. In other words, any blood donor has to stay at the blood center to follow a series of precautious steps including registration, physical examination, and evaluation before the formal blood donation. Nevertheless, not all blood donors are willing to follow such tedious procedures. As a matter of fact, there are over 50 blood donors aborting their donation annually without reasons at Macau Blood Transfusion Center. Thus, it is a consistent challenge for any blood center to recruit and retain voluntary blood donors.

Many investigators and developers attempt to solve this topic through the psychology and behaviors of blood donors. However, no matter from which perspective, the quality of service of blood centers plays a key role in determining the willingness of blood donors. For the blood centers, it can be approached with the help of ISO quality controlling system, or by some quantitative indices such as cost-per-unit. However, from the perspective of blood donors, their evaluation is often based on a few subjective feelings including comfortableness, interest, and others.

An electronic donor card system, which not only optimizes the procedure of blood donation but also enhances the satisfaction of blood donors, is recommended in this section (Li and Dong 2006). Firstly, such system facilitates information acquisition and production during donor registration. An additional benefit is that, in the age of information technology, blood donors often prefer to use electronic donor cards. Therefore, the quality of service of a blood center will definitely be enhanced with the introduction of the electronic donor card system.

The whole blood donation service, as shown in Figure 3, commences with the essential steps of donor registration and donation evaluation. In general, such interactive procedure in a blood center can be outlined as follows: at the site of reception, a blood donor should firstly submit his/her donor card, if any, and a piece of donation form including personal data, medical history and health status to a blood center receptor; the receptor will review the submitted donation form as well as donor card, then make a preliminary decision on the blood donor's eligibility; if eligible, this blood donor will be assigned a unique donation number and then transferred to medical areas for physical examination and further screening of donation eligibility. If validated, this blood donor is qualified for blood donation. To the end of these sessions, the blood donor can get back his/her donor card with this time of donation record.

Owing to the deployed blood bank information system, any data in above procedure can be assembled and preserved well in a centralized database. Then, the staffs can review and approve those return donors' eligibility by referring to their former donation records. However, many bottleneck challenges still persist in aforementioned procedure, especially those related to human-machine interaction. For instance, blood donors have to fill many regular items in the donation form, and blood center receptors have to input those data into the blood bank information system once again. Moreover, after each donation, the donor card should be updated manually. Intuitively, it is possible to optimize those tedious procedures by means of automation equipments.

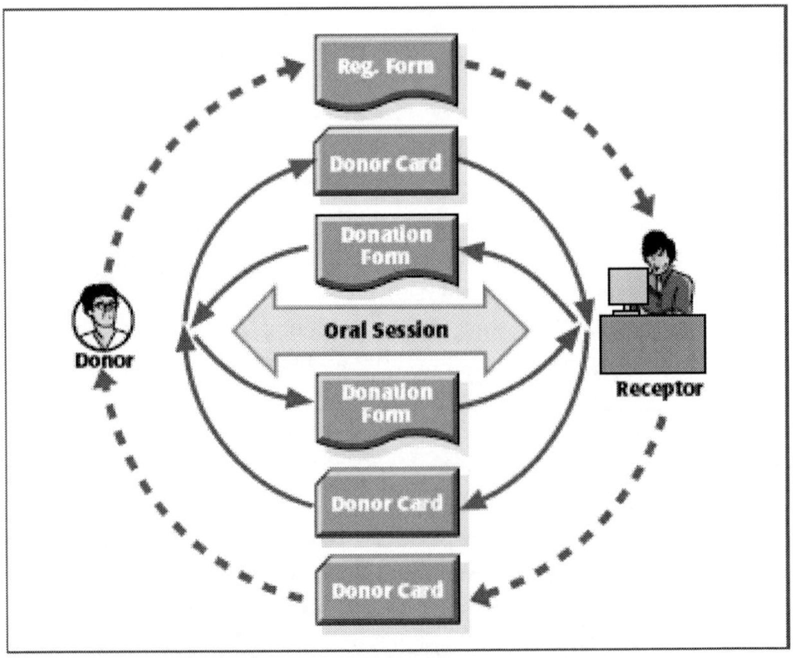

Figure 5. Interactive registration of blood donors.

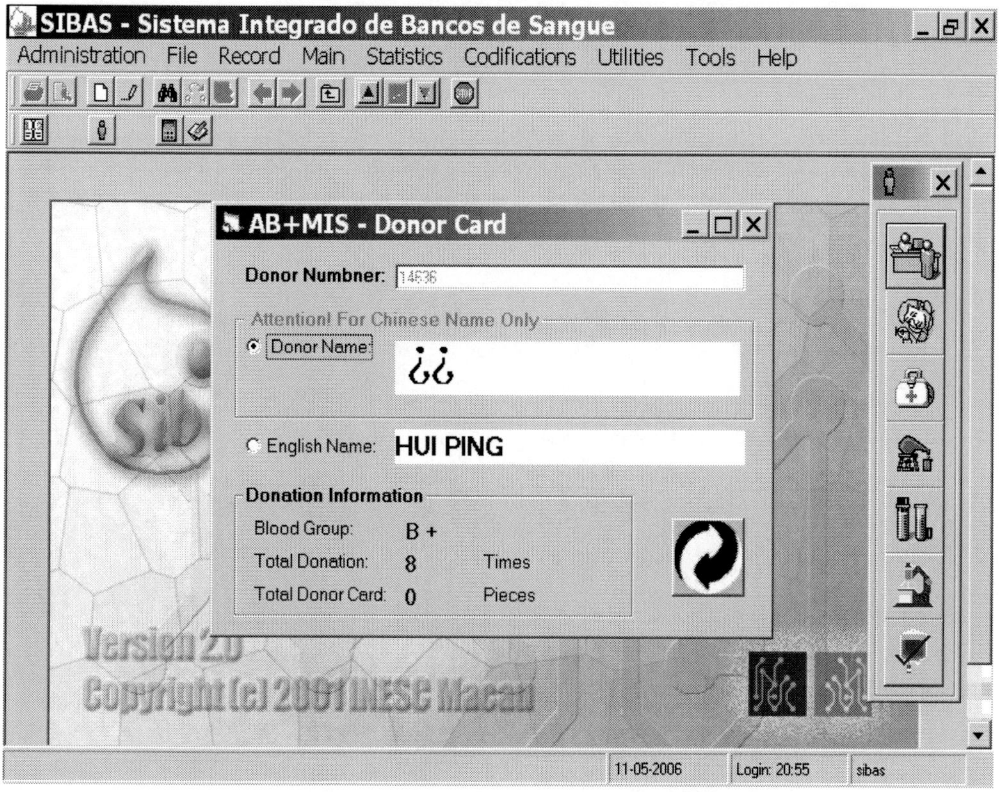

Figure 6. Fully integrated electronic donor card system.

As a matter of fact, with the electronic donor card system, above procedure of donor registration has been redesigned. As shown in Figure 5, only new blood donors involve in the steps of registration forms and card issuing. For those return donors, they just need to present their electronic donor cards to blood bank receptor. With the help of electronic donor cards, the receptor can obtain their former records dynamically from the blood bank information system. For an eligible donor, a piece of donation form will be printed out in site with available information. Further, once a blood donor is qualified by physicians for donation, his/her donor card will be updated automatically with both electronic and printed messages.

Hence, with the introduction of new electronic donor card system, it is predictable that the quality of blood donation service will be enhanced in the following aspects:

- *Operation efficiency*: The processing time can be shortened in that blood donors needn't to fill in many regular items, and the same for blood center receptors. Moreover, the steps with regard to donor cards will be under fully automatic control.
- *Donor satisfaction*: The personalized donation experience, from considerate donation forms to personal electronic donor cards, will definitely contribute to the willingness of blood donors.
- *Public image*: Electronic donor cards, integrated with the blood bank information system, will improve the public image of blood center as a modern institute. The symbols of advanced technology are helpful to enhance the confidence of blood donors.

Coming to the electronic donor card system at Macau Blood Transfusion Center, firstly, the electronic donor cards will be issued to blood donors with their personal records; secondly, the blood bank is equipped with automation equipments for information processing; finally, all data are an integral part of the blood bank information system for centralized management. In a word, the electronic donor card system is complete with electronic donor cards and card readers/writers. In addition, it has been fully integrated into that blood bank information system (Figure 6).

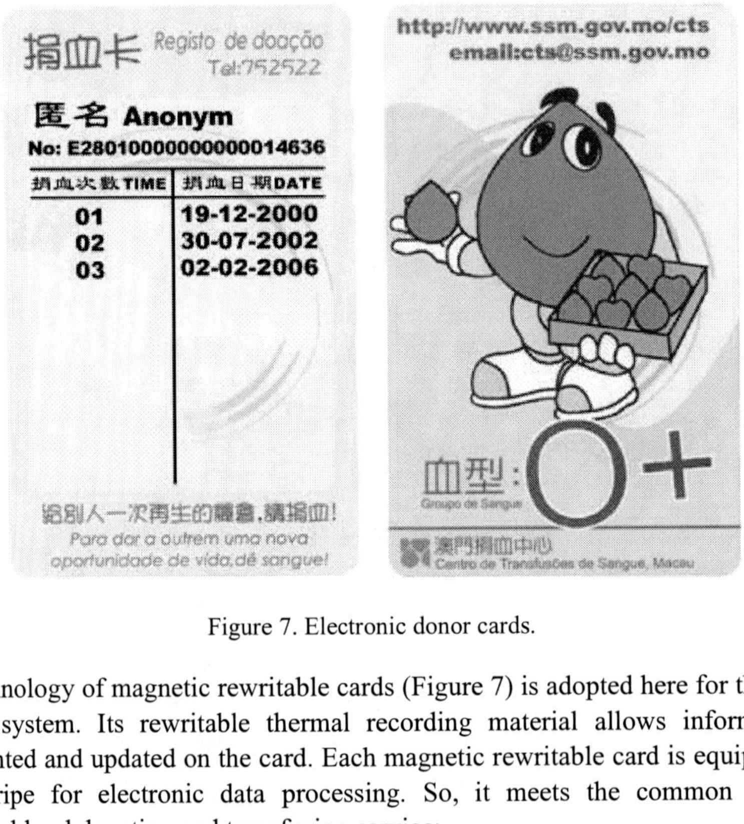

Figure 7. Electronic donor cards.

The technology of magnetic rewritable cards (Figure 7) is adopted here for that electronic donor card system. Its rewritable thermal recording material allows information to be instantly printed and updated on the card. Each magnetic rewritable card is equipped with the magnetic stripe for electronic data processing. So, it meets the common requirements necessary for blood donation and transfusion service:

- *High reliability*: Electronic data processing is contributive to information consistency and validity. Furthermore, the multiform representation, including electronic data in magnetic stripe and printed records in card surface, guarantees the reliability of information circulation.
- *Cost-effectiveness*: In general, a blood bank serves thousands of blood donors, each of whom needs a piece of electronic donor card. In practice, more cards should be ready for possible renewal and damage. Obviously, their costs have to be under rigorous control. Here each piece of magnetic rewritable card costs less than $0.5 dollars.
- *Electronic data interchange*: In view of massive operations in a blood bank, the interface for electronic data interchange is necessary to process data and information accurately. Magnetic rewritable cards provide a magnetic interface for electronic data interchange.

- *Printable card surface*: The donor card with donation records, especially the donation dates, is of great help to remind those regular blood donors. Benefited from the rewritable thermal recording material, visual information can be printed and renewed more than 500 times in a magnetic rewritable card.

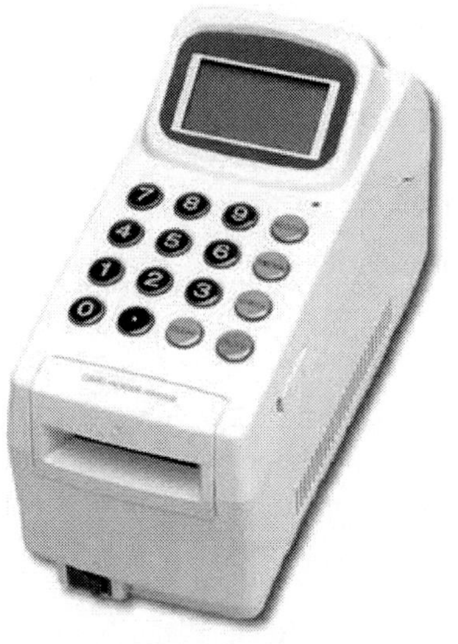

Figure 8. Magnetic card reader/writer wit thermal printer/eraser.

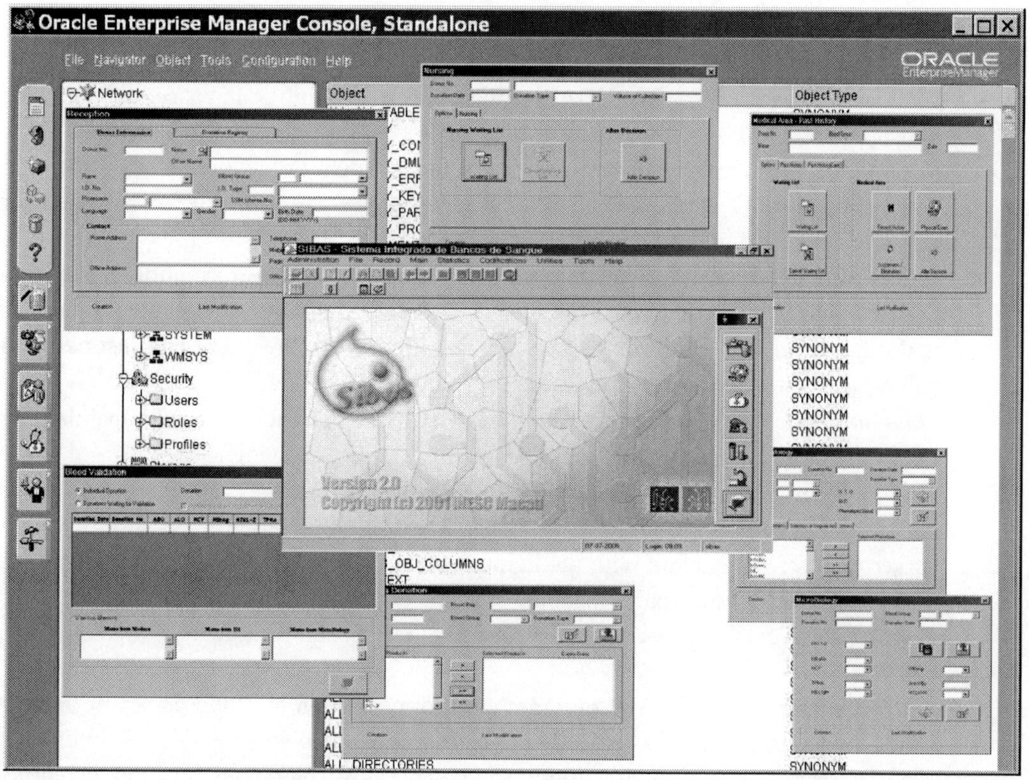

Figure 9. The software framework of SIBAS©.

Finally, the magnetic card reader/writer with thermal printer/eraser (Figure 8) can serve as the data interface between magnetic rewritable cards and the blood bank information system. For magnetic rewritable cards, it firstly accesses the relevant data from the blood bank information system in accordance with the information in electronic donor card; then, it can update the electronic and printed records of blood donation in that electronic donor card. In terms of the blood bank information system, the magnetic card reader/writer with thermal printer/eraser servers as the terminal for information enquiry. Moreover, it accounts for finalizing data and information in electronic donor cards too.

3.5.4. Software Implementation

Based on the preceding discussion, a blood bank information system has to confront with data integrity, workflow synergy, information security, and other challenging issues. The blood bank information system, SIBAS©, follows the client/server infrastructure, namely, implementing data management, user role controlling, electronic data interchange in server end while providing the distributed information processing ability in client end.

In terms of data and information management, SIBAS© provides the all-round Oracle®-based database solution in server end:

- *Data archiving*: The independent Oracle® databases are configured for blood donors, blood donation, laboratory testing, blood producing, and inventory management respectively. In general, there is an optimized database for every blood bank division so that the efficiency of data management and information processing could be improved.
- *User role controlling*: In SIBAS©, there is a specific database for use role management and controlling. Before any operation on real database of blood bank information, the user has to pass a series of independent procedures of identification and verification. On the other hand, any operation by any user will be recorded for possible backward inspection in that database too.
- *Middleware*: To improve the efficiency of data operation and information processing, SIBAS© makes good use of the middleware, namely, packaged procedures, functions and triggers in Oracle® database for data operation. Then the client end just needs to submit its request and receive the desired results.
- *Data mining for decision-making support*: SIBAS© owns the powerful capability for decision making support, which will be detailed in the following sections.
- *Data backup and disaster recovery*: There are independent servers for synchronous and daily backup respectively. Both of them are blind to blood bank staffs. A HP® server runs for hot backup with the help of Oracle® Recovery Manager (RMAN). At the same time, a third-party backup system – Veritas®, combined with the HP® magnetic tape database, accounts for the daily incremental backup.

Coming to the client end, SIBAS© provides a uniform solution of blood bank information system for all divisions. In other words, although it is in essence an integrated environment including donor recruitment, donation monitoring, laboratory blood testing and inventory management, SIBAS© is able to configure itself dynamically according to the group privileges of different user roles, for example, the reception, the nurse, the doctor, the immunohematology, the microbiology, the validation, and the component, etc. In principle, every group of users lives only in its legitimate society. It has the own interface suitable for the specific workflow, and calls different subroutines for data operations and statistic reports (Figure 9). Of course, despite rarely, SIBAS© is powerful enough to permit those users granted more than single user role. For instance, there is a super role "supervisor" allocated for system configuration and maintenance.

4. Computerized Decision Making Support

4.1. Decision Making Support in Blood Banks

It is a challenging task to manage blood banks well: blood products have to be processed and disseminated promptly, meanwhile all operations should be under rigorous controls. On the other hand, a blood center has to take various resource allocations and optimizations into account so as to support the daily operations in blood donation and transfusion service. As a matter of fact, from donor recruitment to blood dissemination, blood bank staffs are involved in various decision making procedures. Hence, other than data management and workflow coordination, an efficient blood bank information system is desired to provide the effective decision making support for customer relationship management (CRM), enterprise resource planning (ERP), supply chain management (SCM), and so on. (Hanson 1996; Kros and Pang 2004)

4.1.1. Customer Relationship Management

Customer satisfaction is a critical indicator in contemporary enterprise evaluation (Roh 2005). No surprise, a blood bank should enhance the blood donors' satisfaction as well as loyalty too. It seems more terrific in face of the double roles of a blood bank: in one side, the blood bank should provide the considerate service in order to attract more and more voluntary blood donors; in the other side, it should guarantee the timely, impartial blood supply to blood transfusion facilities. The implementation of blood bank information systems enables accurate recording and quick responses to various user requirements, which definitely contributes to the enhancement of customer satisfaction. Nevertheless, with its massive records and superior computing capability, a blood bank information system can do more to enhance the customer satisfaction.

The first of all, the nonprofit blood donation and transfusion service makes it a central theme to recruit and retain enough voluntary blood donors. In general, the willingness of blood donors is determined by their impression on blood donation service. It inspires the lasting efforts for the optimization of blood supply chain, blood bank operations, and user information interaction. Moreover, a few accessory strategies have been recommended for blood donation promotion, such as encouraging propaganda, interesting souvenirs, and warm donation reminding, etc. But there is no common strategy for all regions due to the social, economic and cultural diversities. Hence, it has been a long tradition of interest in blood donation and transfusion service to analyze the behavior patterns of local blood donors

(James and Matthews 1996; Brittenham 2001; Glynn 2002; Bosnes 2005; Zaller 2005), including their distributions and variations.

In the other side, the blood bank, as a nonprofit facility, generally depends on the political and financial support of governments and other public organizations, whose decisions are often influenced by the appraisals of blood transfusion facilities. Therefore, a blood bank has to handle its relationship with blood transfusion facilities well (Sime 2005). The relevant affairs include blood dissemination, information exchange, technical cooperation, and so on. Obviously, the blood bank information systems are able to facilitate both blood dissemination and information exchange. Meanwhile, the computerized decision support modules are desired to analyze and predict the needs of blood transfusion facilities.

4.1.2. Enterprise Resource Planning

Enterprise resource planning is recommended to manage and optimize various resource allocations, from raw materials to human resource, within an enterprise (Gregory and Eric 1980; Gupta 2004). There have been two recognized strategies, that is, forecast-based redundant allocation and need-driven reduced allocation. Coming to a blood bank, it usually follows a hybrid paradigm optimized for the procedure of blood collection, processing, storage, and dissemination.

Generally speaking, the arrival of voluntary blood donors has to be characterized as a random event (Bosnes 2005). In other words, any decision in resource allocation for blood collection follows a predicative function in accordance with the historical data and their variations. For example, if there were more blood donors visiting a mobile blood station in a specific period, it is obviously justified to allocate more enterprise resource including materials, apparatus, staffs, and so on. In contrast, once the collected raw blood passes microbiological and immunohematological validations, the needs of blood transfusion facilities often determine the subsequent production and storage of blood components in a blood center. As a consequence, legible statistical reports and appropriate decision supports are of vital importance in blood bank information systems for blood bank resource allocation.

4.1.3. Supply Chain Management

Blood donation and transfusion service is built on an interdependent, distributed network: a few blood banks are in charge of blood collection, processing and dissemination; many blood transfusion facilities request blood products from blood banks and execute the final step of blood infusion. Hence, the researchers are always interested in the optimization of blood supply chain including the number and distribution of blood banks (Sime 2005). However, in terms of a specific blood bank, the central topic turns to guaranteeing the timely, effective blood supply, namely knowing blood consumptions, deciding blood production, optimizing blood dissemination, and so on. Computerized decision making support is just aimed to coordinate the movements of raw materials and final products in order to achieve cost and time minimization (Gardner 1990; Spackman and Beck 1990; Steven and John 2000; Petäjä 2004).

In fact, the researches on inventory management and blood dissemination have a long tradition in blood donation and transfusion service (Gregory and Eric 1980; Raj and Tarun 1991). The first of all, both blood donations and patient needs are a random and dynamic procedure. Therefore, supply chain management in a blood bank, similar as enterprise resource planning, usually follows a hybrid paradigm of predication-based redundant production and need-driven reduced dissemination. In other words, a blood bank has to order various raw materials and accessory equipments in accordance with the local blood donors as well as patient needs. On the other hand, the different therapies often ask for the special treatments of blood components by distinct additive reagents (e.g., anticoagulant and cryoprecipitate, etc) or laboratory processing (e.g., irradiation, deglycerolization and heparinization, etc), which influence their expiry dates too. Thus, it is a topic worthy of exploration to optimize blood production so as to meet patient needs and maximize blood utilization at the same time.

4.2. Computerized Decision Making Support

Thanks to contemporary information and computer technology, there have been quite a few decision support systems proposed for blood bank information systems (Smith 1985; Sielaff 1989; Connelly 1990; Spackman 1990; Hanson 1996; Petäjä 2004; Bosnes 2005; Roh 2005). Generally speaking, they attempt to make use of data and information in a blood bank information system at three different levels: data screening, information analysis, and knowledge discovery.

The conventional blood bank information systems pay more attention on recording and assembling various involved data and resources well. Then it is possible to track and control the huge amount of data and information for decision making support. For instance, any blood bank information system should be able to generate alarming messages once the specific kind of raw materials or accessory equipments exceeds their thresholds. Moreover, blood bank staffs usually take advantage of the historical records in blood bank information systems for blood donor screening.

In contrast, modern blood bank information systems are more interested in data analyses so as to predicatively optimize blood bank resource allocation and management. Take supply chain management in a blood bank for example. More than alarming messages, it is necessary to provide effective requirement analyses and decision evaluations based on a series of quantitative evidences, including the distributions and the variations of blood donors, blood products, and so on.

Furthermore, it has been pointed out that all of quantitative, pseudo-quantitative and qualitative languages are necessary to describe the heterogeneous data and information in blood bank information systems. Thus, it is not a universal strategy of data analysis based on the quantitative statistics. Another popular methodology for computerized decision making support is built on the practical experience from blood bank professionals or the implicit knowledge by data mining or knowledge discovery. Generally speaking, the rule-based expert systems take advantage of the long-term practical expertise while other advanced technologies make decisions with their private knowledge.

Finally, it is worthy to point out that computational intelligence usually takes an assistant role in the decision making procedure of blood bank staffs because any operation in a blood bank should be committed to a specific staff. In other words, it is the blood bank staffs, but not computers, answering for the operations or decisions in that blood bank. Of course, legible analytical reports and effective decision support are definitely contributive to working efficiency as well as working performance of blood bank staffs.

4.2.1. Data-driven Decision Making Support

Decision making is a complicated procedure of brain activities, which can not be characterized by any explicit languages or formula yet. However, computation and tradeoff have been widely accepted as an alternative implementation of human decision making. Therefore, many paradigms of decision making support are in essence a kind of computational methods. In other words, driven by the massive data in blood bank information systems, computational decision making support is to calculate the costs-to-benefits of various solutions and then select out an optimal one (Jelles 1993; Hanson 1996).

4.2.1.1. Hard Computing

In tradition, there is a set of clear formula or models to calculate the final results from raw data. These formula or models are either from the long-term practical expertise or by rigorous numerical analysis. Such kind of solutions is usually termed as hard computing methodology because there is a certain result given the raw data. Up to date, the most popular hard computing methods for decision making support are statistical analysis and the derived probabilistic reasoning.

- *Statistical analysis*: It has been pointed out that human decision making can be essentially regarded as a procedure of computation and tradeoff. Obviously, if there are quantitative indices measuring their costs and benefits, it is not difficult to find out the desired solutions. Statistical analysis is right aimed to calculate various quantitative indices, including means, deviations and expectations, from massive historical records. Those values make the distributions and the variations of past needs and consumptions clear to decision makers. Then, it is possible to make the appropriate decisions for optimal resource allocations.
- *Probabilistic reasoning*: In fact, new situations maybe differ from the historical ones. Consequently, the reasonable conjectures are sometimes necessary. Probabilistic reasoning is hereby developed to find out such conjectures quantitatively by introducing conditional statistics. Then, based on the famous Bayesian theory, it is possible to work out the potential costs or risks of new events. Many contemporary solutions, for example, Monte Carlo Markov Chain analysis, are developed from such idea too.

4.2.1.2. Soft Computing

In computational decision making support, a common challenge is that there is no available formula or models. Namely, the computers have to find out the implicit relations from historical records by themselves. After the long-term exploration, many computational paradigms have been proposed to implement such unknown mapping (James and Matthews 1996; Stephen and John 2000). In general, the users merely provide a set of historical data, and then let the computers learn from them. This kind of solutions is termed as soft computing methodology because there is usually no definitive result but an acceptable region.

There are two kinds of different cases in searching the unknown relations. The first case is that the historical data contain both raw data and related results. Then, it is possible to assume an arbitrary mapping at first, and refine such mapping step by step. Dependent on the historical data and the desired target, the final result may be a modeling function or a mapping network. In another case, there are no known results but raw data only. The procedure of searching unknown relations is hereby a kind of self-organizing mapping. The final results can be regarded as the intrinsic relations of raw data.

As a matter of fact, most of current soft computing methods for decision making support are right aimed to settle above two problems. For instance, Artificial Neural Networks can be trained for the first case by error back-propagation or the second case by self-competition. Support Vector Machines adopt the strategy of expected risk minimization to refine the unknown mapping in stead of empirical risk minimization in Artificial Neural Networks. Those methods in Evolutional Computation attempt to find out the unknown relations by iterative crossover and mutation. It has been approved that all of them can elicit useful knowledge from historical data and support current decision making effectively.

4.2.2. Knowledge-Based Decision Making Support

Other than the data-driven paradigms of computation and comparison, the knowledge-based decision making support is another popular paradigm. Generally speaking, the knowledge here comes into being either from long-term practices or by numerical analyses. In computerized decision making support, the knowledge is often taken advantage to build computing models or linguistic rule bases (Smith 1985; Sielaff 1989; Connelly 1990; Spackman 1990; Hanson 1996; Petäjä 2004; Bosnes 2005; Roh 2005). Then, given the known facts, the procedure of decision making is to find out the corresponding solutions with the help of computing models or rule bases.

However, it is not an easy task to convert the practical expertise into computing models or linguistic rules. In the first side, human knowledge is generally abstract and ambiguous. Human beings can adapt to the real-life situations easily. But it is nearly impossible to settle all situations for computer implementation. In fact, most solutions can be regarded as a tradeoff to the general situations with maximum probability. For instance, the aforementioned soft computing methods are just aimed to find out such models from massive historical records.

In the second side, not all knowledge and experience are quantitative. On the contrary, practical expertise can often be expressed as linguistic statements only. Contemporary knowledge engineering is right devoted to eliciting human knowledge and expressing it as linguistic rules. At present, Fuzzy Logics is a most popular subject in knowledge engineering. It not only makes the linguistic rules more suitable for human languages but also converts the conventional inference strategy, namely searching and matching, into numerical calculation. Obviously, the latter is more suitable for computer implementation.

At last, it is noteworthy that a practical system generally contains more than one strategy of computerized decision making support. Appropriate combination is necessary to cope with different situations and tasks. Here the combination may be in the intra-method level, such as

Fuzzy Neural Networks and Bayesian Neural Networks, or in the inter-method level, such as statistical analyses and rule-based inference systems.

5. Computerized Decision Making Support in Blood Banks: A Case Study

In general, a computerized blood bank information system refers to acquiring, validating, storing, and circulating various data and information electronically in blood donation and transfusion service. Given the top priority of concerns on blood transfusion security, most reported systems are particularly concerned with the issues of data credibility, information consistency, and system reliability, etc. Even official implementation and evaluation guidelines pay little attention on the topics other than security and reliability in blood bank information systems (BCSH 1999; ISBT 2003; CBER 2005).

However, from the perspective of blood bank staffs, they often seek more support from the blood bank information systems other than inputting and retrieving historical data only. At least, such system should be able to assemble the heterogeneous data into legible reports for appropriate decision making support. It is known that any reasonable decision should comply with the objective data and subject to the supervision of knowledge. From effective donor screening to optimal blood dissemination, those electronic data in a blood bank information system indeed can contribute to various blood bank decisions. Thus, for an efficient blood bank information system, it is not a trivial task to develop the effective decision support modules.

A series of literatures have been published with emphases on decision making support in blood bank operations. Generally speaking, those papers tried to approach this subject from the following perspectives: deriving out hypothesized models or parameters for quantitative description or evaluation of blood bank operations; constructing the knowledge-based systems to support local operations in blood banks. For example, in the first side, Gregory (1980; 1984) proposed to measure blood bank operations with two performance indicators as shortage rate and outdate rate, which were revisited by Raj and Tarun (1991) for a model of perishable inventory. Bosnes *et al* (2005) took advantage of logistic regression models to fit the arriving behaviors of blood donors. On the other hand, there have been research groups introducing the expert system for platelet request evaluation (ESPRE) in 1990s, including knowledge acquisition and representation (Sielaff *et al* 1989; Connelly, Sielaff and Scott 1990). And Gardner *et al* (1990) used a HELP computer system, with built-in physician-approved criteria, to assist in criticizing the orders of blood products.

Based on above discussion, it is safe to claim that the necessity and feasibility have been recognized in terms of computerized decision making support for blood bank operations. However, a unified perspective has yet not been well established to implement decision making support in blood bank information systems. Firstly, most of reported methods and systems were merely devoted to the specific procedure such as inventory management (Raj and Tarun 1991) and transfusion decisions (Sielaff *et al* 1989). Secondly, the former exploration was limited in the level of strategies and policies (Gregory 1980; Kros and Pang 2004). Finally, those experts had prepared most data and information by themselves because they were interested in the characteristic parameters or decision models only (Raj and Tarun

1991; Gelles 1993). In other words, few of them were constructed with the full support of computerized blood bank information systems.

Once within the frame of blood bank information systems, the topic of decision making support should be reconsidered for several unique features. The first of all, the blood bank information system means an environment of full automation in the sight of data and information for decision making support. Secondly, decision support modules, embedded in the blood bank information system, are close integrated with blood bank operations from donor screening to blood dissemination. Hence, the major objective of this paper is to address the underlying mechanisms of computerized decision making support, with enough literature review, in blood bank information systems.

As mentioned in above, the exemplified blood bank information system, SIBAS, is specially designed for blood donation service at Macau Blood Transfusion Center, and equipped with many advanced technologies. With respect to computerized decision making support, two kinds of paradigms have been deployed in that blood bank information system: rule-based expert systems and quantitative statistical analyses. It is worthy to point out that the decision making support modules, different from other experimental systems, have been fully integrated into that running blood bank information system. Furthermore, both kinds of decision making support modules are distributed in that blood bank information system so as to support the decentralized affairs. Finally, the decision making support modules merely provide analytical results and operational suggestions. Any decision should be validated and approved by the corresponding blood bank staffs.

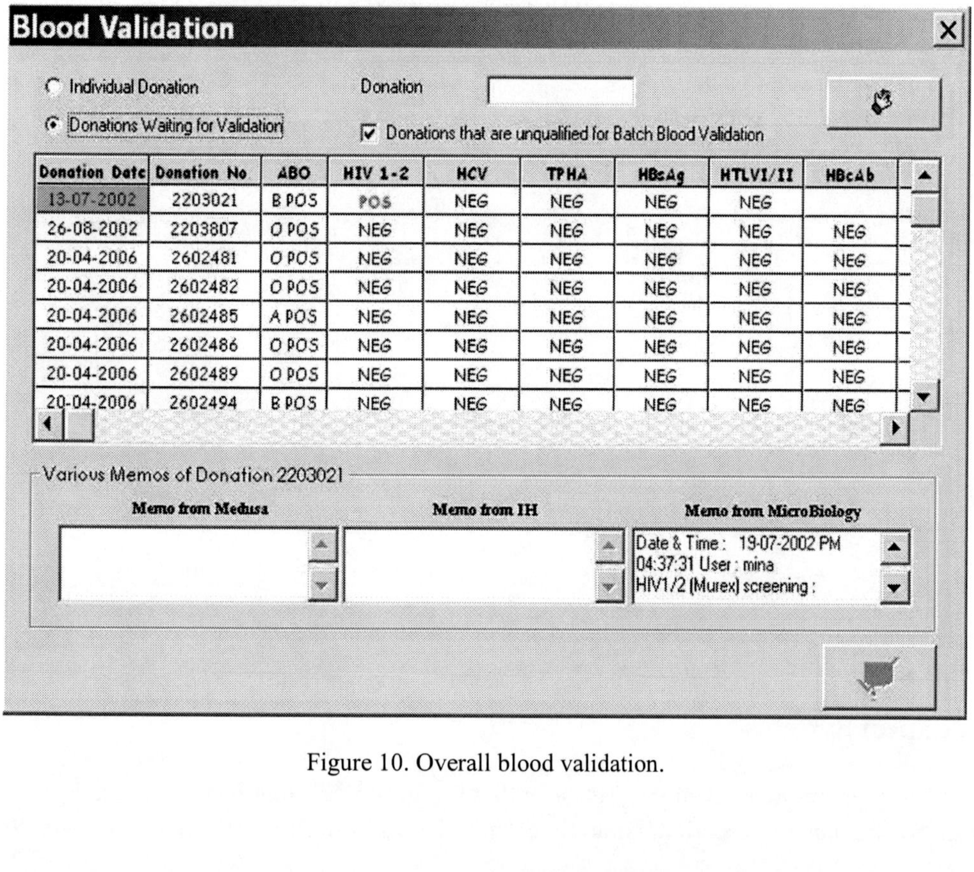

Figure 10. Overall blood validation.

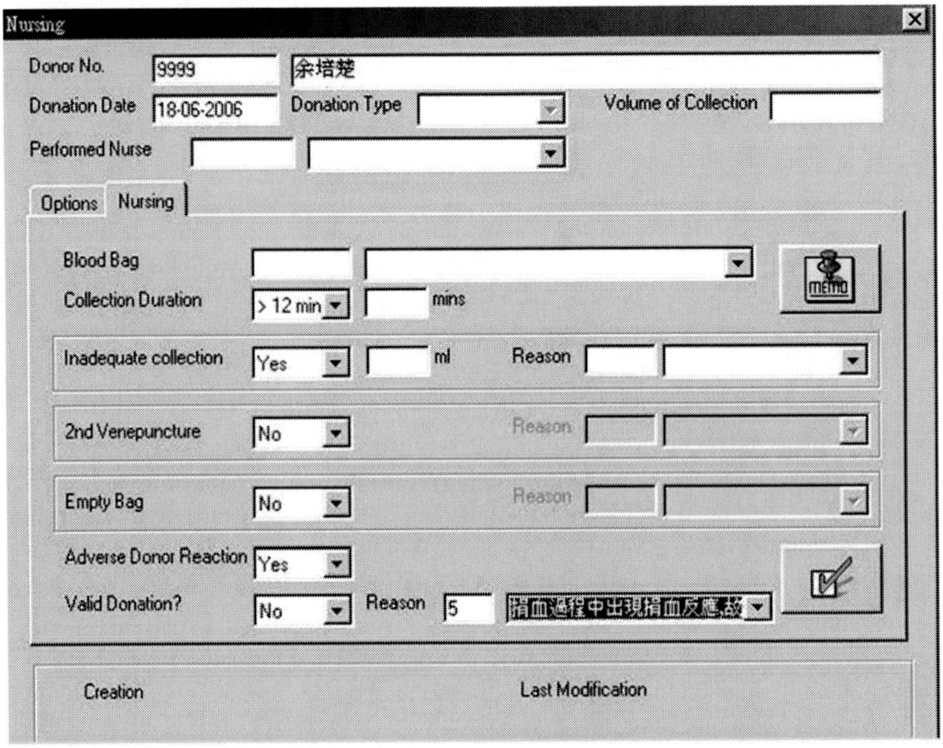

Figure 11. Validating blood collection.

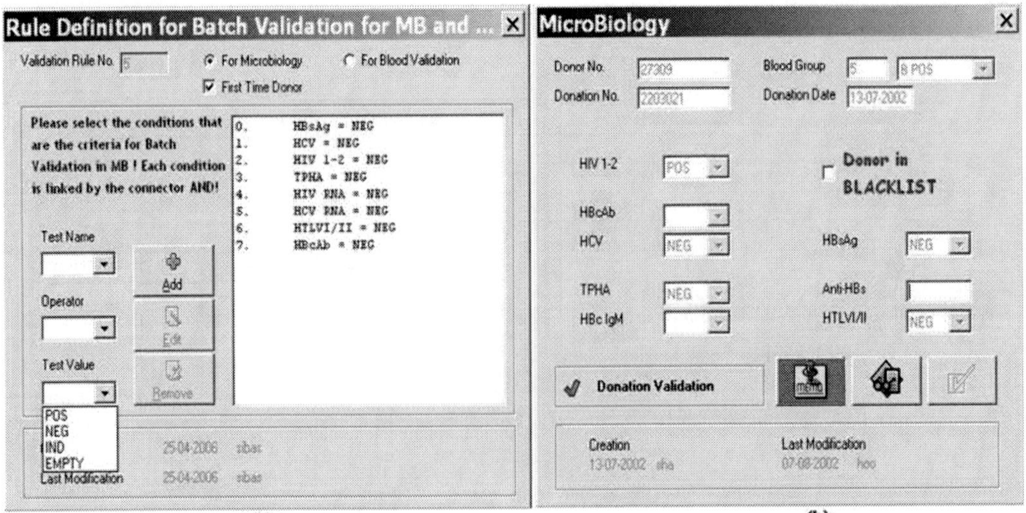

(a) (b)

Figure 12. Rule-based donation validation in Microbiology.

5.1. Expert Systems

It has been pointed out more than once that the life-threatening nature of blood products demands the punctilious administration of blood donation and transfusion service. So a critical challenge is how to efficiently while effectively deal with the huge amount of data and information from the daily operations even within a blood center. In fact, there have been a set of operational guidelines and specifications during the long-term practice of blood donation and transfusion service. For instance, before the formal blood collection, blood donors should undertake a series of physical examinations and eligibility evaluations. Similarly, before the final inventory archiving, the collected blood has to pass a series of microbiological and immunohematological testing. Such kind of practical guidelines and specifications have been taken into account while implementing the computerized decision making support.

In the aforementioned blood bank information system, before the formal blood collection, a blood donor should be evaluated and validated by the receptor, the nurse and the physician respectively. If eligible, the blood donor's data will be submitted to medical area so that the nurses could add the data regarding blood collection, such as blood bag, blood volume, and donation reaction, etc. Otherwise, his/her information will be transferred to suspension or elimination, and this time of blood donation has to be ceased. If that blood donor passes the part of blood collection successfully, the collected blood and samples will be submitted to the divisions of manufacturers and laboratories. In each division, there is an independent evaluation report on its validity. Finally, the blood bank information system is equipped with a special module for overall validation (Figure 10): if passes, to stock; otherwise, to incineration.

Coming to the specific division, no matter physical examination or laboratory testing, there is a set of professional rules for eligibility validation. The blood bank information system adopts two kinds of strategies to organize those expertise and practical experience:

- *Embedded modules*: There are a series of comparatively fixed operational procedure and evaluation rules in the blood bank divisions, including the reception, the nurse, the physician, and the manufacturers. With the consensus of blood bank staffs, the blood bank information system has incorporated that kind of knowledge into its source codes. Take the module of blood collection for example (Figure 11). Various parameters and symptoms during that procedure will be recorded in detail. If an abnormal symptom emerges, it is mandatory to record the related information, for example, the values and the reasons. Then, the system can screen and evaluate the eligibility of that time of blood donation preliminarily.
- *Adjustable modules*: Those comparatively fixed decision making procedures are not suitable for laboratory testing and validations. In accordance with their operational procedure and testing items, the professionals in blood banks have to add or modify the validation rules occasionally. As a consequence, the blood bank information system provides the specially-designed user interfaces for rule definition and updating (Figure 12a). With that customized rule base, the system is hereby able to check the validity of blood donations automatically (Figure 12b). Overall speaking, the inference strategy for those decision making support modules is: one true, then valid; otherwise, invalid.

5.2. Statistical Decision Making Support

Other than the knowledge-based expert systems, that blood bank information system is equipped with the complete statistical reports as listed in Table III. Here the statistical reports are oriented to facilitating the daily operations and supporting the relevant decisions in that blood center.

- *Facilitating blood bank operations*: In many steps of blood donation and transfusion service, blood bank staffs hope to learn the detailed historical records in order to make appropriate decisions. Take the receptors for example. They may screen blood donors better if they have the statistical reports like "Donation History". Similarly, the nurses and the physicians can evaluate the eligibility of blood donors better if they have the statistical reports like "Canceling Waiting List", "Analysis by Donation" and "Suspension and Elimination".
- *Supporting blood bank resource planning*: The effective blood donation and transfusion service depends on the accessory materials, apparatus and professional staffs in that blood center. As mentioned in above, it is not a simple task to allocate those blood bank resources and products optimally. But the requirements in history and their variations are of great help for such kind of decisions. Hence, that blood bank information system provides the complete statistical reports, including "Donation by Blood Group", "Monthly Donation Report", "Blood Components in Stock" and "Unutilized Products", for requirement analysis of blood bank resources and dissemination optimization of blood products.

Table III. The statistical reports

Division	Report	Contents
Reception	Donors by Group and Phenotype	Extracting the blood donors, whose latest donation record is within the specified period, belonging to a specified blood group and having specified phenotypes.
	Donors by Profession	Displaying the blood donors, whose donation was gathered within that period, belonging to the specified profession.
	Distribution by Profession	Calculating the distribution, by percentage of each profession, of blood donors within the specified period.
	Mobile Unit Donors	Displaying the records of blood donations gathered at the specified Mobile Unit within the specified period.
	Donation by Blood Group	Displaying the distribution, by percentage of each blood group, of blood donations.
	Daily Donation Distribution	Listing the daily distribution of blood donations.
	Monthly Donation Report	Listing the figure of blood donations by individual date of the specified month.
	Donation History	Listing the individual blood donor's donations history by specifying the donor number.
	Donor Letter Printing List	Listing the donor letters for the blood donations within the specified period if they are printed or not.
	Award List of Donors	Displaying the blood donors who gain award in the specified period.

Table III. Continued

Division	Report	Contents
Nurses	Canceling Waiting List	Listing the records that the donations have been cancelled in the Nurse area within a specified period.
	Suspension and Elimination	Displaying the suspended or eliminated blood donors whose last donation is within the specified period.
Physicians	Distribution of Anemia	Listing the blood donors with anemia during blood collection within the specified period.
	Distribution of Check/Followup	Displaying the distribution of blood donors according to their check/followup records.
	Distribution of Adverse Reaction	Listing the blood donors with adverse reactions within the specified period.
	Canceling Waiting List	Listing the records that the donations have been cancelled in the Doctor area within a specified period.
	Analysis by Donation	Displaying laboratory testing results of specified donation or blood donations gathered within the specified period.
	Suspension and Elimination	Displaying the suspended or eliminated blood donors whose last donation is within the specified period.
Manufacturers	Donation not Validated	Listing the donations that are not validated yet within the specified period.
Laboratories	Donation not Validated	Listing the donations that are not validated yet within the specified period.
	Donors by Group and Phenotype	Extracting the blood donors, whose latest donation record is within the specified period, belonging to a specified blood group and having specified phenotypes.
	Donors by Special Testing Result	Displaying the distribution of blood donors by their special testing results.
	Analysis by Donation	Listing the testing results of specified donation or blood donations gathered within the specified period.
Stocks	Blood Components in Stock	Displaying the distribution of blood components currently stored in stock.
	Monthly Movement of Blood	Summarizing the general information, within the specified month, of blood donations and blood products.
	Annual Movement of Blood	Summarizing the general information of blood donors, blood donations and blood products by the specified year.
	Unutilized Products	Listing the products of blood donations, gathered within the specified period, that have been transferred to incineration.

5.2.1. Implementation

In that blood bank information system, all statistical analyses and reports are implemented in a distributed infrastructure as shown in Figure 13. Firstly, there is a concise interface through which the user can specify a few limiting conditions. Once that user submits the request, there are two kinds of different paradigms for statistical results. The first one is to call the relevant procedures stored in central Oracle® server, and the statistical results are sent back directly. The alternative paradigm is to synthesize a set of executable SQL statements, submit them to ODBC engine, and obtain the statistical results through ODBC engine. Then, that blood bank information system provides a concise preview interface to display the essential statistical results (Figure 13b). Of course, the user can choose to review and print out the complete statistical results assembled by Crystal Report© engine (Figure 13c).

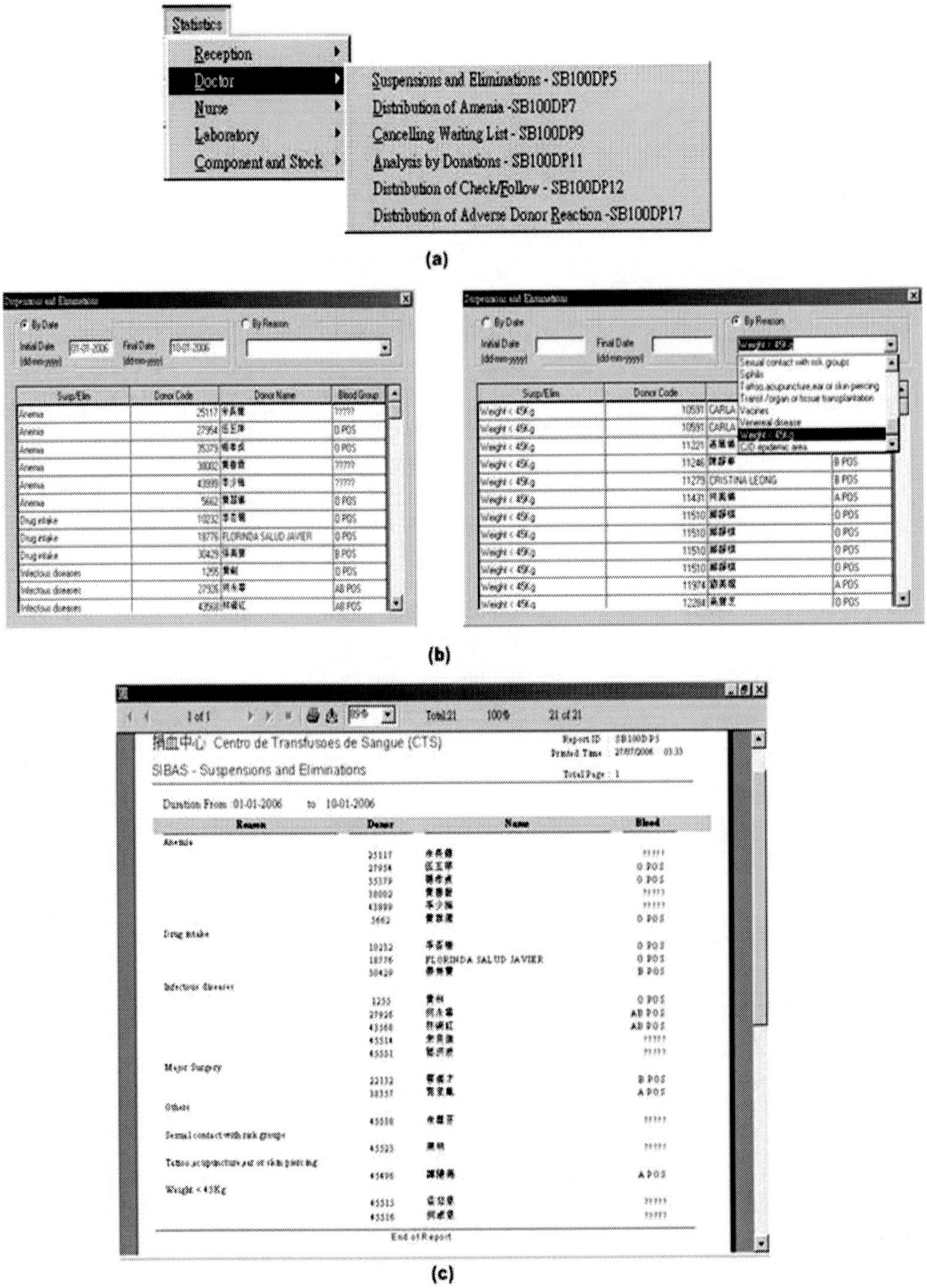

Figure 13. The implementation of statistical reports.

6. Conclusion

The works and results reported in this chapter are based on the initiative of computerized blood bank information management at Macau Blood Transfusion Center. The Institute of Systems and Computer Engineering (INESC-Macau) has fully participated in that initiative since 2000. From the deployment of basic automated apparatus to the release of official SIBAS©, the Institute of Systems and Computer Engineering, cooperating with Macau Blood Transfusion Center, desires to promote the advanced information and computer technology in blood bank automation and information management.

Currently, SIBAS© has been deployed in every division of Macau Blood Transfusion Center, and provides various interfaces for barcode technology, electronic donor cards, blood testing and processing apparatus. It has recorded the complete data and information of blood donation and transfusion service in its Oracle® database since 1999. In spite of the comparatively tiny service at Macau, there are more than one million pieces of records distributed among 3 hundreds of tables in five independent databases. The amount of data is still in increasing with daily thousands of records. While deploying and maintaining such a complicated system, the researchers/developers at the Institute of Systems and Computer Engineering think the following issues and topics worthy of further discussion.

6.1. Implementation Challenges and Solutions

The uninterrupted 24/7 running mode of blood bank information systems requires the quick response of technical support and service, especially for disaster recovery. Owing to reliable planning and designing, SIBAS© has run over 5 years without any serious accidents. In every quarter, the Institute of Systems and Computer Engineering undertakes a regular maintenance. Occasionally, the Macau Blood Transfusion Center requests technical support and service, including technical training and tutorial, abnormal data or inconsistent information, database migration and revision, new user specifications (e.g., user interface revision, workflow revision, and the interfaces for new apparatus, etc). Nevertheless, most of them are in essence related to SIBAS© scalability.

It is an acknowledged fact that any revision or modification will influence the integrity and robustness of blood bank information systems. But three facts make it necessary to update blood bank information systems:

- *The development of blood banks*: Due to the absence of computer knowledge, it is implausible to request blood bank staffs to provide the perfect user specifications on a blood bank information system. Moreover, in face of the advances of blood donation and transfusion service, any blood bank should take workflow optimization as a long-term strategy.
- *The development of blood donation and transfusion service*: As a matter of fact, even the official standards keep evolving to provide safer blood donation and transfusion service (CBER ; BCSH 1995; 1999). For instance, the newly discovered diseases transmitted by blood will no doubt demand the upgraded technologies of blood testing. Moreover, one of the underlying revisions in accordance with official

standards is the upgrade of blood bank barcode technology from Codabar to ISBT 128.

- *The development of information and computer technology*: It is necessary to upgrade the blood bank information system in order to keep up with the technologies of computer hardware and software. Take the operation system for example. Windows 98® was the mainstream system in 1999, whereas it has turned to Windows XP® today. Similarly, Oracle® database has evolved from version 8.0 to version 10g. At the Macau Blood Transfusion Center, the database was upgraded from Oracle® 8.0 to Oracle® 9i in 2004 due to the better operation environment and the better support to Chinese characters.

The Institute of Systems and Computer Engineering is still tracking the latest technologies and products, such as WiFi® and RFID®, and explores their applications in blood bank information management. However, in view of life-threatening blood components, any new technology should pass rigorous testing and assessment before its formal introduction to blood banks.

Acknowledgments

The author is grateful to Ms. Sam Chao, Mr. Wiok Sam, Mr. Jerry Lau and Mr. Bryan from the Institute of Systems and Computer Engineering of Macau (INESC-Macau) and the Macau Blood Transfusion Center (CTS-Macau) for their technical supports as well as comments.

References

American Association of Blood Banks (AABB). (2002) *Standards for Blood Banks and Transfusion Services* (21st ed). Bethesda, MD: American Association of Blood Banks.

Australian and New Zealand Society of Blood Transfusion Inc. (ANZSBTI) (2004). *Guidelines for the Administration of Blood Components*.

Bloodbooks (2008). *The History of Blood Banking and Transfusion Service*. www.bloodbook.com

Bosnes, V., Aldrin, M. and Heier, H.E. (2005). Predicting blood donor arrival. *Transfusion* **45**(2), 162-170.

Brataas, G., Hughes, P.H. and Solvberg, A. (1998). Framework for performance engineering of workflows: a blood bank case study. *Proceedings of 31st Hawaii International Conference on System Sciences*, 230-239.

British Committee for Standards in Hematology (BCSH) (1995). Recommendations for evaluation, validation and implementation of new technologies for blood grouping, antibody screening and cross-matching. *Transfusion Medicine* **5**(2), 145-150.

British Committee for Standards in Hematology (BCSH) (1999). Guidelines for the management of blood and blood components and the management of transfused patients. *Transfusion Medicine* **9**(9), 227-238.

British Committee for Standards in Hematology (BCSH) (2000). Guidelines for blood bank computing. *Transfusion Medicine* **10**(4), 307-314.

Brittenham, G.M., Klein, H.G., Kushner, J.P. and Ajioka, R.S. (2001). Preserving the national blood supply. *Hematology* **1**, 422-432.

Brodheim, E. (1978). Regional blood center automation. *Transfusion* **3**(3), 298-303.

Brodheim, E. (1983). Automated systems in blood banking. *Clinical Lab Medicine* **1**(1), 111-132.

Butch, S.H. (2002). Computerization in the transfusion service. *Vox Sanguinis* **83** (suppl. 1), 105-110.

Center for Biologics Evaluation and Research (CBER) (1988). *Recommendations for Implementation of Computerization in Blood Establishments.* http://www.fda.gov/cber/guidelines.htm

Center for Biologics Evaluation and Research (CBER) (1989). *Requirements for Computerization of Blood Establishments.* http://www.fda.gov/cber/guidelines.htm

Center for Biologics Evaluation and Research (CBER) (1994). *A Letter to Blood Establishment Computer Software Manufacturers.* http://www.fda.gov/cber/guidelines.htm

Center for Biologics Evaluation and Research (CBER) (1997). *Reviewer Guidance for a Premarket Notification Submission for Blood Establishments.* http://www.fda.gov/cber/guidelines.htm

Center for Biologics Evaluation and Research (CBER) (2001). *Guidance for FDA Reviewers: Premarket Submissions for Automated Testing Instruments Used in Blood Establishments* (Draft Guidance). http://www.fda.gov/cber/guidelines.htm

Center for Biologics Evaluation and Research (CBER) (2003). *Guidance for Industry: Streamlining the Donor Interviewing Process: Recommendations for Self-Administered Questionnaires.* http://www.fda.gov/cber/guidelines.htm

Center for Biologics Evaluation and Research (CBER) (2004). *Guidance for Industry: Acceptable Full-Length Donor History Questionnaire and Accompanying Materials for Use in Screening Human Donors of Blood and Blood Components* (Draft Guidance). http://www.fda.gov/cber/ guidelines.htm

Center for Biologics Evaluation and Research (CBER) (2005). *Draft Guideline for the Validation of Blood Establishment Computer Systems.* http://www.fda.gov/cber/guidelines.htm

Center for Biologics Evaluation and Research (CBER) (2006). *Guideline for the Uniform Labeling of Blood and Blood Components.* http://www.fda.gov/cber/guidelines.htm

Center for Biologics Evaluation and Research (CBER) (2008). *510K Blood Establishment Computer Software.* http://www.fda.gov/cber/guidelines.htm

Central Laboratory of the Blood Transfusion Service of the Netherlands (CLBTS) (2001). The use of the computer cross-match. *Vox Sanguinis* **82**(3), 184-184.

Chambers, R.W., Lundy, J.A., Friedman, L.I. and Gordon, S.B. (1975). A computerized donor processing system for a regional blood collection center. *Transfusion* **2**(2), 170-173.

Connelly, D.P., Sielaff, B.H. and Scott, E.P. (1990). ESPRE - expert system for platelet request evaluation. *American Journal of Clinical Pathology* **94**(4), s19-s24.

Department of Health and Human Services in FDA (2004). Bar code label requirements for human drug products and biological products (Final Rule). *Federal Register* **38**, 9119-9171.

Gardner, R.M., Golubjatnikov, O.K., Laub, R.M., Jacobson, J.T. and Evans, R.S. (1990). Computer-critiqued blood ordering using the HELP system. *Journal of Biomedical Informatics* **23**(12), 514–528.

Glynn, S.A., Kleinman, S.H., Schreiber, G.B., Zuck, T., McCombs, S., Bethel, J., Garratty, G. and Williams, A.E. (2002). Motivations to donate blood: demographic comparisons. *Transfusion* **42**(2), 216-225.

Gregory, P.P. and Eric, B. (1980). PBDS: a decision support system for regional blood management. *Management Science* **26**(5), 451-464.

Gupta, O., Priyadarshini, K., Massoud, S. and Agrawal, S.K. (2004). Enterprise resource planning: a case of a blood bank. *Industrial Management and Data Systems* **104**(7), 589-603.

Hanson, M. (1996). Should we do another test? Decision making in blood banking. *Clin Lab Med.* **16**(4), 883-893.

Hirsch, R.L. and Brodheim, E. (1981). Blood distribution systems and the exchange of information between hospital blood banks and regional blood centers. *Vox Sanguinis* **3**(4), 239-244.

International Council for Commonality in Blood Banking Automation (ICCBBA) (2004). *ISBT 128 Standard: Technical Specification* (Version 2.1.0).

International Society of Blood Transfusion (ISBT) (2003). Guidelines for validation and maintaining the validation state of automation systems in blood banks. *Vox Sanguinis* supplement (1), s1-s14.

James, R.C. and Matthews, D.E. (1996). Analysis of blood donor return behavior using survival regression methods. *Transfusion Medicine* **6**(1), 21-30.

Jelles, G.M. (1993). Costs and benefits of HIV-1 antibody testing of donated blood. *Journal of Policy Analysis and Management* **12**(3), 512-531.

Kempf, B. (1967). All French blood donors in a computer: outlook in 1990 or reality in 1970? *Transfusion (Paris) (French)* **10**(1), 59-62.

Kern, D.A. and Bennett, S.T. (1996). Informatics applications in blood banking. *Clinical Lab Medicine* **4**(1), 947-960.

Kros, J.F. and Pang, R.Y. (2004). A decision support system for quantitative measurement of operational efficiency in a blood collection facility. *Computer Methods and Programs in Biomedicine* **74**(1), 77-89.

Larson, P. (1999). *EDI: Electronic Data Interchange*. Pittsburgh, PA: International Council for Commonality in Blood Banking Automation (ICCBBA).

Li, B.N., Chao, S. and Dong, M.C. (2006). Barcode technology in blood bank information systems: upgrade and its impact. *Journal of Medical Systems* **30**(12), 449-457.

Li, B.N. and Dong, M.C. (2006). Banking on blood [Electronic donor cards]. *Computing and Control Engineering* **17**(4), 22-25.

Li, B.N., Chao, S. and Dong, M.C. (2007). SIBAS: a blood bank information system and its 5-year implementation at Macau. *Computers in Biology and Medicine* **37**(4), 588-597.

Li, B.N., Dong, M.C., and Chao, S. (2008). On decision making support in blood bank information systems. *Expert Systems with Applications* **34**(4), 1522-1532.

Linden, J., Paul, B. and Dressler, K.P. (1992). A report of 104 transfusion errors in New York State. *Transfusion* **32**(7), 601-606.

Linden, J. and Kaplan, H. (1994). Transfusion errors: causes and effects. *Transfusion Medicine Reviews* **8**(3), 169-183.

Moncharmont, P., Lacruche, P., Planat, B., Morizur, A. and Subtil, E. (1999). The case for standardization of transfusion medicine practices in French blood banks. *Transfusion Medicine* **9**(1), 81-85.

Moore, R. (1973). A computer-assisted method to retrieve information about blood donors. *Computers in Biology and Medicine* **3**(1), 63-70.

Myhre, B.A. and Ritland, F. (1986). The computer in the blood bank. *Critical Review Clinical Lab Science* **1**(1), 21-42.

Oba, Y., Otani, S., Yasuda, N. and Terada, K. (1971). Management of blood donor examination by computers. *Rinsho Byori (Japanese)* **19** (suppl.), 422-422.

Page, B. (1980). A review of computer systems in blood banks and discussion of the applicability of mathematical decision method. *Methods of Information in Medicine* **2**(1), 75-82.

Petäjä, J., Andersson, S. and Syrjälä, M. (2004). A simple automated audit system for following and managing practices of platelet and plasma transfusions in a neonatal intensive care unit. *Transfusion Medicine* **14**(4), 281-288.

Peyretti, F. (1971). Automation and computer science in blood transfusion. *Minerva Medicine (Italy)* **88**(1), 4363-4363.

Pietersz, R.N.I. (1995) Automation/computerization in blood processing. *Transfusion Science* **16**(3), 235-241.

Raj, J. and Tarun, S. (1991). Storing cross-matched blood: a perishable inventory model with prior allocation. *Management Science* **37**(3), 251-266.

Roh, T.H., Ahn, C.K. and Han, I. (2005). The priority factor model for customer relationship management system success. *Expert Systems with Applications* **28**(4), 641-654.

Sapountzis, C. (1984). Allocating blood to hospitals from a central blood bank. *European Journal of Operational Research* **16**(2), 157-162.

Sazama, K. (1990). Reports of 355 transfusion-associated deaths: 1976 through 1985. *Transfusion* **30**(7), 583-590.

Sielaff, B.H., Connelly, D.P. and Scott, E.P. (1989). ESPRE: a knowledge-based system to support platelet transfusion decisions. *IEEE Transactions on Biomedical Engineering* **36**(5), 541-546.

Sime, S.L. (2005). Strengthening the service continuum between transfusion providers and suppliers: enhancing the blood services network. *Transfusion* **45**(s4), 206S-223S.

Singman, D., Catassi, C.A., Smiley, C.R., Wattenburg, W.H. and Peterson, E.L. (1965). Computerized blood bank control, *Journal of American Medical Association* **194**(6), 583-586.

Smith, J.W., Svirbely, J.R., Evans, C.A., Strohm, P., Josephson, J.R. and Tanner, M. (1985). RED: a red-cell antibody identification expert module. *Journal of Medical Systems* **9**(3), 121-138.

Spackman, K.A. and Beck, J.R. (1990). A knowledge-based system for transfusion advice. *American Journal of Clinical Pathology* **94**(4), s25-s29.

Steven, W. and John, E.S. (2000). Reducing surgical patient costs through use of an artificial neural network to predict transfusion requirements. *Decision Support Systems* **30**(2), 125-138.

Turner, C.L., Casbard, A.C. and Murphy, M.F. (2003). Barcode technology: its role in increasing the safety of blood transfusion. *Transfusion* **43**(9), 1200-1200.

Weisshaar, D. (1991). Electronic data transfer from computer to computer in blood banks using HL7. *Beitr Infusionsther* **28**(5), 370-372.

Zaller, N., Nelson, K.E., Ness, P., Wen, G., Bai, X. and Shan, H. (2005). Knowledge, attitude and practice survey regarding blood donation in a Northwestern Chinese city. *Transfusion Medicine* **15**(4), 277-286.

In: Progress in Management Engineering
Editors: L.P. Gragg and J.M. Cassell, pp. 173-197
ISBN: 978-1-60741-310-3

Chapter 7

RISK MANAGEMENT ADOPTED BY FOREIGN FIRMS IN VIETNAM: CASE STUDY OF A CONSTRUCTION PROJECT

***Florence Yean Yng Ling*[1,a] *and Vivian To Phuong Hoang*[2,b]**
1. Department of Building, School of Design and Environment, National University of Singapore, Singapore
2. Rider Hunt Levett and Bailey, Singapore

Abstract

Vietnam's economic growth has led to a demand for infrastructure facilities, residential and commercial buildings, and hi-tech parks. This has resulted in a high volume of construction activities. With Vietnam's membership in the World Trade Organization, foreign architectural, engineering and construction (AEC) firms now have the opportunity to operate in Vietnam. However, undertaking overseas construction projects is usually considered a high risk business due to a lack of information and overseas experience. Risk management is thus an important aspect of international construction.

To investigate risks associated with managing construction projects in Vietnam and to examine how *foreign firms* manage those risks, a case study was conducted. The main objectives of this case study were to find out the different type of risks encountered by *foreign* players and various risk response strategies adopted by them. These include political and legal risks, financial and economic risks, design risk, construction related risk and cultural risk. The case study relates to the development of a yeast factory in the southern part of Vietnam. Data for the case study were obtained by interviewing experts from different firms that undertook important parts in this project.

The research revealed that Vietnam has a complex government administrative system. Foreigners overcome the political risk by transferring it to a local joint venture partner who is in a better position to deal with local government officials and to obtain the necessary approvals. Negotiation is found to be the best way to settle disputes instead of suing each other in the court of law because Vietnam's legal framework is not robust. Prequalification of bidders is found to be the most effective and practical way to ensure that the contractor

[a] E-mail address: bdglyy@nus.edu.sg. 4 Architecture Drive, Singapore 117566.
[b] 150 Beach Road, #09-01 Gateway West, Singapore 189720.

engaged to carry out the work is financially sound and competent, thereby reducing financial risk. Design risk was severe in this project and it caused many disputes among project team members. Design risk was mitigated through negotiation and by having many coordination meetings. The project faced some construction related risks, such as low quality of workmanship, low safety consciousness, unavailability of sophisticated materials, plant and equipment. These were solved by engaging a safety supervisor, training workmen, and changing specifications to locally available products. The project faced many cultural risks due to different mindsets between foreigners and Vietnamese, and different working styles. This risk can be overcome if foreigners strive to adapt to the local environment, and be mindful and watchful of how locals behave.

Introduction

On January 11, 2007, Vietnam became the 150th member country of the World Trade Organization (WTO). WTO member countries cannot normally discriminate between their trading partners, and imported and locally-produced goods and services should be treated equally following the 'national treatment' principle of giving others the same treatment as one's own nationals (WTO, 2005). As a WTO member, Vietnam must allow WTO member-country's architectural, engineering and construction (AEC) firms to operate in its construction industry, albeit in a controlled way.

In 2007, Vietnam attracted a record of US$20 billion foreign direct investment (FDI), rising by 70% over the previous year (Blume, 2007). Foreign AEC firms are expected to enter Vietnam to develop the facilities which are needed by foreign investors and also public infrastructure projects, in line with Vietnam's WTO commitments to open up its construction market to foreign businesses. However, many of these firms are not familiar with Vietnam's construction environment and the risks that they will face.

The aim of this study is to investigate risks associated with managing construction projects in Vietnam and to examine how *foreign firms* manage those risks. The main objectives of this study are to find out the different type of risks encountered by *foreign* AEC firms and to recommend effective risk response strategies. The scope of this study covers political risk, legal risk, financial and economic risks, design risk, construction related risk, cultural risk and nature risk faced by foreign firms in Vietnam's construction industry.

Literature Review

Construction projects have an abundance of risks. The construction industry is subject to more risk and uncertainty than many other industries (Flanagan and Norman, 1993). International construction involves all the uncertainties common in domestic construction projects as well as risks specific to international transactions (Han et al., 2005). Contracting overseas construction projects is usually considered a high risk business, mostly because of the lack of adequate overseas environmental information and overseas construction experience (He, 1995).

Risk Definition

Risk is perceived as 'the potential for unwanted or negative consequences of an event or activity, a combination of hazard and exposure (Chicken and Posner, 1998). Porter (1980) expressed risk as an exposure to economic loss or gain arising from involvement in the construction process. Risk is a variable in the process of constructing a project and may result in uncertainty to the final cost, duration and quality of the project (Akintoye and MacLeod, 1997). Construction projects are one-off endeavors with many unique features such as long period, complicated processes, abominable environment, financial intensity and dynamic organization structures and such organizational and technological complexity generates enormous risks (Zou et.al, 2007).

Risk Management and Identification

International projects are generally more difficult to manage due to conditions such as multiple ownership, elaborate financial provisions and different political ideologies (Gunhan and Arditi, 2005). Thus, international projects face even more risks than domestic projects. Risk management then becomes more complicated and crucial for overseas construction projects. An effective risk management method can help in understanding not only what kinds of risks are faced, but also how to manage these risks at the stages of design, contracting and construction (He, 1995).

Risk management is a management discipline with the goal of protecting assets, reputation and profits of a company by reducing the possible losses before the risks occur (Bing et. al., 1999). Optimal risk management should aim to minimize the total cost of risk to a project and not necessarily the costs to each party separately (Rahman and Kumaraswamy, 2002). It is important to adopt risk management techniques when projects are large in size, complex and the potential for delay and cost overruns are high (Burchett et al, 1999). Kim and Bajaj (2001) discovered that Korean contractors' unfamiliarity with risk management techniques caused them to manage risks based on intuition, judgment and past experience. In Australia too, Uher and Toakley (1999) found the lack of knowledge to be the main barrier to the implementation of risk management.

Risk management is an entire series of activities related to identification, evaluation and the control or mitigation of risk. While risks cannot be entirely eliminated, successful projects are those where risks are effectively managed, hence early and effective identification and assessment of risks is essential (Smith, 1999).

There are many types of risk management procedures adopted in the construction industry. The stages of risk management and assessment are given below (Construction Industry Institute, 2004):

i. Early identification of hazards and opportunities
ii. Communication of risk between project participants
iii. Identification and management of uncertainty
iv. Acknowledgement of risk issues and mitigation actions
v. Enhance risk-based decision making.

Hampton (1993) used a multiple-step process chart to explain risk management process, which involves: set objectives, identify risks, evaluate risks, design a comprehensive program, implement the program, and monitor results. Bing et al. (1999) proposed that the risk factors be categorized into three main groups: (1) internal; (2) project-specific; and (3) external. The internal risk group represents the risks developed from the nature of the internal operation of that firm when doing business overseas. The project-specific risk group refers to unexpected developments during the construction period that lead to time and cost overruns or in shortfalls in performance parameters of the completed project. A high capital outlay and a relatively long construction period would make project costs particularly susceptible to delays and cost overruns. The external risk group represents the risks that emanate from the competitive macro-environment that the company operates in.

Risks can also be categorized into several levels: country; market; project and client (Wang et al., 2000; 2004). Edwards and Bowen (1998) grouped risks into: natural risks (caused by weather and geological systems) and human risks (comprising social, political, economic, financial, legal, health, managerial, technical and cultural risks). Some of these risks are now reviewed.

Political Risk

Political risk in international business exists when discontinuities occur in the business environment, when they are difficult to anticipate and when they result from political change (Robock and Simmonds, 1983). To constitute a 'risk', these changes in the business environment must have the potential for significantly affecting the profit or other goals of a particular enterprise. Kapila and Hendrickson (2001) defined political risk as the likelihood that political forces will cause drastic changes in a country's business environment, which would hurt the profit and other goals of a business enterprise. It is very important to consider political risks in overseas projects from the national/regional macroeconomic and political standpoints, simply because these risks are unfamiliar compared with those of the domestic environment, and they are significant, particularly for large projects (He, 1995).

Political risk may be classified as follows (Root, 1987):

i. General instability risk: uncertainty about the future viability of a host country's political system.
ii. Ownership/control risk: uncertainty about host government actions that would destroy or limit the investor's ownership or effective control of his affiliate in the host country.
iii. Operations risk: uncertainty about host government policies or acts sanctioned by the host government that would constrain the investor's operations in the host country.
iv. Transfer risk: uncertainty about government acts that would: (a) restrict the transference of profits out of the host country, or (b) lead to currency depreciation.

Robock and Simmonds (1983) distinguished between macro and micro political risks. Macro political risk occurs when all foreign enterprises are affected in much the same way by politically motivated discontinuities in the business environment. Micro risks occur when changes affect only selected industries, firms or even projects. Examples of macro risks are political force majeure events, revolutions, civil wars, nation-wide strikes, protests, riots and mass expropriations. Examples of micro risks are elective expropriations, discriminatory taxes and import restrictions directed at specific firms.

Political risk may also involve inconsistency in policies, changes in law and regulations, restriction on fund repatriations, and import restrictions. In less extreme cases, political changes may result in increased tax rates, the imposition of exchange controls that limit or block a subsidiary's ability to remit earnings to its parent company, the imposition of price controls, and government interference in existing contracts (Kapila and Hendrickson, 2001).

East Asian countries present a wide variety of ruling political systems: democratic, authoritarian, socialist, communist and dictatorships. Governments in developing nations can face serious problems, as seen in the Asian financial crisis that could jeopardize their stability and continuity. Some governments in East Asian countries directly influence the public construction sector by setting the rules for development and contractual relationships. Their influence is also felt in the private sector through policies and legislation regarding licenses and permits, building codes, minimum wage rates, corporate taxes and discriminatory taxation, rules on the importation of material and spare parts (World Bank, 1984). Because of the bureaucratic system, many of the operational decisions of the company have to be approved by the relevant government officials such as approval for land acquisition, permission for starting construction which may cause the development of the project to be very complicated and inefficient. In addition, governments can intervene in the operations of foreign-owned firms by restricting ownership and control, regulating financial flows and employment of foreign management. More drastic measures would be nationalization, expropriation or confiscation of assets (Mortanges and Allers, 1996).

Economic and Financial Risks

The types of financial risks in international construction include: fluctuation in foreign exchange rates, interest rates, labor and material prices; inflation; default by contractors/subcontractors; import/export restrictions; delayed or non receipt of payment; financial failures; and restriction on repatriation of funds (Ling and Lim, 2007). These are now discussed.

Fluctuation in Foreign Exchange Rates

An unfavorable change in exchange rates can result in a loss when the revenue received is in one currency but production costs are in another currency (Xenidis and Angelides, 2005). In build-operate-transfer (BOT) projects, foreign exchange fluctuation risk is moderately critical during the pre-investment stage and slightly critical during other BOT stages (Lam and Chow, 1999). Chua et al. (2003) found that fluctuation of foreign exchange rates is one of the most critical factors causing budget overrun in East Asia. Other studies have found risk arising from fluctuation in foreign exchange rates to be of varying importance to joint ventures (Bing et al., 1999; Shen et al., 2001; Wang et al., 2004).

Interest Rate Fluctuation

Interest rate is a key factor in determining the intensity of a debt and internal rate of return, which consequently affects the feasibility, construction and operation of a project (Lam and Chow, 1999). Loss due to fluctuation of interest rate is moderately critical (Shen et al., 2001), especially during pre-investment stage and slightly critical in all other stages (Lam and Chow, 1999).

Inflation

Inflation fluctuation in a country affects various financial indices such as the interest rate, rate of return and currency exchange rate (Lam and Chow, 1999). Several studies have found that rise in inflation have some bearing on construction projects (Lam and Chow, 1999; Bing et al., 1999; Fang et al., 2004; Shen et al., 2001; Wang et al., 2004).

Labor and Material Price Fluctuations

The economic conditions in the host country may lead to an increase of production costs (Xenidis and Angelides, 2005). The increase in demand for construction work will result in shortages of resources, which leads to higher prices (Chen, 1997). Smith et al. (2004) found labor and material costs to be volatile when a country is undergoing economic reforms.

Import/Export Restrictions

A deficit trade balance of the host country may be the reason for the imposition of several restrictions concerning imports and exports. It is common for a host government to implement policies such as increasing tariffs for imported products or requiring special permission to import certain products (Xenidis and Angelides, 2005). This leads to an increase in the prices of goods and services.

Delayed or Non Receipt of Payment

In some developing countries such as China, established banks only provide project financing to large national projects and this lack of construction credit is a major constraint in the construction industry (Chen, 1998). It may lead to owners of smaller projects not making regular payments to contractors. Disputes have been found to arise from the shortage of necessary capital, due to the lack of construction financial credit facilities (Smith et al., 2004).

Financial Failures

Foreign AEC firms may face the risk of financial failures of their own firms or their business partners. Companies that face financial failures have a serious impact on the project's progress. Nevertheless, potential bankruptcy is not necessarily connected to the project but could be related to other business activities (Xenidis and Angelides, 2005).

Restriction on Repatriation of Funds

Restriction on the repatriation of funds occurs when a host country forces foreign companies to spend their earnings in the host market (Chua et al., 2003). This results in loss of profit either by preventing exploitation of foreign bank account privileges or by additional convertibility costs to lift restrictions. Furthermore, the enforcement of such restrictions may not be predictable (Xenidis and Angelides, 2005).

Legal Risk

Legal risk entails issues such as breach of contract by project partners or other participants, lack of enforcement of legal judgment when problems arise, insufficient law for joint ventures, and uncertainty and unfairness of court justice (Shen et al. 2001; Ling and Low, 2007) and insolvency of a partner (Wang et al., 2000). The components of legal risks are now reviewed.

Laws and Regulations

The legal risks relating to laws and regulations include non compliance with them and inability to keep up with frequent changes. The first significant law that AEC firms face upon entering a foreign country relates to licensing. Vietnam's key WTO commitments in services (of which AEC belongs to) are provided in WTO (2006). Vietnam must allow foreign enterprises to establish commercial presence in the form of business cooperation contracts, joint venture enterprises and 100% foreign-invested enterprises.

To give time to domestic firms to adjust, for a period of two years from the date of WTO ascension, 100% foreign-invested enterprises may only provide construction services to foreign-invested enterprises in Vietnam (WTO, 2006). Construction services include architectural services, engineering services, integrated engineering services, urban planning, urban landscape architectural services and construction work. In addition, for construction and related engineering services, after three years from the date of ascension, foreign firms would be allowed to set up branch offices, but the chief of the branch has to be a resident in Vietnam. For urban planning and urban landscape architectural services, the service must be authenticated by an architect who has a Vietnamese practicing certificate, and who works in a Vietnamese architectural organization. Foreign architects working in foreign-invested enterprises must possess professional practicing certificate granted or recognized by the Government of Vietnam.

Vietnam is expected to continue her economic reform policies, aiming to achieve improvement in her laws and regulations in order to attract more investors. Many of the reforms are new or at the experimental stage and there is still room for enhancement and modification. Hence, changes in laws are inevitable because of constant reviews and updating as reform progresses. In China, for example, Shen et al. (2001) found that 'cost increase due to policy changes' is the most significant risk to foreigners operating in its construction industry.

Contract Formation and Performance

For a contract to come into existence, foreign AEC firms are used to a written contract being signed, to signify that the parties have agreed to all its terms. An agreement is said to be completed, and the contract formed, when there is acceptance of the offer, which must be unequivocal and must be communicated to the offeror (Wallace, 1995).

When an offer is accepted and the fact is notified to the offeror, the parties usually consider themselves bound, unless the acceptance is subject to some overriding condition, the parties have not come to a common understanding on an important term of the contract, or the tender could not be accepted due to the passage of time (Wallace, 1995). Generally, the formation of construction contracts is more complicated because of negotiations between the parties which continue for a substantial period after work has commended (Wallace, 1995).

Appropriate, clear and equitable conditions of contract are commonly considered to be invaluable for successful projects (Rahman and Kumaraswamy, 2002). Contract conditions must be clear in order to define the rights and duties of project players and explicitly allocate risks to the different contracting parties. Apart from merely following the often adopted principle of assigning the risks to those best prepared to deal with them, contract conditions are also expected to be equitable, so as to allocate these risks in a 'just' way.

It is generally accepted that when a contract is entered into, parties would perform their obligations in the contract. Contract performance include obligation to complete the work, maintenance and defects issues, time for completion, prices and damages (Wallace, 1995). Parties can only be discharged from their liability for further performance if some special situations take place such as frustration, discharge by breach, illegality and limitation (Wallace, 1995).

Foreign AEC firms need to evaluate their Vietnamese business associates carefully, to ensure that they have the capability to perform the contract. This is not an easy task because besides checking on their ability to perform, the reliability and creditworthiness of Vietnamese entities are difficult to ascertain.

Dispute Resolution

The common methods of dispute resolution include: lawsuits/litigation; arbitration; mediation; conciliation; and negotiation. Foreign AEC firms are used to having the dispute resolution method stated in the contract conditions. Cheung et al. (2000) found that it is easier to settle a dispute through negotiation when there is plenty of work available in the market. In addition, there should be strong commitment and a strong desire by the client to settle the dispute. Clients' involvement in the management of the project helps settle disputes at the job site level.

Design Risk

Designers generate risks such as defective design, deficiency in drawings, changes in design and documents not issued on time (El-Sayegh, 2008). Design risk arises when there are design errors and omissions and design change. This source of risk includes defective or incomplete design documents, inconsistencies and flaws among the plans and specifications

(Petrov, 2006). Design errors could be due to defective design, and unavailable or inappropriate design detail. These happen due to: incomplete design scope; incomplete or erroneous geological and geotechnical exploration; and inadequate interaction of design with methods of construction.

Late design changes may happen due to change in clients' requirements arising from changing needs. Late design changes can impact already procured and performed work and require the contractor to change its execution plan, and thus may impact its productivity and cost (Petrov, 2006).

Andi and Minato (2003) found that the problems of defective design are complex and deep rooted. Some of the causes include: insufficient overall design time; low fee for designers; insufficient project budget; and client's needs change as the design develops.

Construction Related Risk

Contractors generate risks such as construction accidents, poor quality, low productivity, technical problems, incompetency and lack of or departure of qualified staff (El-Sayegh, 2008). Although these risks can adversely affect the progress of construction, they are not easily grasped and controlled.

Some construction related risks include difficulty to access the site and unforeseen soil conditions. In many countries, the construction industry also has the worst accident record. In developing countries, the problem is compounded by a lack of safety management system and safety awareness. In India, for example, proper safety equipment is not always provided or used (Ling and Hoi, 2006).

Bing et. al. (1999) found that incompetence of subcontractors and suppliers is a major risk factor for contractors. In China, for example, poor management, technology and quality of materials are significant risks (Fang et al., 2004). Other associated risks are: unexpected delay in delivery of materials; subcontractors' breach of contract; and disputes between main and sub contractors. It is therefore important to select contractors/subcontractors carefully, paying attention to their construction and management ability (Fang et. al., 2004).

Cultural Risk

International construction projects are those in which contractors, lead consultants or employers are not of the same domicile, and at least one of them is working outside his country of origin. The cross cultural encounter and cultural differences are expected to contribute to conflicts among parties to an international project and increase difficulties in management of the project (Fellow and Hancock, 1994) such as low productivity, lack of management capability, bureaucracy in work procedures, dispute settlement methods and miscommunication. For successful outcomes of international construction projects, AEC practitioners should understand the culture of the host country, and even if they do not know what the similarities between cultures of home and host countries are, they should at least know the differences (Low and Shi, 2002).

National cultures may be categorized into the following dimensions: power distance; individualism vs collectivism; masculinity vs femininity; uncertainty avoidance; and long vs.

short term orientation (Hofstede and Hofstede, 2005). For example, the Chinese (race) is characterized by importance of preserving face, building relationships (guan xi), trust and friendship (Ang and Ofori, 2001).

Cultural differences are usually represented by dissimilar language, background, perceptions and mentalities (Swierczek, 1994). Difficulties encountered in international projects often find their genesis in the differences between the cultures, language, religion and custom background of foreigners and the locals in the host country. Other differences include factors such as educational background, beliefs, arts, moral, customs and laws (Evans et al., 1989). Cultural differences are also associated with individual characteristics such as gender, age, job experience and race (Earley and Mosakowski, 2004). Ankrah and Langford (2005) found organizational cultural differences between architects and contractors. Loosemore and Chau (2002) discovered that Asian operatives perceive significant levels of blatant racial discrimination and harassment in the Australian construction industry. Cultural differences are not effectively managed; and Asian operatives are expected to assimilate and integrate into mainstream white society.

These differences lead to problems in communication and working together (Swierczek, 1994). Chua *et al.* (2003) found that problems arising from cultural differences in international construction are usually related to low productivity and skill levels, poor management, bureaucratic work procedures and disputes. Ngowi (1997) found that in construction projects in which team members are from different cultural backgrounds, there are inhibitions to innovation compared to the ones in which team members have similar cultural backgrounds.

The need to recognize, and deal with cultural issues is critical in developing countries where sizable projects often involve foreign companies and/or professionals (Ofori, 2007). Doing business across national boundaries requires interaction with people and their institutions and organizations nurtured in different cultural environments. Values that are important to one group of people may mean little to another. In brief, there exists among nations striking and significant differences of attitude, belief, ritual, motivation, perception, morality, truth, superstition, and an almost endless list of other cultural characteristics (Jain,1996).

In the context of joint ventures, the members can be brought in from different territories or countries that are culturally, historically and economically different. Possible differences (e.g. 'ethnocentrism' vs. 'polycentrism') may significantly affect their approaches to problem solving (Lai and Truong, 2005). Different approaches to the problems arising within the organization can cause misunderstanding among the members. A particularly bad situation may arise when the partners do not possess the necessary skills to cope with the conflict. In this situation, communication within the organization becomes poor and its members are reluctant to cooperate and work together, knowledge is less likely to be shared and trust between the members is low (Dodgson, 1993).

Natural Risks

As stated earlier, natural risks include risks caused by weather and geological systems (Edwards and Bowen, 1998). Natural risks consist of the environmental force majeure risk such as landslide, earthquake, flood and volcano eruption could cause the destruction of facilities, equipment, material and give rise to an unsafe working condition.

For most projects, inclement weather is the most significant natural risk. Chan and Au's (2007) studied Hong Kong building contractors' risk-pricing behaviors. They found that smaller contractors are more willing to absorb weather risks in their tenders. Their study revealed that it is generally more cost efficient for employers to delete the contractual provision for extension of time due to inclement weather as contractors' allowance of weather-caused delays (in monetary terms) in tenders is less than the actual number of days of inclement weather obtained from weather records.

While local firms may be familiar with natural risks, foreign AEC firms could face severe challenges if these risks are not taken into consideration in project planning and execution.

Risk Response

There are several risk response techniques such as risk elimination, risk transfer, risk retention and risk reduction (Carter and Doherty, 1974; Thompson and Perry, 1992; and Flanagan and Norman, 1993). Risk reduction technique is the most frequently utilized, followed by risk transfer, risk retention and risk elimination (Baker et. al 1999). Risk response techniques are now reviewed.

Risk Elimination

Thompson and Perry (1992) found that risk management is most valuable at an early stage in a project, for example at the proposal stage, where there is still some flexibility available in design and planning to consider how the serious risk might be avoided or eliminated. Some methods to eliminate risks are to design out of the risk, and coordinate closely with all project team members such as main contractors, subcontractors, clients and consultants.

Risk Transfer

Risks can be transferred to other parties to help a firm reduce loss. To do this, the transfer must be stated clearly in the contract terms and conditions (Wang and Chou, 2003). Kartam and Kartam (2001) found that transferring risks to others is not an effective way to reduce construction delay risk in Kuwait.

Insurers provide financial support to projects by accepting risks that are either outside the main participants' control or beyond their financial capacity (Orman, 1991). Risks are transferred to insurance companies for an insurance premium to cover the cost of the 'gamble' and other fees (Orman, 1991). Some of the types of insurance available are fire insurance, public liability insurance, workmen's compensation insurance and completion risk insurance covering delay causes.

Risk Retention

Project participants may decide that they can tolerate some risks, and hence make a decision to retain the risk. The risk of unforeseen circumstances arising is sometimes retained as well. If risk retention strategy is adopted, some form of contingency sum or markup must be added to the budget to cover the retained risk (Burcu and Martin, 1998). The amount of contingency sum varies between projects and is mostly dependent on the attendant risk and decision makers' risk attitude. For contractors, adding a contingency sum or markup would increase the bid price and consequently decrease the probability of winning the bid due to the highly competitive construction market (Kartam and Kartam, 2001).

Risk Reduction

Project managers should endeavor to reduce risks that cannot be eliminated, retained or transferred. Construction delay risks can be reduced by: producing a high quality schedule by getting updated project information; produce a proper program; refer to previous and ongoing similar projects for accurate program; coordinate closely with subcontractors; increase manpower and equipment on site and close supervision (Kartam and Kartam, 2001).

Gap in Knowledge

Operating in foreign countries is generally perceived to be more risky than domestic operations. Hitherto, not many studies of foreign AEC firms in Vietnam or Vietnam's construction industry have been conducted. Long et al. (2004) investigated the problems faced in large construction projects in Vietnam, and found that these are: incompetent designers and contractors; poor estimation and change management; social and technological issues; site related issues; and improper techniques and tools. Their study did not cover the risks faced by foreign AEC firms in Vietnam. Luu et al. (2008) proposed a framework that integrates balanced scorecard and SWOT matrix to evaluate the strategic performance of large contractors in Vietnam. The study focused on one large construction firm, AnGiang Construction Company, and hence the findings are not applicable for foreign AEC firms. Past studies on Vietnam did not investigate the risks that foreign AEC firms face in Vietnam, particularly with regard to political and legal systems, economic and financial risk conditions, design risks, construction related risk, cultural risk and natural risk. They also did not identify what strategies foreign AEC firms could adopt to respond to these risks. In the fieldwork, the risks faced by foreign AEC firms are investigated, and effective risk response techniques are examined.

Research Method

The case study research design was employed for this study. A case study is an empirical inquiry that: (a) investigates a contemporary phenomenon within its real-life context; (b) is appropriate when the boundaries between phenomenon and context are not clearly evident; and (c) incorporates multiple sources of evidence (Yin, 2003). Many researchers have also adopted the case study approach in their research (e.g. Rowlinson, 2001; Awakul and Ogunlana, 2002; Ling and Lau, 2002). The purpose of the case study was to investigate the risks faced by project team members from different countries and scrutinize into how these were handled and managed.

The case study relates to the development and construction of a foreign-invested factory in the southern part of Vietnam. The clients, project manager, consultant architects and engineers are from France. The main contractor is from Vietnam, while the water treatment and architectural finishes contractor is from Singapore. The data collection instrument comprised a list of open-ended questions, which were derived from the literature review. Open ended questions were used so that interviewees could share their experience and opinions without any influence from prefixed alternatives. The interviewees were allowed to raise other issues.

The sampling frame comprised project team members who worked for foreign firms (subject matter experts). All the senior project team members were contacted for the study. After an expert has agreed to be interviewed, a soft copy of the questions was sent to him. An in depth face-to-face interview was then carried out at the most convenient time and date for the interviewee.

After much persuasion through the telephone and follow up emails, three project team members working for foreign firms agreed to be interviewed – the client's representative, consultant project manager and consultant engineer. The main contractor's construction manager (a Vietnamese) was also interviewed to get multiple views of the same issue. Multiple interviewing of key participants provided rich sources of data. Each interview took between 50 and 70 minutes. The interviews were carried out between July and November 2007.

The face-to-face interview method was preferred by the interviewees because they could seek clarifications on the questions. As the questions were open ended, face-to-face interviews meant that interviewees need to only give verbal comments rather than fill in the questionnaire with long answers. The face-to-face meeting also allowed interviewees and interviewers to discuss and exchange points of view. All the interviews were recorded on paper. Follow-up interviews were conducted through emails. Besides interviews, archival research was carried out to obtain non-confidential project details and company reports.

Results: Case Study

The main objective of the case study was to find out different type of risks encountered by different players within the same project, and the risk response measures which were undertaken. The data for the case study was obtained by interviewing four experts from different firms that undertook important parts in this project.

Background of the Project

This case study relates to the development of a factory to produce yeast in the southern part of Vietnam. The client is a joint venture based on a 50:50 share between a France company and a Vietnamese company. The site area is about 40,000m^2 and the gross constructed floor area about 6,200 m^2. The contract sum for construction works is approximately US$10 million.

The client informed that a few consultants and project management firms were invited to submit fee proposals. French architects, engineers and project managers were selected as they came from the same country as the foreign client.

Table 1. Project team members

Team member	Country of origin	Contractual arrangement
Client/developer	France and Vietnamese	Equity joint venture
Project management	French	Consultancy contract
Consultant architect	French	Consultancy contract
Consultant engineers	French	Consultancy contract
Main contractor	Vietnam	Construction management
Contractor for water treatment and architecture finishes	Singapore	Turnkey contract
Subcontractors	Vietnam	Nominated subcontract

The construction management contractual arrangement was adopted. The project was divided into many work packages. Each work package was separately awarded when its design was ready. The client chose this method because it wanted to have the flexibility of making changes at the later stages of the project without paying exorbitant costs for variations. The first contractor, who also won the largest package for civil construction, became the main contractor and the construction manager. The successful bidder of other work packages became subcontractors.

The client revealed that the tendering procedure to select the construction management contractor was based on the two-envelope system. The first envelope comprised information on contractor's competency - track record, financial and manpower resources and quality of previous projects. The second envelope contained the bid price. The evaluation team ranked the bidders based their competency, as provided in the first envelope. Contractors who did not meet the base line competency were excluded from the next phase of evaluation. With only competent contractors, the second envelopes were opened. The contract was then awarded to the lowest competent contractor.

The main package was awarded to a Vietnamese contractor. Other subcontract packages were also awarded to Vietnamese firms. Another large package for water treatment and architecture finish was awarded to a Singaporean firm who came on board as a turnkey contractor. Details of project team members are shown in Table 1.

As regards project performance, this project exceeded budget by more than 5% and was completed 2 years behind schedule. The client felt that the quality of the built facility and workmanship quality did not meet his expectations but it was acceptable by Vietnamese standards.

Planning and design started in late 2001. Due to some problems within the joint venture and technical issues with the yeast manufacturing method and process of operating the factory, construction only started in early 2005 and was expected to be completed in late 2006. However, the project was only completed in 2008 due to changes in design and technology, late delivery of imported materials, slow progress of work, and main contractor's inadequate level of manpower. While the client brought up the issue of liquidated damages, it was not imposed. The main reasons were the uncertainty surrounding adequacy of the local law to support the imposition of liquidated damages and fairness of the judicial system.

Profile of Interviewees and Their Firms

The profile of the interviewees and their firms are shown in Table 2. The experts who were interviewed were the foreign client (coded as CL), consultant engineer (ER) and project manager (PM) and the local main contractor (CM).

The client's representative (CL) is a regional director who is in charge of developing the factory. CL had 8 years of experience in the construction industry, and had worked in Vietnam for 3 years. CL's parent company manufactures yeast in France, and this was its first project in Vietnam. He was tasked with transferring the technology of manufacturing yeast to Asia.

Table 2. Characteristics of interviewees

Interviewee (Code)	Role in project	Work experience (years)	Experience in Vietnam (years)	Firm's country of origin	Firm's years in Vietnam (years)
Regional director (CL)	Client's representative	8	3	France	3
Project Manager (PM)	Consultant PM	7	3	France	14
Chief Engineer (ER)	Consultant electrical engineer	13	13	France	14
Construction manager (CM)	Main contractor	15	15	Vietnam	20

The consultant project manager (PM) and the chief consultant engineer (ER) came from a multi-disciplinary AE consultancy firm in France. It formed a joint venture with a local firm to undertake this project in Vietnam. The foreign firm had operated in Vietnam for 14 years and undertaken more than 40 projects of differing types. Interviewees PM and ER had 7 and 13 years of experience in the construction industry respectively. ER is a Vietnamese working for this French company, and thus had the benefit of knowing the risks from local and foreigners' points of view.

The main contractor's Construction Manager (CM) was also interviewed. CM is a Vietnamese and had worked for 15 years in Vietnam's construction industry. The main contractor is a local firm which had been operating in Vietnam for more than 20 years and undertaken 45 projects.

The foreign client and consultants revealed that their companies ventured into Vietnam to increase sales volume and profitability, to gain competitive advantage and to create a new market for their products and services. These foreign firms did not engage any risk analyst to give a preliminary assessment of the risk they would face in Vietnam's construction market, but identified and assessed the risks by themselves.

Findings and Discussion

The results of the interviews showed that foreigners faced many risks in this project – some were more serious than others. These risks and how they were managed are now discussed.

Political Risk

The Vietnamese government is constantly trying to improve the legal system. It also makes new policies to attract foreign investors and tries to fulfill its commitments to the international community. According to the interviewees, these progressive steps have successfully increased the level of confidence of foreign investors in making foreign direct investments in Vietnam. They agreed that the government's policies for foreign companies are quite encouraging. For instance, the foreign client was exempted from paying taxes when importing equipment and machines that were to be used in the factory.

The interviewees complained that the government's administrative system is complicated, and administrative procedures are unclear. All applications for permits must go through many government departments for approval. The requirements of central, provincial and local governments also need to be met.

To respond to political risk, the common strategy adopted by foreigners was to transfer administrative applications to their local partners who are more familiar with Vietnam's administrative systems and regulations.

Legal Risk

One of the causes of the 2-year delay was the main contractor's slow progress and insufficient manpower. According to the project manager, the main contractor had too many ongoing projects and was unable to handle all of them adequately. The main contractor even withdrew its skilled workers from this project and assigned them to other projects. Due to the instability of resources, substantial delays were experienced. For example, 15 workers were allocated to foundation work, later on 10 workers were pulled out before the work was finished. The 5 left could not comply with the schedule. Delay in the foundation work had a massive impact on other parts of the project because when the foundation was not completed, other superstructure works could not proceed.

Despite several warnings from the client, the main contractor did not seem to try to resolve the problem. Even though liquidated damages were stated clearly in the contract, the client could not impose it because it was not a common practice. The client was close to bringing the dispute to the courts but did not proceed with this course of action because it perceived that the judge may be biased, and the legal framework may be inadequate. The French client was also not familiar with Vietnam's legal system.

To respond to the legal risk, the client entered into serious negotiations with the main contractor. It had no choice but to implore the main contractor to increase manpower in order to keep to the schedule.

Financial and Economic Risk

Vietnam has high inflation rates, and with the project delayed, higher material cost, staff salary and other cost were incurred. To respond to the risk of rising prices due to high inflation, the client set aside 10% of the contract sum as contingency sum. At the planning stage, adequate floats were built-in the schedule. When prices of materials rose, the consultant engineer allowed the contractor to use cheaper materials of the same quality. For example, the price of cable imported from France increased drastically, and he allowed cables produced in Vietnam to be used instead. The main contractor shared that due to rising prices, it bought building materials early. The two year delay in this project led to the foreign client and main contractor suffering huge financial losses. The lesson learned is that foreign clients should engage contractors who have a sound financial background and could weather the increase in prices of resources due to high inflation rate in Vietnam. Large contingency sums should be set aside. The schedule should be logical, enable activities to be carried out efficiently, have adequate duration for each activity, all important activities accounted for, and adequate float provided.

Fluctuation of foreign exchange rate is another risk faced in international construction. Vietnam's Central Bank controls the official exchange rate. It imposes a limit (eg 2%) on the daily change in the tightly controlled official rate for trading Vietnamese dongs into U.S. dollars and vice versa. The client felt that this relatively stable exchange rate made the risk not severe. The consultants received 60% of their fees upon finishing the design, and 30% upon the award of main contract, with 10% as retention sum. Hence, within the relatively short design period, they had received the bulk of the fees, making foreign exchange rate fluctuation an insignificant risk. The lesson learned is that consultants should sign consultancy contracts that allow significant payments when the design is completed, so that they are subjected to less risk in foreign exchange rate fluctuation.

Financial failure is one of the risks that may be encountered. To prevent this risk from arising, the foreign client and consultants selected contractors and nominated subcontractors carefully using the two envelope system. As stated earlier, the first envelope examined bidders' financial viability, technical know-how and management competence. The consultants also took steps to investigate the status of bidders' other on-going projects to lessen the chance of financial failure in the middle of the project. During construction, a give-and-take attitude should be adopted when a firm is facing financial difficulties. For example, one subcontractor had cash flow problems. The strategy adopted to prevent his financial failure was to make some advance payments to ease the company's cash flow problem.

While the main contractor and nominated subcontractors did not face finance failures, the client found out that the main contractor had sublet some of the work to domestic subcontractors without going through careful selection and assessment of their credit worthiness. The substandard domestic subcontractors lacked competence and financial capacity. To mitigate this risk, the client demanded that the main contractor submit the names of domestic subcontractors to the project manager, who then monitored them closely.

Design Risk

In this project, changes were made to the design many times to: fit actual site conditions; be in sync with current technological level in Vietnam; and fit the tight budget. The consultants advised that designers should make provisions for change and prepare a flexible design during the outline design stage, instead of being caught off guard later.

One major design risk faced in this project was revision of the design to fit into a tight budget, to the point that the completed facility was not fit for its purpose. The original design by the consultant engineer called for underground electrical cable and a medium voltage electrical substation. The client approved the design. When bids were invited for electrical works, the lowest bid exceeded the estimated cost significantly. In order to cut cost and speed up the process, the client asked the contractor to redesign to fit into the budget. The contractor changed the design to overhead electrical line and proposing using low quality materials. He also proposed transformer cables ran inside the substation wall to reduce the length of cables, and reduce the size of emergency oil pit.

The consultant engineer refused to approve the contractor's design since there were many of serious mistakes, the design was outdated, and there was insufficient provision for future expansion to the factory. Due to shortage of funds, the client approved the contractor's design and allowed it to proceed with construction. Dispute arose among parties mainly because the consultants knew that the contractor's design was inadequate and would not be accepted by the local authority.

As expected, the completed electrical system failed the inspection by Industrial Park Management Authority, the main reasons being the overhead cables may endanger the national power grid and fire safety was compromised. The reworks insisted by the Authority were similar to the consultant's original design: install underground cables; build a medium voltage substation; move transformer away from the wall; and enlarge the emergency oil pit. As a result, cost increased significantly. The client paid for the reworks, much to its chagrin and disappointment. To solve this problem, all project team members attended a coordination meeting. In the meeting, all requirements were clarified, and advice from other project members was sought.

Construction Related Risk

The main construction related risk faced in this project was substandard quality of materials and workmanship. The client's representative had to ensure that the factory achieved similar quality as the client's other yeast production factories throughout the world. However, in this project, the client's representative had to accept a lower quality standard because that was what the Vietnamese contractors were capable of. The low quality workmanship was because some of the main contractor's workers were unskilled. They also had low productivity. The main contractor's excuse was that he needed the skilled manpower for his other ongoing projects. The lesson learned is that foreign firms should enter Vietnam knowing that their built facility would not be of the quality and standard that they are used to in developed countries. Hence, quality control must be enforced strictly.

The project manager recounted that the contractors changed their site management staff frequently. In response to this risk, the consultants required that any change of personnel in the main contracting, nominated and domestic subcontracting firms would need the project manager's approval.

The interviewees shared that run-of-the-mill equipment and materials are easily found in Vietnam. Specialized items however had to be imported. The local main contractor had to contend with long delivery time, delay in customs clearance and payment of tax. There was a long lead time for equipment to arrive from overseas, and the main contractor overcame this risk by placing orders early. The delay in customs clearance is something that foreign AEC firms should be aware of and build this into their schedule. Import tax is complicated, and the rate depends on the type of material and equipment, and when the items arrive in Vietnam. Firms should therefore build in a large contingency sum for paying import tax on materials and equipment.

Another construction risk faced by the project was unsafe work practices. The project manager found that most of the contractors did not have safety management systems in place. To manage this risk, the main contractor was required to engage a safety specialist, who must ensure that safe work practices were adopted, and the working condition on site was safe.

Cultural Risk

The foreigners shared that working style in Vietnam is totally different from French practice which they were used to. Several examples are given below, and where possible, the local contractor's response was sought.

The first is that local staff were more concerned about big items that were visible, ignoring the 'nuts and bolts'. The client, however, was very concerned that small details must also be perfectly executed, so that the factory could function efficiently to produce yeast. On the other hand, the main contractor felt that the client was too strict, micro-managed them, and tried to control every detail, even minor issues.

The second example was the quality of schedule. The foreigners wanted a detailed weekly work schedule and a comprehensive weekly report to be submitted by the main contractor. Instead, Vietnamese managers gave only the milestones to be achieved in different months. To exacerbate the problem, they did not meet their own deadlines. The main contractor complained that there were so many things to take care of, so it did not want to waste time recording them. Since there might be minor changes to the schedule, the construction manager preferred not to give a weekly work plan to the client, so that he could keep his schedule flexible. To overcome this problem, after many rounds of negotiation, both the client and main contractor reached a compromise – the main contractor would submit work schedules fortnightly, and revisions can be made if there were unforeseen changes.

The foreigners also feedback that Vietnamese do not seem to have a sense of urgency about milestones and deadlines. They complained that the locals were not punctual – so meetings and appointments could not start on time. The local construction manager shared that the locals are influenced by the agrarian practices where time is not of the essence.

To mitigate cultural risk, it is recommended that foreigners adapt to the local practices – and not expect it to be the other way around, ie locals adapt to foreigners. As guests in the host country, foreigners should be mindful (involves paying attention, being watchful and attentive of locals), and also acquire more knowledge about how locals function (Ling et al., 2007).

Natural Risk

Construction work in Vietnam may be affected by heavy rain. The construction manager explained that in southern Vietnam, the wet season lasts from May to October. In central Vietnam, flooding may occur from October to December. In northern Vietnam, the rainy months are August to November, with February and March having persistent, light, drizzling rain. The construction manager said that he would undertake construction work that would be affected by bad weather condition outside of the rainy season. Apart from planning work to avoid heavy rain, the foreign interviewees did not face any other severe natural risk that affected construction progress.

Conclusion

Risks may exist throughout the project lifetime from inception to design, bidding, construction and commission. Using the case study approach of the development of a foreign invested manufacturing plant in southern Vietnam, this study identified the types of risk that foreign AEC firms face in Vietnam. This study uncovered that the main types of risks faced are political risk, legal risk, financial and economic risks, design risk, construction related risk and cultural risk. Bureaucracy, inadequate legal framework, high inflation rate, low credit worthiness of local partners and cross culture issues are significant factors affecting foreign firms' venture into Vietnam.

Effective risk response techniques must be adopted to manage the risks. Foreign investors should make a risk assessment of Vietnam's construction market before venturing in. They should familiarize themselves with Vietnam's legislation, laws and regulations before undertaking projects there. Due to Vietnam's high inflation rate, foreign firms should set aside a large contingency sum to pay for escalating cost of resources.

Foreign firms should also form joint ventures with local firms, and transfer administrative work such as getting approvals and licenses to local partners, since they were more familiar with Vietnam's administrative system, regulations and policies. It is important to select local partners, contractors and subcontractors who are reputable and financially sound. Stringent selection criteria should be set.

Projects in Vietnam may face delays, but whether the liquidated damages clause can be effectively put into operation is untested. As such, foreign AEC firms should pay close attention to key activities on the critical path to avoid project delay. They should also build relationships with project team members to foster co-operation.

To minimize design risks, designers are advised to check whether the materials and equipment that they specify are available locally and if the quality is acceptable. This is to avoid design changes or having to import them, which may delay the project. Designers should also endeavor to have a flexible design, because clients' needs change as Vietnam's economic and political landscape changes.

As the quality of work and safety could be compromised in projects, it is recommended that foreign firms provide Vietnamese workers with some training so that they can achieve certain skills level and safety awareness. Safety management systems should also be implemented on site.

Another effective risk response technique is to use negotiation to solve problems. It is found to be the most practical and effective way to overcome difficulties faced in Vietnam. The implication is that practitioners should be well-equipped with negotiation skills when undertaking projects in Vietnam. Foreign AEC practitioners should also familiarize themselves with Vietnam's working environment and culture, which are significantly different from developed countries. To overcome difficulties in cross cultural encounters, it is proposed that frequent coordination meetings among project members be held. Foreigners should try to adapt to local working culture, be flexible in achieving their aims and not to rigidly follow the terms of the contract. Team building activities should also be carried out. Foreigners should spend time interacting and understanding the locals. By understanding their culture and needs, foreigners can build up relationship with locals, win their trust and get them to commit to projects goals.

References

Akintoye, A., and MacLeod, M.J.(1997). Risk analysis and management in construction. *International Journal of Management in Engineering*, **15**(1), 31-38.

Andi, and Minato, T. (2003). Design documents quality in the Japanese construction industry: factors influencing and impacts on construction process. *International Journal of Project Management,* **21**(7), 537-546.

Ang, Y.K., and Ofori, G. (2001). Chinese culture and successful implementation of partnering in Singapore's construction industry. *Construction Management and Economics*, **19**, 619-632.

Ankrah, N.A., and Langford, D.A. (2005). Architects and contractors: a comparative study of organizational cultures. *Construction Management and Economics*, **23**, 595-607.

Awakul, P., and Ogunlana, S.O. (2002). The effect of attitudinal differences on interface conflicts in large scale construction projects: a case study. *Construction Management and Economics*, **20**, 365-77.

Baker, S., Ponniah, D., and Smith, S. (1999). Risk response techniques employed currently for major projects. *Construction Management and Economics*, **17**, 205 - 213.

Bing, L., Tiong, R.L.K., Wong, W.F., and Chew, D. (1999). Risk management in international construction joint ventures. *Journal of Construction Engineering and Management*, **125**(4), 277-284.

Bing, L., and Tiong, L.K. (1999). Risk management model for international construction joint ventures. *Journal of Construction Engineering and Management*, **125**(5), 377-384.

Blume, C. (2007, December 31). Vietnam attracts record amount of foreign direct investment. Washington D.C.: Voice of America. Retrieve 23 July 2008 from http://www.voanews.com/english/archive/2007-12/2007-12-31-voa7.cfm?CFID=16008504andCFTOKEN=97455739

Burchett, J.F., Tummala, V.M.R., and Leung, H.M. (1999). A world-wide survey of current practices in the management of risk within electrical supply projects. *Construction Management and Economics*, **17**, 77 – 90.

Burcu A., and Martin F. (1998). Factors affecting contractors risk of cost overburden. *Journal of Management in Engineering*, **14**(1), 67–76.

Carter, R.L., and Doherty, N.A. (1974). *Handbook of Risk Management.* London: Kluwer-Harrap.

Chan, E.H.W., and Au, M.C.Y. (2007). Building contractors' behavioural pattern in pricing weather risks. *International Journal of Project Management,* **25**(6), 615-626.

Chen, J.J. (1997). The impact of Chinese economic reforms upon the construction industry. *Building Research and Information*, **25**(4), 239-245.

Chen, J.J. (1998). The characteristics and current status of China's construction industry. *Construction Management and Economics*, **16**, 711-719.

Cheung S.O., Tam, C.M., Ndekugri, I., and Harris, F.C. (2000). Factors affecting clients' project dispute resolution satisfaction in Hong Kong. *Construction Management and Economics*, **18**, 281-294.

Chicken J.C., and Posner, T. (1998). *The philosophy of risk*. London: Thomas Telford.

Chua, D.K.H., Wang, Y., and Tan, W.T. (2003). Impacts of obstacles in east Asian cross border construction. *Journal of Construction Engineering and Management,* **129**(2), 131-141.

Construction Industry Institute (2004). *Risk assessment of international projects: a management approach*. Austin: University of Texas.

Dodgson, M. (1993). Organizational learning: a review of some literature. *Organization Studies*, **14**(3), 375–394.

Earley P.C., and Mosakowski, E. (2004). Toward Cultural Intelligence: turning cultural differences into a workplace advantage. *Academy of Management Executive,* **18**(3), 151-157.

Edwards, P.J., and Bowen, P.A. (1998). Risk and risk management in construction: a review and future directions for research. *Engineering, Construction and Architectural Management*, **5**(4), 339-349.

El-Sayegh, S.M. (2008). Risk assessment and allocation in the UAE construction industry. *International Journal of Project Management,* **26**(4), 431-438.

Evans, W.A., Hau, K.C., and Sculli, D. (1989). A cross-cultural comparison of managerial styles. *Journal of Management Development,* **15**(3/4), 28-32.

Fang, D., Li, M., Fong, S., and Shen, L. (2004). Risks in Chinese construction market – contractors' perspective. *Journal of Construction Engineering and Management,* **130**(6), 853-64.

Fellow, R.F., and Hancock, R.(1994). Conflict resulting from cultural differentiation: An investigation of the new engineering contract. In Proceedings on *Construction Conflict: Management and Resolution*, (pp. 259-267). Rotterdam: CIB.

Flanagan, R., and Norman, G. (1993). *Risk Management and Construction*. Oxford: Blackwell Scientific.

Gunhan, S., and Arditi, D. (2005). Factors Affecting International Construction. *Journal of Construction Engineering and Management*, **131**(3), 273-282.

Hampton, J. J. (1993). *Essentials of risk management and insurance*. New York: AMACOM.

Han, S.H., Diekmann, J.E., and Ock, J.H. (2005). Contractor's risk attitudes in the selection of international construction projects. *Journal of Construction Engineering and Management*, **131** (3), 283-292.

He, Z. (1995). Risk management for overseas construction projects. *International Journal of Project Management*, **13**(4), 231-237.

Hofstede, G., and Hofstede, G.J. (2005). *Culture and organizations: Software of the mind* (2nd ed). New York: McGraw-Hill.

Jain, S.C. (1996). *International marketing management*. Cincinnati: South Western College Press.

Kapila, P. and Hendrickson, C. (2001). Exchange rate risk management in international construction ventures. *Journal of Management in Engineering*, **17**(4), 186-191.

Kartam, N.A., and Kartam, S. (2001). Risk and its management in the Kuwaiti construction industry: a contractors' perspective. *International Journal of Project Management*, **19**(6), 325-335.

Kim, S., and Bajaj, D. (2001). Risk management in construction: an approach for contractors in South Korea. *Cost Engineering*, **42**(1), 38 – 44.

Lai, X.T., and Truong, Q. (2005). Relational capital and performance of international joint ventures in Vietnam. *Asia Pacific Business Review*, **11**(3), 389 – 410.

Lam, K.C., and Chow, W.S. (1999). The significance of financial risks in BOT procurement. *Building Research and Information*, **27**(2), 84-95.

Ling, Y.Y., and Hoi, L. (2006). Risk faced by Singapore firms when undertaking construction project in India. *International Journal of Project Management*, **24**(3), 261-270.

Ling, Y. Y., and Lau, B.S.Y. (2002). A case study on the management of the development of a large-scale power plant project in East Asia based on design-build arrangement. *International Journal of Project Management*, **20**(6), 413-423.

Ling, Y.Y., and Lim, H.K. (2007). Foreign firms' financial and economic risk in China. *Engineering, Construction and Architectural Management*, **14**(4), 346-362.

Ling, Y.Y., and Low, S.P. (2007). Legal risks faced by foreign architectural, engineering and construction firms in China. *Journal of Professional Issues in Engineering Education and Practice*, **133**(3), 238-245.

Ling, F. Y. Y., Ang, A. M. H., and Lim, S. S. Y. (2007). Encounters between foreigners and Chinese: Perception and management of cultural differences. *Engineering, Construction and Architectural Management*, **14**(6), 501-517.

Long, N.D., Ogunlana, S., Quang, T., and Lam, K.C. (2004). Large construction projects in developing countries: a case study from Vietnam. *International Journal of Project Management*, **22**(7), 553-561.

Low, S.P., and Shi, Y.Q. (2002). An exploratory study of Hofstede's cross-cultural dimensions in construction projects. *Management Decision*, **40**(1), 7-16.

Loosemore, M., and Chau, D.W. (2002). Racial discrimination towards Asian operatives in the Australian construction industry. *Construction Management and Economics*, **20**, 91-102.

Luu, T.V., Kim, S.Y., Cao, H.L., and Park, Y.M. (2008). Performance measurement of construction firms in developing countries. *Construction Management and Economics*, **26**, 373-386.

Mortanges, C.P., and Allers, V. (1996). Political risk assessment: Theory and the experience of Dutch firms. *International Business Review,* **5**(3), 303-318.

Ngowi, A.B. (1997). Impact of culture on construction procurement. *Journal of Construction Procurement*, **3**(1), 3-15.

Ofori, G. (2007). Construction in developing countries. *Construction Management and Economics*, **25**(1), 1– 6.

Orman, G.A.E. (1991). New applications of risk analysis in project insurances. *International Journal of Project Management,* **21**(7), 537-546.

Petrov, M. (2006). Managing project risk with proper scheduling**.** Nielsen-Wurster Communique, 1.5. Retrieved on 30 July 2008 from http://www.nielsen-wurster.com/Email_Announcements/NW_Communique/NW_Communique_2006_NOV.html

Porter, M.E. (1980). *Competitive strategy: techniques for analyzing industries and competitors*. Boston: Free Press.

Rahman, M.M., and Kumaraswamy, M.M. (2002). Risk management trends in the construction industry: moving towards joint risk management. *Engineering, Construction and Architectural Management*, **9**(2), 131-151.

Robock, S. H., and Simmonds, K. (1983). *International business and multinational enterprises*. Homewood, IL: Irwin.

Root, F. (1987). *Entry strategies for international markets*. Lexington, MA: Lexington Books.

Rowlinson, S. (2001). Matrix organizational structure, culture and commitment: a Hong Kong public sector case study of change. *Construction Management and Economics*, **19**(7), 669-73.

Shen, L.Y. (1997). Project risk management in Hong Kong. *International Journal of Project Management,* **15**(2), 101–105.

Shen, L. Y., Wu, W. C., and Ng, S. K. (2001). Risk assessment for construction joint ventures in China. *Journal of Construction Engineering and Management,* **127**(1), 76–81.

Smith, J., Zheng, B., Love, P.E.D., and Edwards, D.J. (2004). Procurement of construction facilities in Guangdong province, China: factors influencing the choice of procurement method. *Facilities,* **22**(5/6),141-148.

Smith, N.J. (1999). *Managing risk in construction projects*. Oxford: Blackwell.

Swierczek, F.W. (1994). Culture and conflict in joint ventures in Asia. *International Journal of Project Management*, **12**(1), 39-47.

Thompson, P., and Perry, J. (1992). Engineering construction risks: a guide to project risk analysis and risk management. London: Thomas Telford.

Uher, T.E., and Toakley, A.R. (1999). Risk management in the conceptual phase of a project. *International Journal of Project Management*, **17**(3), 161 – 169.

Wallace, I. N. D. (1995). *Hudson's building and engineering contracts* (Vol.1). London: Sweet and Maxwell.

Wang, M.T., and Chou, H.Y. (2003). Risk allocation and risk handling of highway projects in Taiwan. *Journal of Management in Engineering*, **19**(2), 60-68.

Wang, S.Q., Tiong, L.K., Ting, S.K., and Ashley, D. (2000). Evaluation and management of foreign exchange and revenue risks in China's BOT projects. *Construction Management and Economics*, **18**, 197-207.

Wang, S.Q., Dulaimi, M.F., and Aguria M.Y. (2004). Risk management framework for construction projects in developing countries. *Construction Management and Economics*, **22**, 237-252.

World Bank. (1984). *The construction industry: Issue and strategies in developing countries.* Washington, D.C.: World Bank.

World Trade Organization (WTO). (2005). *Understanding the WTO*. Geneva. Retrieved on 13 January 2007 from http://www.wto.org/english/thewto_e/whatis_e/tif_e/understanding_e.pdf.

World Trade Organization (WTO). (2006). *Working Party on the Accession of Vietnam, Part II - Schedule of Specific Commitments in Services.* Retrieved on 10 February 2008 from http://docsonline.wto.org/imrd/directdoc.asp?DDFDocuments/t/wt/acc/vnm48a2.doc.

Xenidis, Y., and Angelides, D. (2005). The financial risks in build-operate-transfer projects. *Construction Management and Economics*, **23**, 431-441.

Yin, R.K. (2003). *Case study research: design and methods* (3rd ed). Thousand Oaks, CA: Sage Publications.

Zou, P.X.W., Zhang, G., and Wang, J. (2007). Understanding the key risks in construction projects in China. *International Journal of Project Management*, **25**(6), 601-614.

In: Progress in Management Engineering
Editors: L.P. Gragg and J.M. Cassell, pp. 199-236
ISBN: 978-1-60741-310-3

Chapter 8

EVALUATION OF COOLING, HEATING, AND POWER SYSTEMS BASED ON PRIMARY ENERGY OPERATIONAL STRATEGY

Pedro J. Mago[1], Louay M. Chamra[2] and Nelson Fumo[3*]
Department of Mechanical Engineering,
Mississippi State University, USA

Abstract

Cooling, Heating and Power (CHP) systems have been recognized as a key alternative for thermal energy and electricity generation at or near end-user sites. CHP systems are a form of distributed generation that can provide electricity while recovering waste heat to be used for space and water heating, and for space cooling by means of an absorption chiller. Although CHP technology seems to be economically feasible, due to the constant fluctuations in energy prices, CHP systems cannot always guarantee economic savings. However, a well-designed CHP system can guarantee energy reduction. This energy reduction could be increased depending on the CHP system operational strategy employed. CHP systems operational strategy defines the goal of the system's response to the energy demand, which is one of the factors that characterize the energy performance of the system. CHP systems are usually operated using a cost-oriented operational strategy. However, an operational strategy based on primary energy would yield better energy performance. In this chapter the CHP system energy performance is evaluated based on primary energy consumption and a primary energy operational strategy is implemented to optimize energy consumption. To determine the energy performance, a model has been developed and implemented to simulate CHP systems in order to estimate the building-CHP system energy consumption. The novel characteristic of the developed model is the introduction of the Building Primary Energy Ratio (BPER) as a parameter to implement a primary energy operational strategy, which allows obtaining the best energy performance from the building-CHP system. Results show that the BPER operational strategy always guarantees energy savings. In addition, the BPER operational

[1] E-mail address: mago@me.msstate.edu.
[2] E-mail address: chamra@me.msstate.edu.
[3] E-mail address: fumo@me.msstate.edu.
[*] 210 Carpenter Engineering Building, P.O. Box ME, Mississippi State, MS 39762-5925, Phone: (662) 325-6602 - Fax:(662) 325-7223.

strategy is compared with a cost oriented operational strategy based on energy cost. Results from a cost-oriented operational strategy show that for some operation conditions, high economic savings can be obtained with unacceptable increment of the energy consumption. This chapter also considers how the application of the BPER operational strategy can improve the Energy Star Rating and the Leadership in Energy and Environmental Design (LEED) Rating, as well as reduce the emission of pollutants.

Nomenclature

BPER	Building primary energy ratio
CH	Absorption chiller
CHP	Cooling, heating, and power
CCHP	Combined cooling, heating, and power
COP	Coefficient of performance
CCS	Cooling coil system
cutoff	Fraction of PGU nominal power bellow which the unit not operate
DD	Degree-days
E	Electricity
EG	Electric grid
ECF	Site-to-primary energy conversion factor for electricity
ECR	Energy cost ratio
EIA	Energy Information Administration
F	Fuel
FCF	Site-to-primary energy conversion factor for fuel (natural gas)
HCS	Heating coil system
HVAC	Heating, ventilating, and air conditioning
I	Increment on size (PGU and CH)
MEC	Monthly energy consumption
P	Power (nominal power of the PGU)
PEC	Primary energy consumption
PES	Primary energy savings
PER	Primary energy ratio
PGU	Power generation unit
Q	Thermal energy (cooling or heating)
SEC	Site energy consumption
VC	Vapor compression system

Symbols

η	Efficiency level (ratio between useful output and input amount)
f	Fraction

Subscripts

b	Boiler
c	Cooling
ch	Chiller
chp	Cooling, heating, and power
f	Furnace
h	Heating (space and water); heating system (furnace, boiler)
hs	Space heating
hw	Water heating
np	Nominal power (of the PGU)
m	Meter
rec	Heat recovery system
p	Parasitic electricity
pgu	Power generation unit
R	Recovered (thermal energy)
Ra	Recovered and available (thermal energy for heating)
s	Space (heating or cooling)
vc	Vapor compression
yr	Year

1. Introduction

Dependence on imported energy, reliability and efficiency of energy systems, environmental concerns, and energy costs for end users, are factors that continually press for the improvement and development of new technologies, and new energy and environmental legislations (policies and regulations).

Cooling, Heating and Power (CHP) systems have been recognized as a key alternative for thermal energy and electricity generation at or near end-user sites. CHP systems are a form of distributed generation that can provide electricity while recovering waste heat to be used for space and water heating, and for space cooling by means of an absorption chiller. Since CHP systems generate the electricity on-site, losses due to transmission and distribution are considerably reduced compared with the electricity supplied by distant central power plants. While central power plants have a total efficiency between 30% - 51%, CHP systems are potentially 70% - 85% efficient in utilizing fuels [1].

General accepted benefits from the use of CHP systems are: increased energy efficiency, improved air quality, lower energy cost, increased power quality, and increased power reliability. Beyond these benefits, a non-conventional evaluation of CHP systems will show additional benefits such as building energy performance, fuel source flexibility, brand and marketing benefits, protection from electric rate hikes, and benefits from promoting energy management practices.

The design and analysis of CHP systems is simplified by using models, which can be used to develop computer software for simulation purpose, allowing the reduction of time analysis. CHP system analysis involves variables such as type and size of the components, individual component efficiencies, system operating mode, operational strategy, and building demand for power, heating, and cooling loads [2-7]. These seem to be the most relevant variables to consider when designing and estimating the performance of CHP systems.

Among these variables, the system operating mode and system operational strategy are crucial for the design and analysis of CHP systems. For a CHP system, common operation modes are electric load following, electrically sized, or thermally sized [8]. For the electric load following operation mode the power generation unit (PGU) is able to handle the variations on the electric demand. The electrically sized operation mode is a "base loaded" operation. The thermally sized operation mode is a "following thermal demand" operation. For this chapter, the model is based on the electric load following operation mode. This operation mode was chosen for two reasons. First, in order to optimize the energy performance of the CHP system, the model was implemented to account for different PGU and the absorption chiller (CH) sizes. Second, because other operation modes frequently results in more production of power than needed by the building, which require either selling electricity back to the grid or electric power storage. CHP operational strategy defines the goal of the system's response to the energy demand, which is one of the factors that characterize the energy performance of the system. The most frequent operational strategy is cost-oriented, although a primary energy operational strategy yields a better energy performance. Cardona and Piacentino [9] presented a summary of the most common evaluation criteria for combined heat and power plants and combined cooling, heating, and power plants (CCHP). They reported that the primary energy saving management strategy is the operational strategy that allows achieving maximum energy savings during the plant life cycle. Other studies [10-12] also consider primary energy as the appropriate criterion for evaluation of CHP systems. Sun et al. [10] compared the thermal efficiency of separated cooling and heating system versus the combined system. They stated that "to compare systems with different types of driving and produced energy, the primary energy rate (PER) is a satisfactory criterion." The PER is defined as the ratio of the required output to the primary energy demand. For estimating the PER for electrical equipment, they considered the efficiency of generation and distribution of electricity, which can be compared to the inverse of site-to-primary energy conversion factor used in this study. Sun et al. [10] and Li et al. [12] compared the energy utilization evaluation of separated systems versus combined cooling, heating and power, but using the fuel energy ratio which considers total primary energy use. As suggested by Li et al. [12], when comparing energy performance, primary energy savings "... is not mainly resulted from the performance of CCHP systems but the difference of primary energy."

In this chapter, CHP system energy performance is evaluated based on primary energy consumption and a primary energy operational strategy is implemented to optimize energy consumption. Therefore, a model is developed for the analysis of CHP systems. The model accounts for the variables that govern the exchange and use of energy for the CHP system components and other components of the building HVAC system. A new parameter called Building Primary Energy Ratio (BPER) is introduced to evaluate the CHP system energy performance under a primary energy operational strategy. Although the results presented in this chapter can be applied for different type of buildings, this study concentrates on the use of CHP systems for office buildings.

1.2. Site Energy

The Energy Information Administration (EIA) [13] defines Site Energy as "The Btu value of energy at the point it enters the home, sometimes referred to as "delivered" energy. The site value of energy is used for all fuels, including electricity." Building energy use is mainly a consequence of the building characteristics, use, operation, and climate conditions. The combination of these factors will give a unique amount and type of energy consumption for each building. Site Energy Consumption (SEC) is referred to the energy consumed at the building doors, that is, the energy use registered by the utility meters.

CHP system design requires knowing the building energy consumption profiles or patterns for accurately sizing the components and modeling the system [1, 14-17]. Hourly energy consumption profiles are commonly used as a good reference for energy evaluation. For new buildings, or when the energy use is unknown, simulation software such as EnergyPlus™ [18] can be employed to estimate the building energy consumption. For this study only electricity and natural gas are considered as site energy sources, although other energy sources such as fuels (propane, biofuels, fuel oil, etc.) or secondary energy (steam, hot water, etc.) can also be utilized. When a CHP system is incorporated to the building, it changes the site energy consumption profiles mainly because: (a) the electric energy consumed by the cooling system is substituted by fuel consumption; (b) the electric energy from the grid is substituted by electric energy from the power generation unit; and (c) the fuel consumption for heating is substituted by heat recovered from the power generation unit. For economic evaluations the site energy plays an important role because the energy consumption is billed based on the SEC.

1.3. Primary Energy

The EIA [13] defines Primary Energy as "All energy consumed by end users, excluding electricity but including the energy consumed at electric utilities to generate electricity. (In estimating energy expenditures, there are no fuel-associated expenditures for hydroelectric power, geothermal energy, solar energy, or wind energy, and the quantifiable expenditures for process fuel and intermediate products are excluded.)", and Primary Energy Consumption (PEC) is defined as "is the amount of site consumption, plus losses that occur in the generation, transmission, and distribution of energy." Primary energy reduction is important because it is related to the energy resources and environmental impact. In fact, Energy Star [19], a government-backed program uses primary or source energy as the basis for benchmarking buildings energy performance. In concordance with Energy Star, the standard site-to-primary energy or site-to-source energy conversion factors are applied as national averages and it is stated that the application of these national averages is consistent with the objective of comparing the total annual energy consumption among buildings with similar operations. In this study, the site-to-primary energy conversion factors, presented in Table 1.1, correspond to those obtained from Target Finder [20] for office type commercial buildings.

Table 1.1. Site-to-Primary Energy Conversion Factors

Fuel Type	Conversion Factor [a]
Electricity	3.343
Natural Gas	1.047
Propane	1.010
Fuel Oil (No. 2)	1.010
Diesel (No. 2)	1.010
Wood	1.000

[a] Values obtained in January 2008.

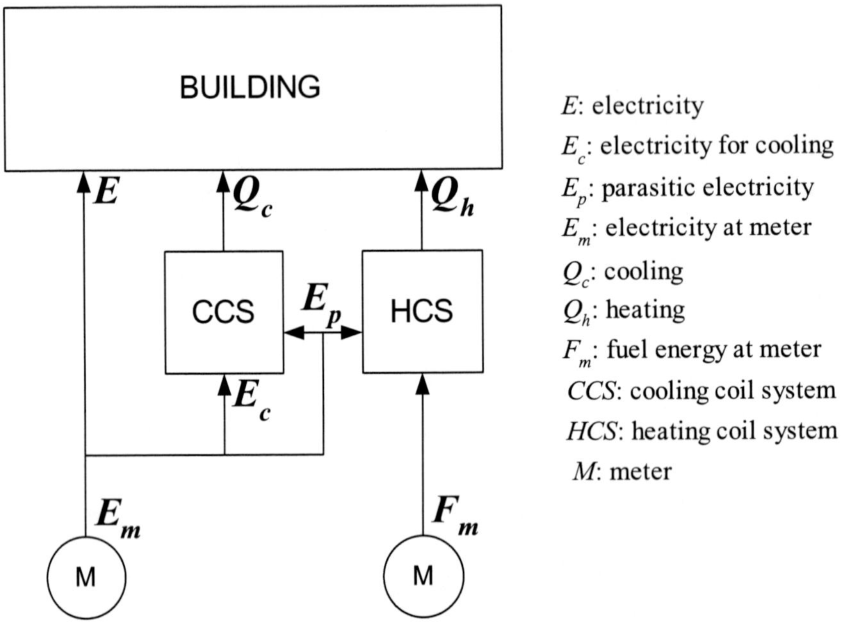

Figure 2.1. Building-HVAC System Arrangement.

In the evaluation of any energy system, primary energy has more significance than site energy because it is related to the energy resources and the environment. For example, electricity as site source does not show that more than three times of the energy is being used at the origin; and while electricity as site source has zero emissions at the origin, it has significant amount of pollutants released into the environment.

2. Building-CHP System Simulation

The building-CHP system site energy consumption is computed based on the energy consumption measured at the utility meters. The model uses the actual building energy consumption to estimate the energy consumption for the case when a CHP system is incorporated. The model is derived based on the building-HVAC system and the building-CHP system sketched in Figures 2.1 and 2.2, respectively.

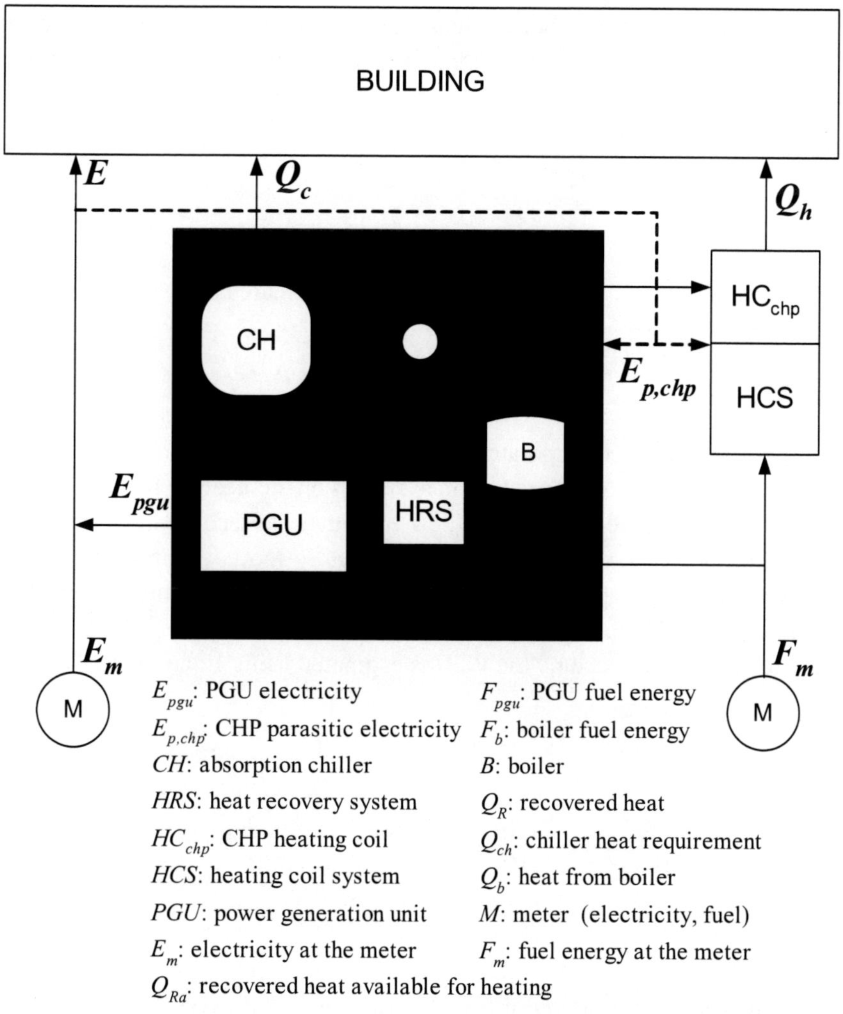

Figure 2.2. Building-CHP System Arrangement.

The efficiency of the power generation unit (PGU), η_{pgu}, is considered as the fuel to electricity conversion efficiency, and the efficiency of the boiler, η_b, is considered as the thermal efficiency. The efficiency of the heat recovery system, η_{rec}, and the HVAC heating system (furnace or boiler), η_h, are considered as the relation between the thermal energy gain by the working fluid and the available thermal energy for heat transfer from the source. All fuel use is considered as a thermal energy source at the fuel lower heating value.

The grid electric energy use at the meter, E_m, can be determined as

$$E_m = E + E_{p,chp} - E_{pgu} \tag{2.1}$$

where E is the building electric energy consumption (electric equipment, lights, etc), E_{pgu} is the electric energy generated by the PGU, and $E_{p,chp}$ is the CHP parasitic electricity. For the hour time step simulation, the electric energy demand from the PGU is assumed to be equal to the energy consumption for the specific hour. The actual HVAC system parasitic electric energy, E_p, is increased by a factor $F_{p,c}$ when cooling is required, and by a factor $F_{p,h}$ when heating is required. Then, for cooling demand the CHP system parasitic electricity is estimated as

$$E_{p,chp} = E_p \cdot F_{p,c} \tag{2.2}$$

and for heating demand the CHP parasitic electricity is estimated as

$$E_{p,chp} = E_p \cdot F_{p,h} \tag{2.3}$$

When a CHP system is incorporated, most of the original parasitic electricity demand remains as part of the HVAC air distribution system. For the heating mode of CHP systems, additional electric energy is required by the new equipment to recover the waste heat from the prime mover. For the cooling mode of CHP systems, more electric energy is required compared with the heating condition because of the additional equipment associated with the absorption chiller (CH). Therefore, in general, $F_{p,c}$ is greater than $F_{p,h}$.

The electric energy produced by the PGU is estimated using Equation (2.4):

$$E_{pgu} = 0 \text{ if } E + E_{p,chp} < cutoff \cdot E_{np,pgu} \tag{2.4a}$$

$$E_{pgu} = E + E_{p,chp} \text{ if } E + E_{p,chp} < E_{np,pgu} \tag{2.4b}$$

$$E_{pgu} = E_{np,pgu} \text{ if } E + E_{p,chp} > E_{np,pgu} \tag{2.4c}$$

where $cutoff$ refers to the fraction of the nominal power below which the PGU should not operate; and $E_{np,pgu}$ is the energy produced by the PGU during an hour at the nominal energy rate (nominal power, P_{pgu}). Numerically $E_{np,pgu}$ corresponds to $E_{np,pgu} = P_{pgu} \cdot 1hr$.

The PGU fuel energy consumption is estimated as

$$F_{pgu} = \frac{E_{pgu}}{\eta_{pgu}} \tag{2.5}$$

where η_{pgu} is the PGU thermal efficiency. The efficiency of the power generation unit is assumed to be constant independently of the electric demand. Then, the ratio between electricity and fuel remains constant for any demand higher than the $cutoff$ fraction of the nominal power of the PGU.

The heat required by the absorption chiller to handle the cooling load is estimated as

$$Q_{ch} = \frac{COP_{vc}}{COP_{ch}} E_c \tag{2.6}$$

where COP_{ch} and COP_{vc} represent the coefficient of performance of the absorption chiller and vapor compression system, respectively; and E_c is the electric energy consumption for cooling from the vapor compression system. Equation (2.6) defines the amount of heat required by the absorption chiller to provide the same cooling as the vapor compression system for any specific time of analysis.

The recovered waste heat from the prime mover is estimated as

$$Q_R = F_{pgu} \cdot \eta_{rec} \left(1 - \eta_{pgu}\right) \tag{2.7}$$

where Q_R is the recovered thermal energy and η_{rec} is the heat recovery system efficiency. The recovered thermal energy corresponds only to the useful energy, that is, the heat required by the absorption chiller, and heat required for space heating. For the cases when the recovered thermal energy is greater than the total heat required, Q_R is set equal to the total heat required.

The priority for the use of the recovered thermal energy is the heat required by the absorption chiller. Then, the thermal energy recovered and available for space heating, Q_{Ra}, will exist only when the recovered thermal energy is greater than the chiller heat consumption

$$Q_{Ra}^{+} = Q_R - Q_{ch} \tag{2.8}$$

The fuel energy saving from the waste thermal energy recovered is estimated using the efficiency of the heating system, η_h, and is determined as

$$F_{Ra} = \frac{Q_{Ra}}{\eta_h} \tag{2.9}$$

When the recovered thermal energy does not satisfy the requirement of the absorption chiller, additional heat is provided by the boiler of the CHP system. The boiler fuel energy consumption is computed as

$$F_b = \frac{Q_{ch} - Q_R}{\eta_b} \tag{2.10}$$

where η_b is the boiler thermal efficiency.

The fuel energy consumption required to provide the heat needed by the building is

$$F_h = \frac{Q_h}{\eta_h} \tag{2.11}$$

Then, the fuel energy consumption registered at the meter is estimated as

$$F_m = F_h + F_{pgu} + F_b - F_{Ra} \tag{2.12}$$

2.1. Primary Energy Operational Strategy

Thermal energy efficiency from the use of CHP systems has to be assessed through primary energy consumption. Therefore, the building primary energy consumption (*PEC*) is determined, for the actual building energy consumption (subscript 1) and building-CHP system energy consumption (subscript 2), as

$$PEC_1 = E_{m1} \cdot ECF + F_{m1} \cdot FCF \tag{2.13}$$

$$PEC_2 = E_{m2} \cdot ECF + F_{m2} \cdot FCF \tag{2.14}$$

where *ECF* and *FCF* are the site-to-primary energy conversion factors for electricity and fuel, respectively. In this chapter, the site-to-primary energy conversion factor correspond to those used by the Energy Star program (Table 1.1); however, more specific conversion factors for electricity could be used based on the fuel mix of the power plant feeding the grid.

In this study, the definition of Building Primary Energy Ratio (*BPER*) is introduced as a new parameter to evaluate CHP systems energy performance under a primary energy operational strategy. The *BPER* parameter is defined as

$$BPER = \frac{PEC_1}{PEC_2} \tag{2.15}$$

The primary energy strategy is based on the *BPER* values. For values of *BPER* higher than 1, the use of a CHP system reduces the primary energy consumption, and for *BPER* values lower than 1, the use of a CHP system causes an increase of the primary energy consumption.

2.2. Cost Oriented Operational Strategy

CHP systems are commonly designed based on a cost-oriented operational strategy. A cost-oriented operational strategy let the CHP system operates in order to obtain the maximum economic benefit from the energy consumption. However, designers must be aware that the energy price could lead to misleading results when the analysis considers economic feasibility without quantifying the energy consumption.

The variation in energy cost can be evaluated by comparing the energy cost (*EC*) for the actual building (subscript 1) and building-CHP system (subscript 2). The energy cost can be determined as

$$EC_1 = E_{m1} \cdot EP + F_{m1} \cdot FP \tag{2.16}$$

$$EC_2 = E_{m2} \cdot EP + F_{m2} \cdot FP \tag{2.17}$$

where *EP* and *FP* are the energy price for electricity and fuel, respectively.

The results for the energy cost operational strategy were obtained by implementing the Energy Cost Ratio (ECR) parameter. The ECR parameter is defined as

$$ECR = \frac{EC_1}{EC_2} \tag{2.18}$$

Then, a cost-oriented operational strategy can be implemented based on *ECR* values. For values of *ECR* higher than 1, the use of a CHP system reduces the cost of the energy consumption, and for *ECR* values lower than 1, the use of a CHP system causes an increase of the cost of the energy consumption. The *ECR* operational strategy is equivalent to the *BPER* operational strategy, but accounting for energy cost in lieu of energy consumption.

2.3. Simulation Program

Based on the developed model for an hour time step energy use analysis, a simulation program was developed according with the logic of the flowchart presented in Appendix A. The sub-index “1” represents the actual building energy consumptions, that is, without CHP system. The sub-index “2” represents the case for the building-CHP system operating without the BPER operational strategy. The sub-index “3” represents the case for the building-CHP system operating with the BPER operational strategy.

The inputs for the simulation program are:

- Heat recovery system efficiency, η_{rec}
- Vapor compression coefficient of performance, COP_{vc}
- Absorption chiller coefficient of performance, CPO_{ch}
- CHP boiler efficiency, η_b
- Heating system efficiency, η_h
- Increasing factors for parasitic electricity, $F_{p.c}$ and $F_{p.h}$
- PGU cutoff fraction, $cutoff$
- PGU efficiency, η_{pgu}
- Site-to-primary energy conversion factors, ECF and FCF

- Excel file with the following energy consumption information:
  - Building electric energy consumption, E
  - HVAC parasitic electricity, E_p
  - Vapor compression electricity for cooling, E_c
  - Fuel energy consumption for heating, F_h
  - Building fuel energy consumption for not heating use, F

With the exception of the building energy consumption, the input values used in this study are presented in Table 2.1.

The simulation program allows the analysis for different PGU and CH sizes in order to define the condition for best energy performance. Therefore, some equations of the model must be adjusted to account for this condition. The PGU and CH sizes are varied from zero to the maximum required capacity. PGU and CH sizes of zero represent the case when the CHP system does not exist. The maximum PGU size is calculated based on the maximum building electric energy consumption because no electricity is sold to the grid. While the maximum CH size is calculated based on the maximum vapor compression electric energy consumption for cooling. The size increments for the PGU (I_{pgu}) and CH (I_{ch}) were defined as 5 and 2, respectively. For each CH size the proportion of the cooling load to be handled by the absorption chiller and vapor compression system are defined. The model gives priority to the absorption chiller to handle the cooling load. If the chiller is providing the maximum cooling, but the cooling load is not satisfied, the difference is handled by the vapor compression system.

Table 2.1. Input Values for CHP System Simulation Program

Variable	Symbol	Value
Heat recovery system efficiency	η_{rec}	0.8
Vapor compression coefficient of performance	COP_{vc}	3
Absorption chiller coefficient of performance	CPO_{ch}	0.7
CHP boiler efficiency	η_b	0.8
Heating coil system efficiency	η_h	0.8
Cooling factor for parasitic electricity	$F_{p.c}$	1.4
Heating factor for parasitic electricity	$F_{p.h}$	1.2
PGU cutoff fraction	$cutoff$	0.25
PGU efficiency	η_{pgu}	0.25, 0.30, 0.35
Electricity energy conversion factor	ECF	3.343
Fuel energy conversion factor	FCF	1.047

To incorporate the building primary energy operational strategy two steps must be followed. The first step is to compute the building primary energy ratio parameter (BPER, see

Section 2.1) using the primary energy consumption for the building without CHP system (PEC_1), and the primary energy consumption for the building-CHP system (PEC_2). Based on the sub-indices assigned to identify the energy consumption condition, the computed BPER is identified as $BPER_{12}$. If the $BPER_{12}$ is higher or equal to 1, the grid electric energy use at the meter, and the fuel energy consumption registered at the meter, remain as calculated for the building-CHP system assuming that the CHP system was operating as prescribed. However, if the $BPER_{12}$ is lower than 1, the actual electric energy consumption (E_{m1}) is set as the building-CHP system energy consumption assuming that the PGU was not operating. When the PGU does not operate, no heat is recovered from the prime mover. Therefore, when cooling demand exists, the boiler of the CHP system must supply the heat required by the absorption chiller. For this condition, the fuel energy consumption is the actual fuel energy consumption plus the boiler fuel energy consumption ($F_{m1}+F_b$). However, because of the low energy conversion efficiency from fuel energy to cooling through the absorption chiller, more primary energy could be consumed if the PGU is not operating. Therefore, the next step is to define if the PGU must operate even though the $BPER_{12}$ is lower than 1. To accomplish this, the BPER is now calculated using the PEC_2 and the primary energy consumption for the case when the PGU of the CHP system is not operating (PEC_3). Then, based on the sub-indices assigned to identify the new energy consumption condition, the new BPER is identified as $BPER_{23}$. If the $BPER_{23}$ is higher or equal to 1, the energy consumption is set as those values calculated if the PGU is not operating (E_{m3} and F_{m3}). If the $BPER_{23}$ is lower than 1, the energy consumption is set as those values calculated for the CHP system (E_{m2} and F_{m2}).

A block diagram for general description of the structure for the simulation of the building-CHP system is presented in Figure 2.3.

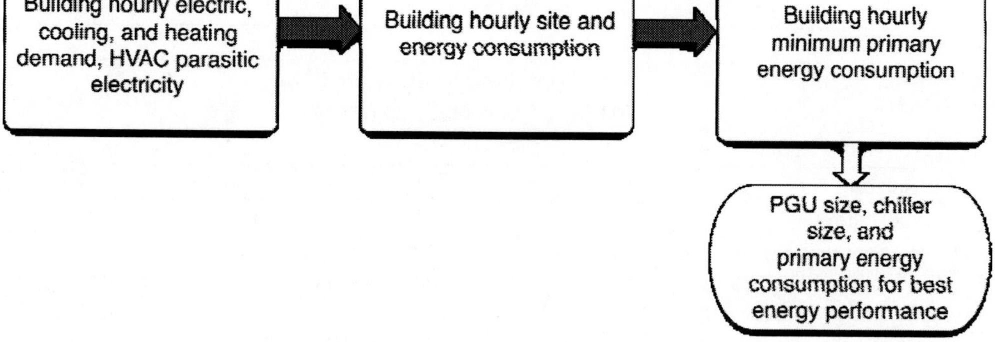

Figure 2.3. Block Diagram for the Building-CHP System Simulation.

3. Results and Analysis

For CHP system analysis, a reference building was defined in order to compare the energy consumption for the cases without and with the implementation of a CHP system. To obtain hourly site energy consumption data a hypothetical building was simulated using the software EnergyPlus™ [18]. General description of the building is presented in Table 3.1. The hourly energy consumption from the EnergyPlus™ simulations were used in the CHP system simulation model presented in Section 2.

The energy consumption profile of a building is highly dependent on the climate conditions. Climate is one of the variables that define the energy consumption profiles (electric and thermal energy demand) of a building. To analyze the effect of the energy consumption profiles on CHP systems energy performance, the same building was simulated using weather data for the cities presented in the map of climate zones of the U.S.A. (based on the 2003 commercial building energy consumption survey) shown in Figure 3.1.

In this study, CHP systems are considered as distributed generation systems with the advantage that waste thermal energy from the prime mover is recovered for space cooling and heating. Therefore, to account for the effect of the power generation unit (PGU) on the CHP system energy performance, three efficiency values were considered: 0.25, 0.30, and 0.35. These values were chosen as representative of general efficiencies for common commercially available PGU.

Table 3.1. General Description of the Simulated Building Using EnergyPlus

Orientation	Aligned with North
Building type	General Offices
Area	1156 m^2 (34 m x 34 m)
Glass area	30% in each wall (windows and door)
People	115 for weekdays, 0 for weekend
Occupancy schedule	Until (fraction): 6 (0), 7 (0.1), 8 (0.5), 12 (1), 13(0.5), 16(1), 17 (0.5), 18 (0.1), 24 (0)
Electric equipment	15000 W
Equipment schedule	Same as for occupancy
Lights	45,000 W
Lights schedule	Until[a] (fraction)[b]: 6 (0.05), 7 (0.2), 17 (1), 18 (0.5), 24 (0.05); for weekends 24 (0.05)
Thermostat schedule:	
For heating	Until[a] (set point, °C)[c]: 6 (18), 22 (22), 24 (18)
For cooling	Until[a] (set point, °C)[c]: 6 (28), 22 (24), 24 (28)

a. Until: indicates the hour of the day until the specified fraction is considered.

b. Fraction: indicates the fraction of the total value of the variable that is considered in the calculation for that specific period of time.

c. Set point: indicates the temperature to be considered as the thermostat set point for that specific period of time.

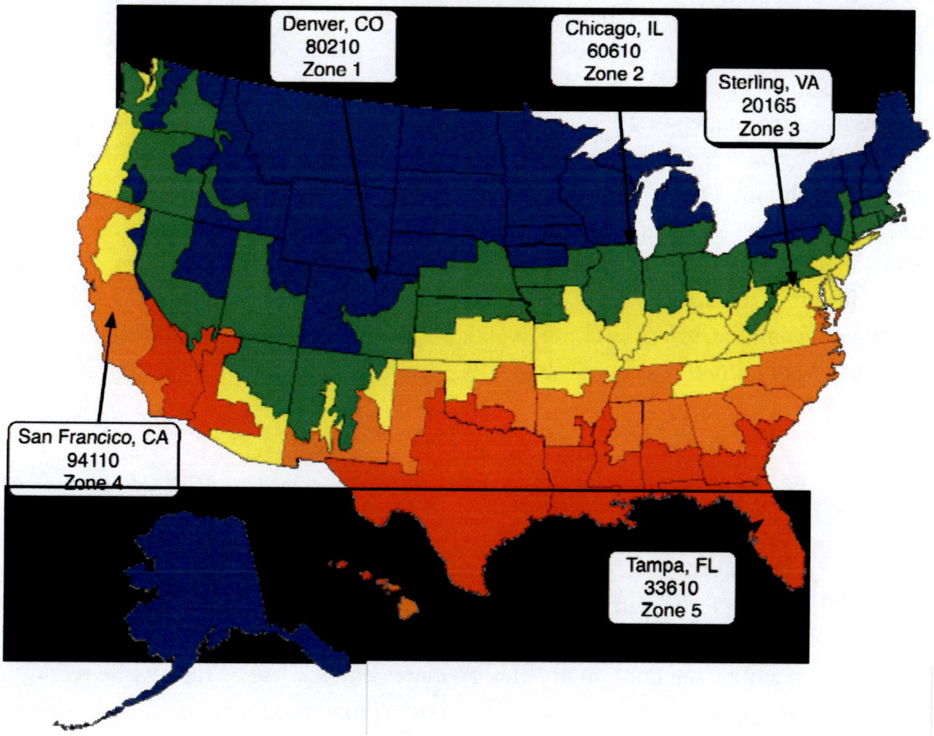

Figure 3.1. Cities Representing the Climate Zones of the U.S.A.

3.1. CHP System Energy Performance

CHP system analysis involves variables related to the components of the system, operation of the system, and building characteristics. The interrelation among all variables will define the system performance. Adequate designs must yield economical savings but, more importantly, they must yield real energy savings based on the best energy performance. In this chapter site energy consumption, primary energy consumption, and system efficiency are the variables considered to evaluate the CHP system energy performance. Since, generally, the economic analysis prevails in the feasibility of CHP systems, the energy cost is also considered to show that economic decisions could yield misleading results.

The results are presented and compared for the cases when the CHP runs without and with the building primary energy ratio (BPER) operational strategy. These two cases are identified as CHP and CHP-BPER, respectively. Based on the nomenclature of the simulation software, sub-index 2 and 3 corresponds to CHP and CHP-BPER, respectively. As mentioned previously, when the PGU and CH sizes are zero, the results correspond to the case when the CHP system does not exist.

The simulation software varies the power generation unit (PGU) and absorption chiller (CH) sizes to find the sizes that yield the best energy performance for each particular case of inputs. The additional electric energy and cooling energy required by the building for any particular hour of analysis are provided by the electric grid (EG) and vapor compression system (VC), respectively. Table 3.1 summarizes the PGU, CH, and VC sizes, and the maximum electric power demanded from the grid for the best energy performance. The optimized parameters in Table 3.1 are used to obtain the results presented in this chapter.

Table 3.1 shows that for PGU and CH size increments of 5 and 2 respectively, the design for best energy performance is the same for the cases when the system run without (CHP) and with the BPER operational strategy (CHP-BPER). However, since different primary energy consumption is obtained for CHP and CHP-BPER, the design will not necessarily be always the same for both cases. In order to understand Table 3.1, the case for Tampa with PGU efficiency of 0.25 is explained. Based on the inputs required by the developed simulation model, the design is specified with a CHP size of 15 kW and a CH size of 8 kW. The required electric power from the grid (EG) to match the maximum electric demand is 100 kW. Similarly, the vapor compression system (VC) capacity to match the maximum cooling demand is 42 kW.

3.2. Site Energy Consumption

The most common criterion to implement any energy conservation system is economic. However, a CHP system economic evaluation could yield misleading results when the main goal is energy resources conservation and environmental protection. CHP systems changes the building site energy consumption profiles by increasing the use of fuel, while reducing the use of electricity from the grid. However, CHP systems increase the total building site energy consumption.

Table 3.1. PGU, CH, and VC Sizes, and EG Demand for Best Energy Performance

	Size (kW)								
	PGU	CHP				CHP-BPER			
City/Zone	Efficiency	PGU	EG	CH	VC	PGU	EG	CH	VC
Denver Zone 1	0.25	15	85	8	28	15	85	8	28
	0.30	50	50	22	14	50	50	22	14
	0.35	75	25	22	14	75	25	22	14
Chicago Zone 2	0.25	15	90	8	34	15	90	8	34
	0.30	50	55	22	20	50	55	22	20
	0.35	80	25	22	20	80	25	22	20
Sterling Zone 3	0.25	15	100	8	40	15	100	8	40
	0.30	50	65	22	26	50	65	22	26
	0.35	80	35	22	26	80	35	22	26
San Francisco Zone 4	0.25	10	75	6	20	10	75	6	20
	0.30	35	50	16	10	35	50	16	10
	0.35	65	20	20	6	65	20	20	6
Tampa Zone 5	0.25	15	100	8	42	15	100	8	42
	0.30	25	90	10	40	25	90	10	40
	0.35	75	40	22	28	75	40	22	28

3.2.1. CHP System Simulation: Site Energy Consumption

Table 3.2 and 3.3 present the SEC for the cases of CHP and CHP-BPER, respectively. Figure 3.1 and 3.2 illustrate the variation of the SEC for the cases of CHP and CHP-BPER, respectively. When the results are presented as percentage of variation, positive and negative values means more or less energy consumption with respect to the reference value (actual building SEC).

Table 3.2. Site Energy Consumption for CHP System without BPER Strategy

	PGU	Site Energy Consumption (kWh)		
City/Zone	**Efficiency**	**Building**	**CHP**	**Variation %**
Denver Zone 1	0.25	511766	620129	21.2
	0.30		791451	54.7
	0.35		763929	49.3
Chicago Zone 2	0.25	587123	685414	16.7
	0.30		844171	43.8
	0.35		821177	39.9
Sterling Zone 3	0.25	494217	609140	23.3
	0.30		788825	59.6
	0.35		767962	55.4
San Francisco Zone 4	0.25	290703	370339	27.4
	0.30		499600	71.9
	0.35		548166	88.6
Tampa Zone 5	0.25	303476	497609	64.0
	0.30		544835	79.5
	0.35		738664	143.4

Table 3.3. Site Energy Consumption for CHP System with BPER Strategy

	PGU	Site Energy Consumption (kWh)		
City/Zone	**Efficiency**	**Building**	**CHP-BPER**	**Variation %**
Denver Zone 1	0.25	511766	594782	16.2
	0.30		738815	44.4
	0.35		763929	49.3
Chicago Zone 2	0.25	587123	666525	13.5
	0.30		804218	37.0
	0.35		821177	39.9
Sterling Zone 3	0.25	494217	586987	18.8
	0.30		742273	50.2
	0.35		767962	55.4
San Francisco Zone 4	0.25	290703	336252	15.7
	0.30		420636	44.7
	0.35		548166	88.6
Tampa Zone 5	0.25	303476	481871	58.8
	0.30		528591	74.2
	0.35		738664	143.4

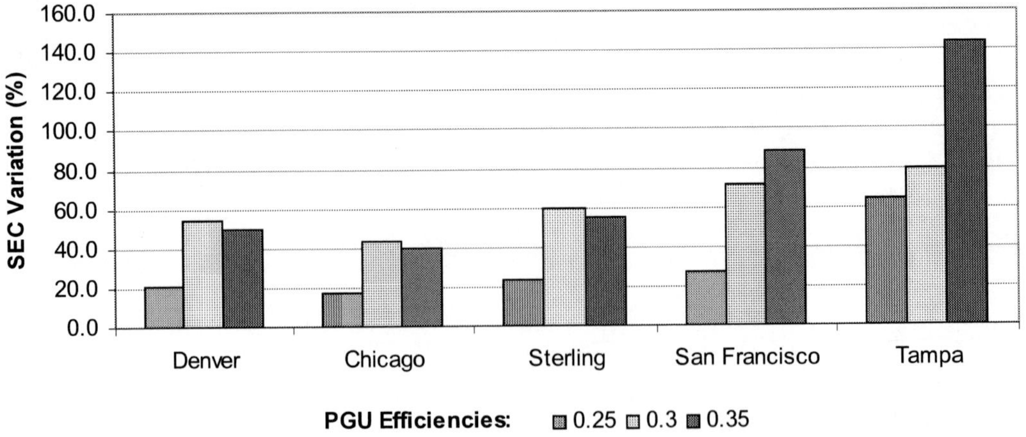

Figure 3.1. SEC Variation for CHP System without BPER Strategy.

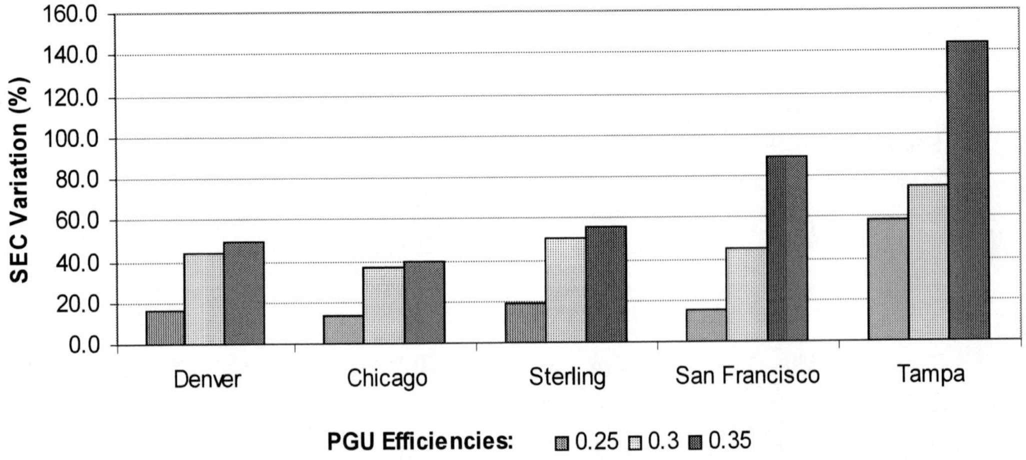

Figure 3.2. SEC Variation for CHP System with BPER Strategy.

The results show the increment on SEC from the use of CHP systems. The increase occurs for all the cities and PGU efficiencies. For the case when the CHP system runs under the BPER operational strategy, the SEC can be reduced compared with the case without the BPER operational strategy. Based on the conditions of this study, the results presented in Figure 3.1 and 3.2 illustrate that if the CHP system operates under the BPER strategy the increment on the SEC can be reduced as much as 19.1%, 6.4%, 9.4%, 27.2%, and 5.4%, for Denver, Chicago, Sterling, San Francisco, and Tampa, respectively. Since economic evaluation is computed based on site energy prices, these results suggest that less energy cost should be achieved when BPER operational strategy is implemented.

3.3. CHP System Efficiency

For any energy system, efficiency is a way to determine how much energy in the required form is generated, added, or removed from the system for a given input. Energy conservation efficiency for the CHP system sketched in Figure 2.2 can be written as

$$\eta_{chp} = \frac{\left(E_{pgu} - E_{p,chp}\right) + \left(Q_{ch} \cdot COP_{ch}\right) + \frac{Q_{Ra}}{\eta_h}}{F_{pgu} + F_b} \tag{3.1}$$

where the term ($Q_{ch} \cdot COP_{ch}$) corresponds to the cooling load handled by the absorption chiller.

Since CHP system increases the site energy consumption, it seems that the use of the first law efficiency alone is not appropriate to evaluate the energy performance of CHP systems. A similar statement was proposed by Zogg [4] "...some CHP promoters report "total efficiency" of CHP systems based on a first-law definition that simply sums electric and thermal outputs. Meaningful efficiency definitions, however, account for the relative values of the electric and thermal outputs." Therefore, an evaluation of CHP systems based on primary energy consumption, such as BPER, would be more adequate.

Examples of the false impression that could be derived from the use of the first law efficiency are presented in Figures 3.3 to 3.5. The figures show the comparison of the CHP system efficiency and the BPER. Figures 3.3, 3.4, and 3.5 show the results for the city of Chicago on October 21st, April 21st, and December 27th, respectively. In these figures values of zero efficiency implies that the CHP system is not operating, and consequently the BPER is 1.

Figure 3.3 illustrates that the CHP system performance based on BPER can follow the CHP system efficiency. Logically, better the efficiency better is the energy performance. However, Figure 3.4 and 3.5 illustrates that BPER as a measure of the CHP system energy performance does not necessarily follows the efficiency.

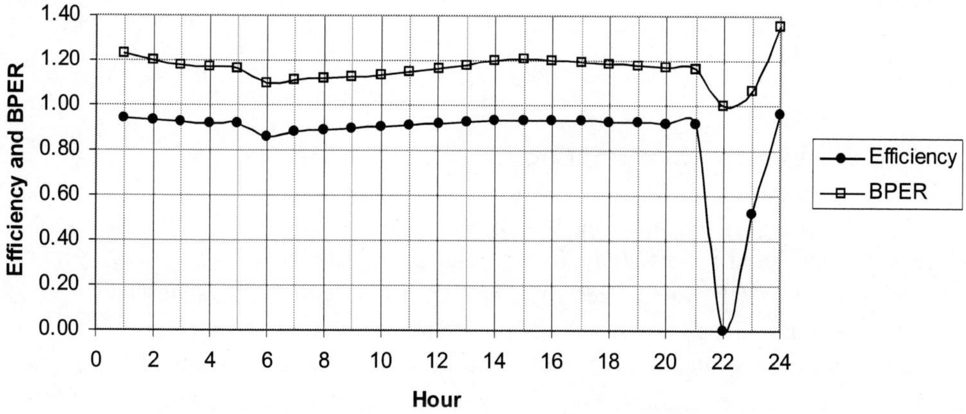

Figure 3.3. CHP System Efficiency and BPER for Chicago (IL) on October 21st.

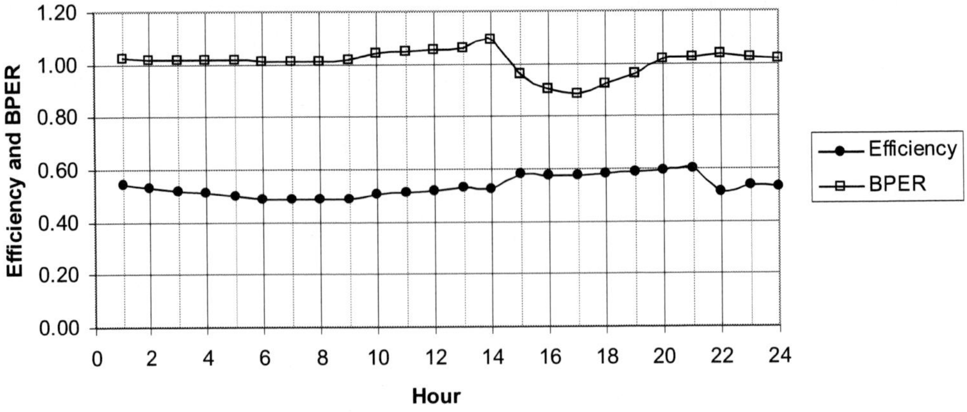

Figure 3.4. CHP System Efficiency and BPER for Chicago (IL) on April 21st.

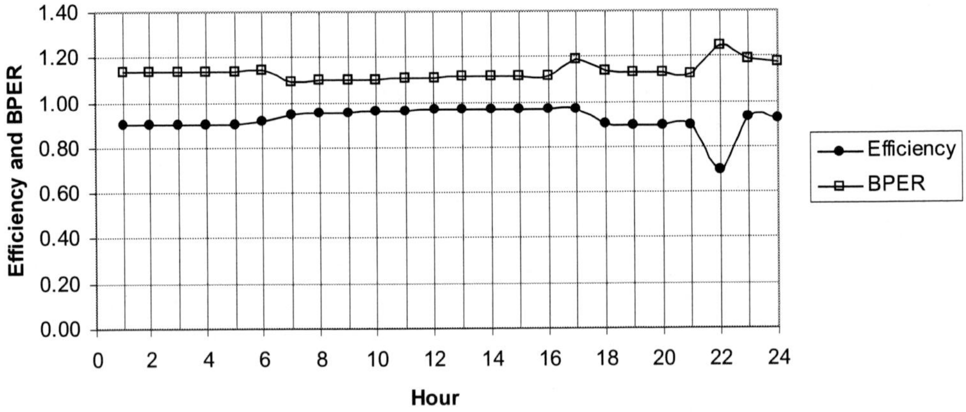

Figure 3.5. CHP System Efficiency and BPER for Chicago (IL) on December 27th.

As an example consider the two particular points at 3:00 p.m. (hour 15) in Figure 3.4, and at 10:00 p.m. (hour 22) in Figure 3.5. In the first point the BPER decreases while the efficiency increases, but in the second point the BPER increases while the efficiency decreases. Therefore, CHP system efficiency is not used in this study to evaluate the system energy performance.

3.4. Primary Energy Consumption

Table 3.4 presents the building primary energy consumption for the cities and PGU efficiencies considered in this study. The results show that for the best energy performance design (Table 3.1), CHP systems decrease the PEC for all cases. The lowest reduction, 2.3%, occurs for Tampa; and the highest reduction, 16.5%, occurs for Chicago.

Table 3.4 demonstrates that for the same PGU efficiency different PEC variations are obtained, which verifies the influence of the building energy consumption profiles on the CHP energy performance. This table also shows that the incremental decrease of the PEC variation is higher for the efficiencies between 0.30 and 0.35 than for the efficiencies of 0.25 and 0.30. This can be explained because a PGU efficiency of 0.35 is higher than the efficiency (generation – transmission – distribution) for the utility power plant which for this study is 0.30 obtained as the inverse of the ECF (1/3.343). Logically, when the PGU efficiency is higher than the power plant efficiency, even if the waste thermal energy is not recovered, better energy performance is achieved.

3.5. Primary Energy Consumption for BPER Strategy

For the cities and PGU efficiencies considered in this study, Table 3.5 shows the building primary energy consumption for BPER operational strategy, while Figures 3.6 to 3.8 illustrate the variation in the PEC.

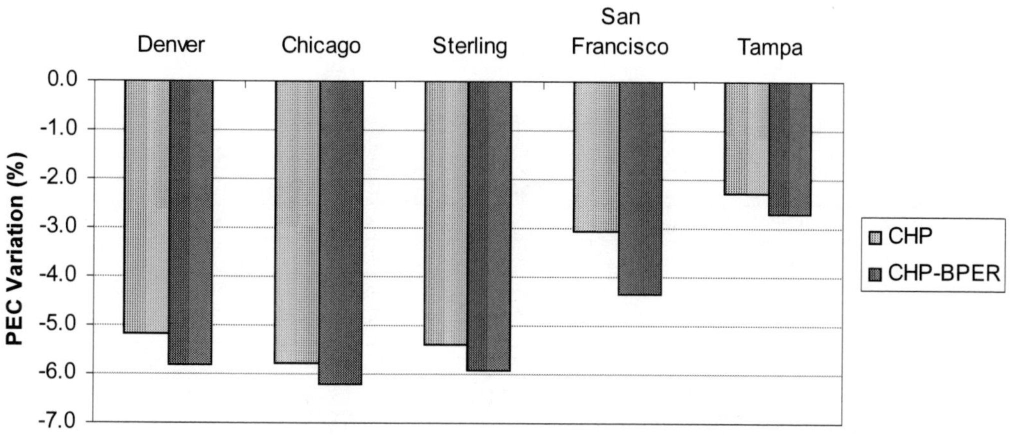

Figure 3.6. PEC Variation for CHP and CHP-BPER Cases, for $\eta_{pgu} = 0.25$.

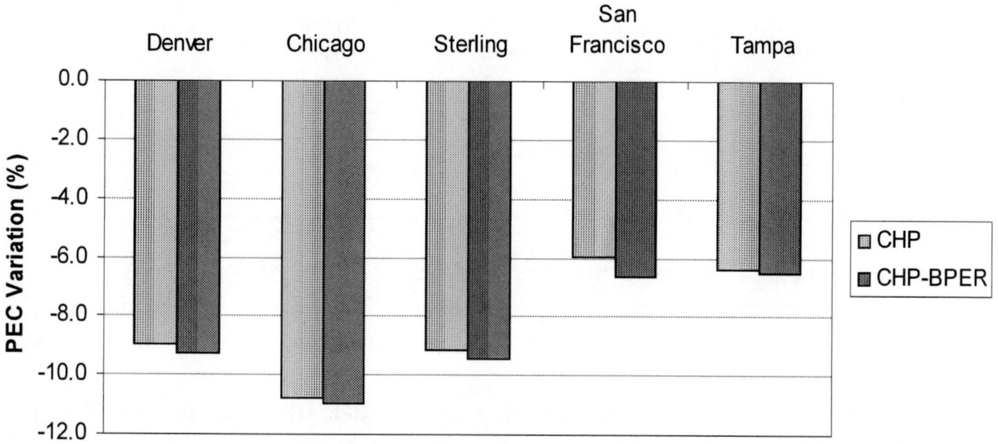

Figure 3.7. PEC Variation for CHP and CHP-BPER Cases, for $\eta_{pgu} = 0.30$.

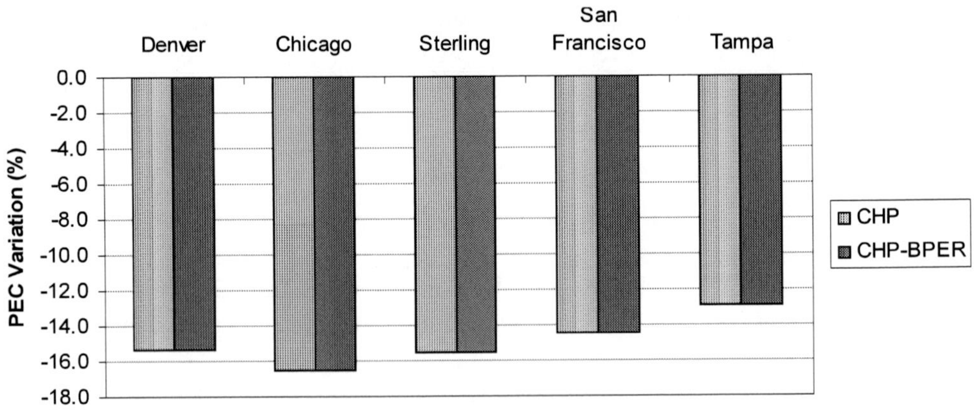

Figure 3.8. PEC Variation for CHP and CHP-BPER Cases, for $\eta_{pgu} = 0.35$

Table 3.4. CHP System Primary Energy Consumption

	PGU	Primary Energy Consumption (kWh)		
City/Zone	**Efficiency**	**Building**	**CHP**	**Variation %**
Denver Zone 1	0.25	1006639	954584	-5.2
	0.30		916336	-9.0
	0.35		851704	-15.4
Chicago Zone 2	0.25	1093141	1030020	-5.8
	0.30		975374	-10.8
	0.35		912650	-16.5
Sterling Zone 3	0.25	1016587	961855	-5.4
	0.30		923577	-9.1
	0.35		858174	-15.6
San Francisco Zone 4	0.25	718772	696790	-3.1
	0.30		675979	-6.0
	0.35		614689	-14.5
Tampa Zone 5	0.25	960771	938926	-2.3
	0.30		899242	-6.4
	0.35		836461	-12.9

Results in Table 3.5 are similar to those obtained for CHP systems running without the BPER operational strategy. However, Figures 3.6 and 3.7 illustrate that with the BPER operational strategy more PEC reduction can be achieved. As discussed in the previous section, the closeness of the PGU efficiency to the utility power plant efficiency has implications on the CHP system energy performance. When BPER operational strategy is applied, the benefits are more significant for lower PGU efficiencies. For higher PGU efficiencies, the results are the same when compared to the case without the BPER strategy as shown in Figure 3.8. When the PGU efficiency is higher than the utility power plant efficiency, better performance is obtained when the PGU operates. Therefore, the BPER operational strategy will not require that the PGU stop, and consequently the results are the same.

Table 3.5. CHP System Primary Energy Consumption for BPER Strategy

	PGU	Primary Energy Consumption (kWh)		
City/Zone	Efficiency	Building	CHP-BPER	Variation %
Denver Zone 1	0.25	1006639	948096	-5.8
	0.30		913236	-9.3
	0.35		851704	-15.4
Chicago Zone 2	0.25	1093141	1025099	-6.2
	0.30		973033	-11.0
	0.35		912650	-16.5
Sterling Zone 3	0.25	1016587	956181	-5.9
	0.30		920910	-9.4
	0.35		858174	-15.6
San Francisco Zone 4	0.25	718772	687412	-4.4
	0.30		671139	-6.6
	0.35		614689	-14.5
Tampa Zone 5	0.25	960771	934843	-2.7
	0.30		898208	-6.5
	0.35		836461	-12.9

Table 3.6. Electricity and Natural Gas Price

	Price of Energy ($/kWh)[a]	
City	Electricity	Natural Gas
Denver	0.0670	0.0236
Chicago	0.0744	0.0293
Sterling	0.0588	0.0327
San Francisco	0.1188	0.0276
Tampa	0.0755	0.0375

a. Values obtained in January 2008.

3.6. Economic Considerations

The results presented in this section correspond to those given by Target Finder [20] of the Energy Star program [19]. The site energy consumption for electricity and natural gas used in Target Finder are summarized in Appendix B. To better understand the results presented in this section, Table 3.6 show the estimated energy price used by Target Finder to compute the energy cost. For the PGU efficiencies considered in this study, Tables 3.7 and 3.8 present the energy cost for the cases when the CHP systems operates without and with the BPER operational strategy, respectively, while Figures 3.9 to 3.11 illustrate the variation in energy cost. The results confirm that CHP systems economic feasibility relies on energy price. This can be explained because as previously demonstrated, while increasing SEC, CHP system changes the building energy consumption profiles. Tables 3.7 and 3.8 show that for the cities of Sterling and Tampa, contrary to common belief, CHP systems increased energy cost. In general, with the exception of the city of San Francisco, Figures 3.9 and 3.10 show that with the BPER operational strategy better economic results can be obtained. However, even with the BPER operational strategy, as consequence of the variation in energy consumption profiles, for the cities of Chicago and Sterling, a peculiar case arises for the PGU efficiency of 0.30. For this PGU efficiency, the energy cost is higher than the cost for a PGU efficiency of 0.25. Figure 3.11 shows that for high PGU efficiency the energy cost is the same for the CHP and CHP-BPER. This is explained based on the same energy consumption for both cases at this PGU efficiency.

Table 3.7. CHP System Energy Cost

	PGU	Cost of Energy ($)		
City/Zone	**Efficiency**	**Building**	**CHP**	**Variation %**
Denver Zone 1	0.25	20969	20395	-2.7
	0.30		20318	-3.1
	0.35		18989	-9.4
Chicago Zone 2	0.25	26617	26241	-1.4
	0.30		26565	-0.2
	0.35		25132	-5.6
Sterling Zone 3	0.25	21843	23613	8.1
	0.30		26917	23.2
	0.35		25744	17.9
San Francisco Zone 4	0.25	24489	22504	-8.1
	0.30		19875	-18.8
	0.35		16770	-31.5
Tampa Zone 5	0.25	22026	25570	16.1
	0.30		25856	17.4
	0.35		28723	30.4

Table 3.8. CHP System Energy Cost for BPER Strategy

	PGU	Cost of Energy ($)		
City/Zone	**Efficiency**	**Building**	**CHP-BPER**	**Variation %**
Denver Zone 1	0.25	20969	20395	-2.7
	0.30		20318	-3.1
	0.35		18989	-9.4
Chicago Zone 2	0.25	26617	25983	-2.4
	0.30		26168	-1.7
	0.35		25132	-5.6
Sterling Zone 3	0.25	21843	23084	5.7
	0.30		25920	18.7
	0.35		25744	17.9
San Francisco Zone 4	0.25	24489	22610	-7.7
	0.30		20784	-15.1
	0.35		16770	-31.5
Tampa Zone 5	0.25	22026	25187	14.4
	0.30		25514	15.8
	0.35		28723	30.4

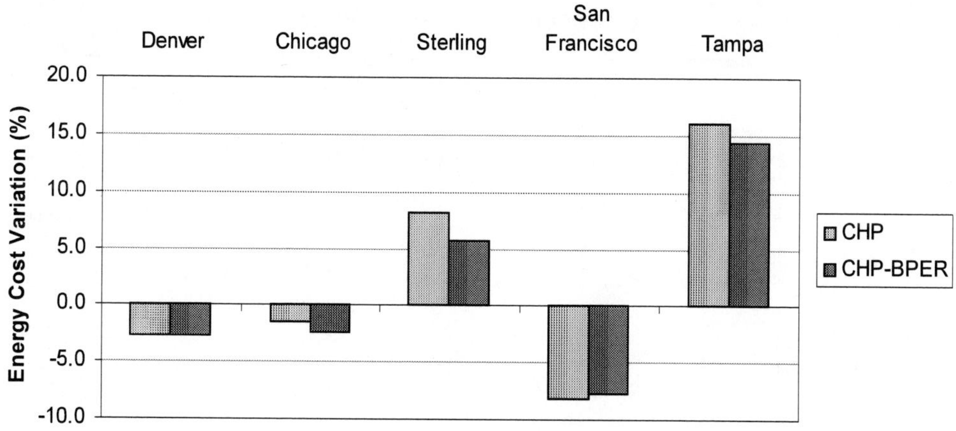

Figure 3.9. Energy Cost Variation for CHP and CHP-BPER Cases, for $\eta_{pgu} = 0.25$.

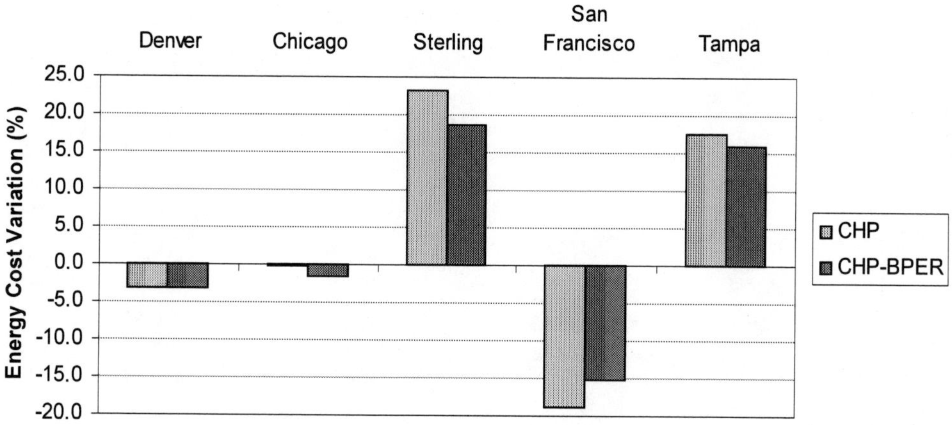

Figure 3.10. Energy Cost Variation for CHP and CHP-BPER Cases, for $\eta_{pgu} = 0.30$.

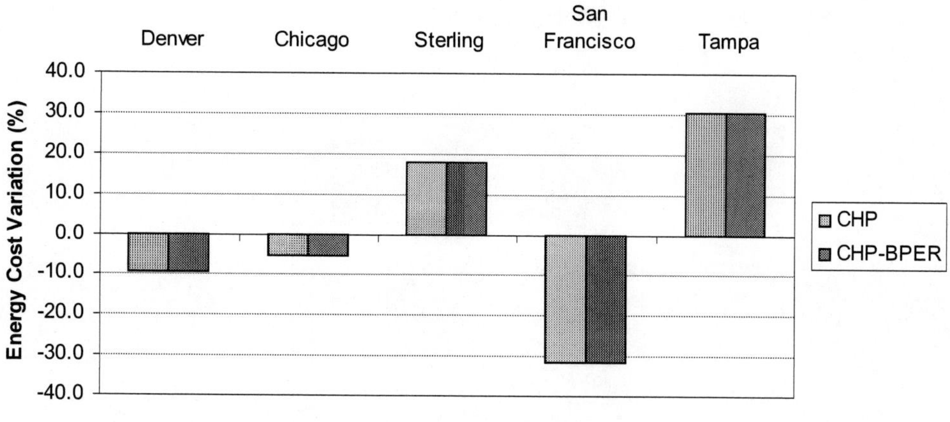

Figure 3.11. Energy Cost Variation for CHP and CHP-BPER Cases, for $\eta_{pgu} = 0.35$.

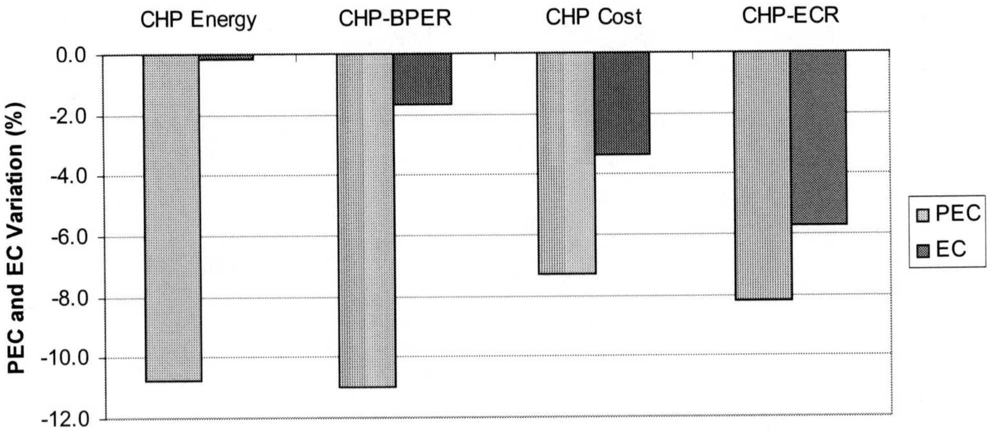

Figure 3.12. Variation of PEC and EC for the City of Chicago, $\eta_{pgu} = 0.30$.

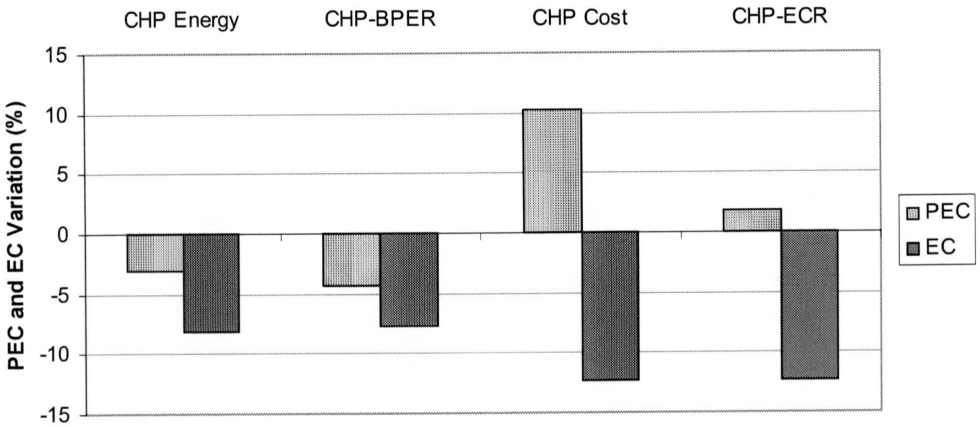

Figure 3.13. Variation of PEC and EC for the City of San Francisco, $\eta_{pgu} = 0.25$.

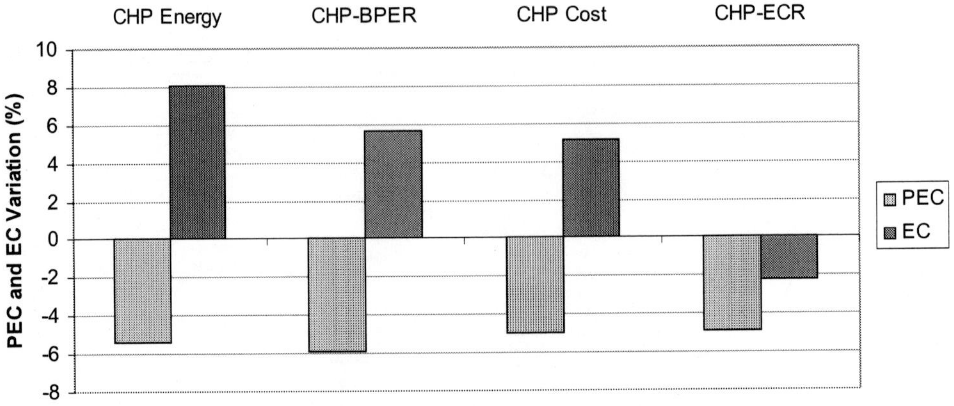

Figure 3.14. Variation of PEC and EC for the City of Sterling, $\eta_{pgu} = 0.25$.

3.6.1. Energy Versus Economics

The main goal of the use of CHP systems is to reduce energy consumption, and to reduce emission of pollutants. One of the main topics that this study wants to show is the misconception that could be derived from the cost-oriented design of CHP system. Figures 3.12 and 3.13 are two examples comparing CHP system design based on primary energy consumption (PEC) and energy cost (EC). In these figures CHP Energy and CHP-BPER are the results discussed in Sections 3.4 and 3.5, respectively. CHP Cost and CHP-ECR refer to the energy cost obtained from the lowest energy cost for the cases without and with the energy cost ratio (ECR) operational strategy (Section 2.2), respectively.

Figure 3.12 illustrates the variation of PEC and EC for the city of Chicago for a PGU efficiency of 0.30. This figure shows that the maximum PEC reduction of 11% is obtained with an energy cost reduction of 1.7% when the BPER operational strategy is used. However, the maximum energy cost reduction of 5.7% is obtained with a lower PEC reduction (8.2%) when the ECR operational strategy is used.

Similarly, Figure 3.13 illustrates the variation of PEC and EC for the city of San Francisco for a PGU efficiency of 0.25. For San Francisco, the maximum PEC reduction of 4.4% is obtained with an energy cost reduction of 7.7% when the BPER operational strategy is used. However, the maximum energy cost reduction of 12.4% is obtained with an increment of PEC of 1.8% when the ECR operational strategy is used. This analysis is more noticeable for the case when the CHP system operates without any operational strategy, shown as CHP Cost Case in Figure 3.13. An energy cost reduction of 12.4% is obtained which could justify the implementation of the CHP system, but with the contradictory result that the PEC would increase 10.2%.

Figure 3.14 illustrates the variation of PEC and EC for the city of Sterling for a PGU efficiency of 0.25. For Sterling, only when the CHP system operates with the ECR operational strategy the EC is reduced, Case shown in Figure 3.14 as CHP-ECR. The BPER operational strategy (Case CHP-BPER) gives the greatest PEC reduction of 5.9%, but with an increase of 5.7% in the EC. Comparing both cases, CHP-BPER and CHP-ECR, the difference in PEC reduction is only 1%, while the difference in EC is 7.9% (CHP-BPER increases the EC in 5.7% and CHP-ECR decreases the EC in 2.2%). This particular result suggests that in a tradeoff between EC and PEC, the ECR operational strategy could be a better option than the BPER operational strategy.

3.7. Non Conventional Evaluation of CHP System

Several researchers have investigated and reported the economic benefits of using CHP systems such as Newborough [21], Keppo and Savola [22], Jablko et al. [23], Tucker [24], De Paepe et al. [25], Zoog [26], and Mago et al. [27]. Although most of the time CHP technology seems to be economically feasible, results from Section 3.6 show that CHP system can not always guarantee economic savings. However, a well designed CHP system can guarantee energy reduction, which makes necessary the quantification of other benefits from this technology in order to offset any economic weakness that can arise as consequence of energy prices.

A non-conventional evaluation of CHP systems, based on non-economical aspects, will show the additional benefits that can be obtained from this technology. As customers, investors, and government continue to be more involved and to develop more understanding about energy choices, a non-conventional evaluation seems to be the solution to offset economic weakness of CHP systems. Besides, as conservation of energy resources and reduction of emissions are guaranteed by CHP technology, economic barriers can be offset through legislation or economic incentives as those given to promote renewable energy technologies.

Some aspects that could be included in a non-conventional evaluation are: building energy rating, emission of pollutants, power reliability, power quality, fuel source flexibility, brand and marketing benefits, protection from electric rate hikes, and benefits from promoting energy management practices. Some of these suggested benefits from a non-conventional evaluation can be factored into an economic evaluation but others would give intangible potential to the technology. This study, for a non-conventional evaluation of CHP systems, focuses on building energy rating and reduction of emissions because both of them are directly related to the CHP energy performance.

3.7.1. Building Energy Ratings

Two building energy ratings are recognized for benchmarking buildings in the U.S.A., Energy Star and Leadership in Energy and Environmental Design (LEED). The methodology to evaluate CHP systems based on Energy Star Rating is described in Mago et al. [27]. For LEED Rating there is one for new constructions (LEED-NC Rating), and other one for existing buildings (LEED-EB Rating). The methodology presented in Mago et al. [27] is described based on the LEED-EB Rating, but can also be applied to the LEED-NC.

3.7.2. Energy Ratings, Energy Star and LEED-EB

Tables 3.9 and 3.10 present the Energy Star Rating and the points for the LEED-EB Rating, respectively. The points for the LEED-EB only consider the points that can be gain from Credit 1 – Optimize Energy Performance. Figure 3.15 illustrates the increment of the Energy Star Rating from the use of CHP Systems.

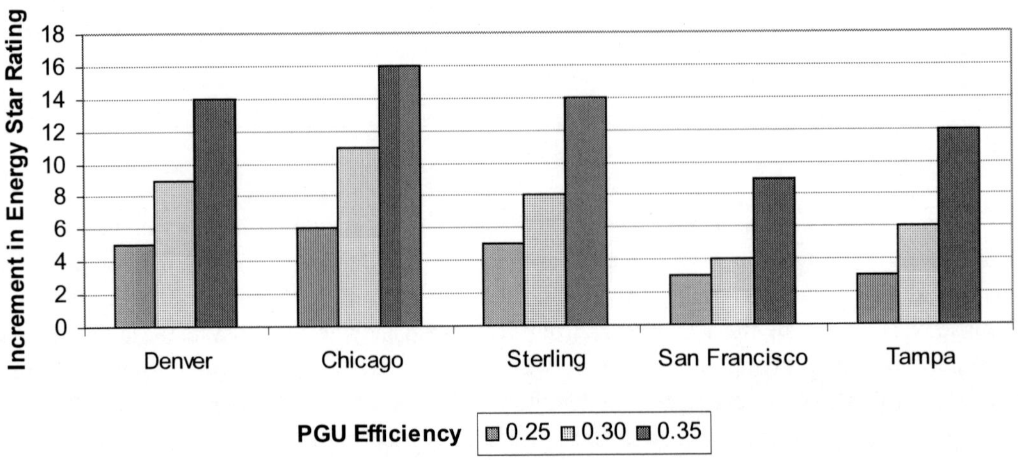

Figure 3.15. Increment of Energy Star Rating from the Use of CHP Systems.

Table 3.9 shows that CHP systems increase the Energy Star. However, the Energy Star Rating is the same when the CHP system operates without and with the BPER operational strategy. This means that the incremental decrease of energy consumption from the use of the BPER operational strategy is not enough to further improve the Energy Star Rating.

Figure 3.15 illustrates that CHP systems increases the Energy Star Rating for all cities. The greatest incremental increase is 16 for city of Chicago, while the lowest incremental increase is 3 for the cities of San Francisco and Tampa. As expected, higher PGU efficiency implies less energy consumption and consequently higher Energy Star Rating.

Table 3.9. Energy Star Rating

	PGU	Energy Star Rating		
City/Zone	**Efficiency**	**Building**	**CHP**	**CHP-BPER**
Denver Zone 1	0.25	56	61	61
	0.30		64	65
	0.35		70	70
Chicago Zone 2	0.25	49	55	55
	0.30		59	60
	0.35		65	65
Sterling Zone 3	0.25	54	59	59
	0.30		62	62
	0.35		68	68
San Francisco Zone 4	0.25	74	76	77
	0.30		78	78
	0.35		83	83
Tampa Zone 5	0.25	58	60	61
	0.30		64	64
	0.35		70	70

Table 3.10. LEED-EB Rating Points from Credit 1

	PGU	LEED-EB Credit 1 Points		
City/Zone	**Efficiency**	**Building**	**CHP**	**CHP-BPER**
Denver Zone 1	0.25	0	0	0
	0.30		0	0
	0.35		2	2
Chicago Zone 2	0.25	0	0	0
	0.30		0	0
	0.35		0	0
Sterling Zone 3	0.25	0	0	0
	0.30		0	0
	0.35		2	2
San Francisco Zone 4	0.25	5	5	6
	0.30		6	6
	0.35		9	9
Tampa Zone 5	0.25	0	0	0
	0.30		0	0
	0.35		2	2

Table 3.11. Emission of Nitrogen Oxides (NO_x)

City/Zone	PGU Efficiency	Kg of NO_x Building	CHP	CHP-BPER
Denver Zone 1	0.25	737	518	544
	0.30		236	303
	0.35		182	182
Chicago Zone 2	0.25	690	493	510
	0.30		237	280
	0.35		184	184
Sterling Zone 3	0.25	489	359	370
	0.30		194	227
	0.35		155	155
San Francisco Zone 4	0.25	173	151	154
	0.30		120	134
	0.35		91	91
Tampa Zone 5	0.25	705	501	512
	0.30		416	430
	0.35		170	170

Table 3.12. Emission of Sulfur Dioxide (SO_2)

City/Zone	PGU Efficiency	Kg of SO_2 Building	CHP	CHP-BPER
Denver Zone 1	0.25	447	291	310
	0.30		84	134
	0.35		50	50
Chicago Zone 2	0.25	2318	1514	1586
	0.30		444	635
	0.35		257	257
Sterling Zone 3	0.25	1398	909	958
	0.30		275	404
	0.35		153	153
San Francisco Zone 4	0.25	118	88	96
	0.30		45	66
	0.35		13	13
Tampa Zone 5	0.25	1099	714	735
	0.30		563	590
	0.35		109	109

Table 3.10 shows that points from Credit 1 for the LEED-EB Rating can not be obtained for all evaluated cities. This is explained because the Energy Star Rating without the CHP system is too low. For the conditions of this study the greatest incremental increase was 4 points for the city of San Francisco, which represents 27% of the total 15 points that can be gained.

3.7.3. Emission of Pollutants

Emission factor for electricity depends on the fuel mix used to generate electricity in the region where the energy has been used. To better understand the results of this section, Appendix C describes the effect of the fuel mix for the electric grid regions associated to the cities considered in this study.

Table 3.13. Emission of Carbon Dioxide (CO_2)

	PGU	Kg of CO_2		
City/Zone	Efficiency	Building	CHP	CHP-BPER
Denver Zone 1	0.25	510612	383321	396563
	0.30		221181	257800
	0.35		184402	184402
Chicago Zone 2	0.25	421992	330244	336622
	0.30		213271	232086
	0.35		183600	183600
Sterling Zone 3	0.25	321685	261077	265216
	0.30		188290	201288
	0.35		164249	164249
San Francisco Zone 4	0.25	192890	171657	174389
	0.30		142216	154268
	0.35		113063	113063
Tampa Zone 5	0.25	409644	320641	324630
	0.30		280031	285904
	0.35		168575	168575

For CHP systems, Tables 3.11, 3.12, and 3.13 present the estimated emission of Nitrogen Oxides (NO_x), Sulfur Dioxide (SO_2), and Carbon Dioxide (CO_2), respectively. Figures 3.16, 3.17, and 3.18 illustrate the percentage of reduction for NO_x, SO_2, and CO_2, respectively. For all the cases considered, CHP systems reduce the emission of pollutants, and high PGU efficiency has a significant impact on emission reductions.

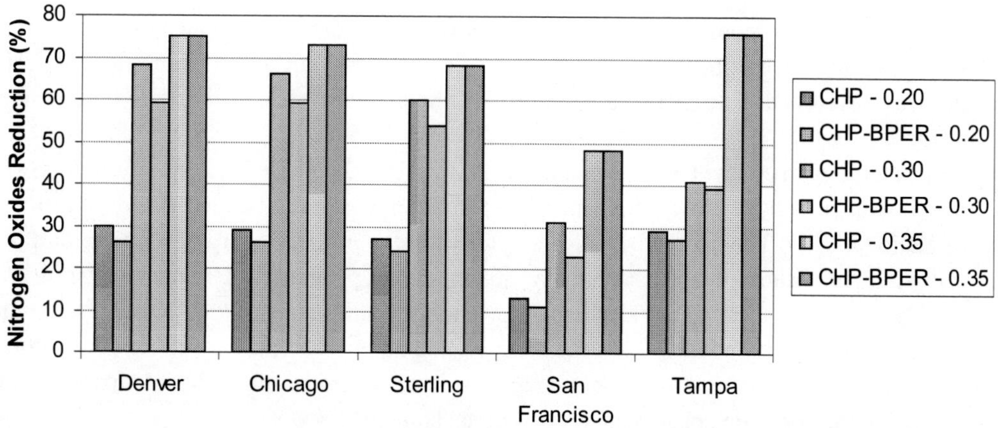

Figure 3.16. Nitrogen Oxides (NO_x) Reduction from the Use of CHP Systems.

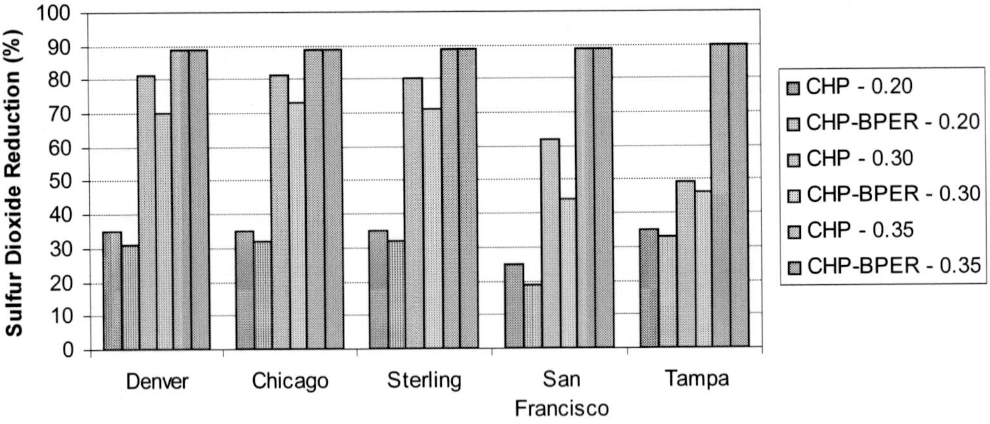

Figure 3.17. Sulfur Dioxide (SO_2) Reduction from the Use of CHP Systems.

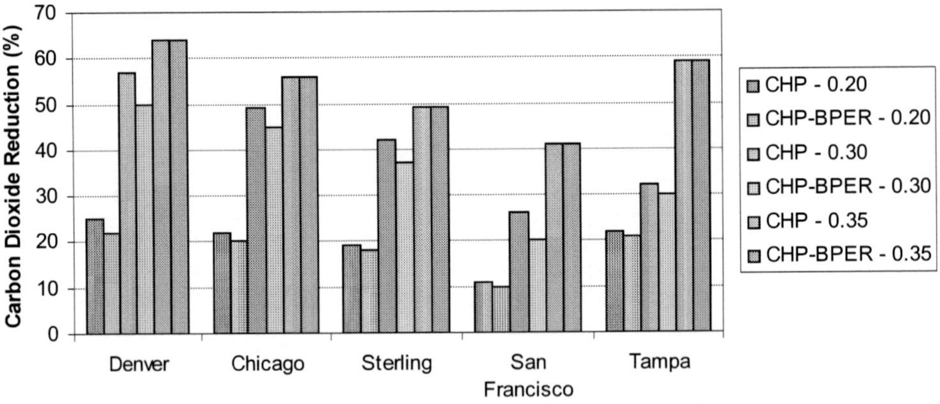

Figure 3.18. Carbon Dioxide (CO_2) Reduction from the Use of CHP Systems.

6. Conclusions

An evaluation of cooling, heating, and power systems based on primary energy operational strategy was presented. A model to estimate the energy consumption of CHP systems was developed. The analysis presented in the chapter leads to the following salient conclusions.

A methodology to evaluate CHP systems energy performance based on primary energy was developed by using a novel parameter introduced in this investigation called Building Primary Energy Ratio (BPER). This parameter allows comparing the primary energy consumption for a building with and without a CHP system. In the simulation program, this parameter allows to simulate the energy performance under a primary energy operational strategy in order to obtain the lowest primary energy consumption for the specified inputs. The primary energy operational strategy introduced in this investigation guarantee primary energy savings which not always is achieved using the common cost-oriented operational strategy.

The use of the primary energy operational strategy can guarantee energy savings from the use of CHP systems, but the economic feasibility is subject to energy prices. When energy prices make economically unfeasible a well designed CHP system, other benefits from this technology must be considered. Some benefits from a non-conventional evaluation could be quantified and transferred into the economic evaluation, while others would give intangible potential to the technology.

Results proved that applying the BPER operational strategy improve the Energy Star Rating and the Leadership in Energy and Environmental Design (LEED) Rating, as well as reduce the emission of pollutants.

Appendix A. Flowchart for the CHP System Simulation Program

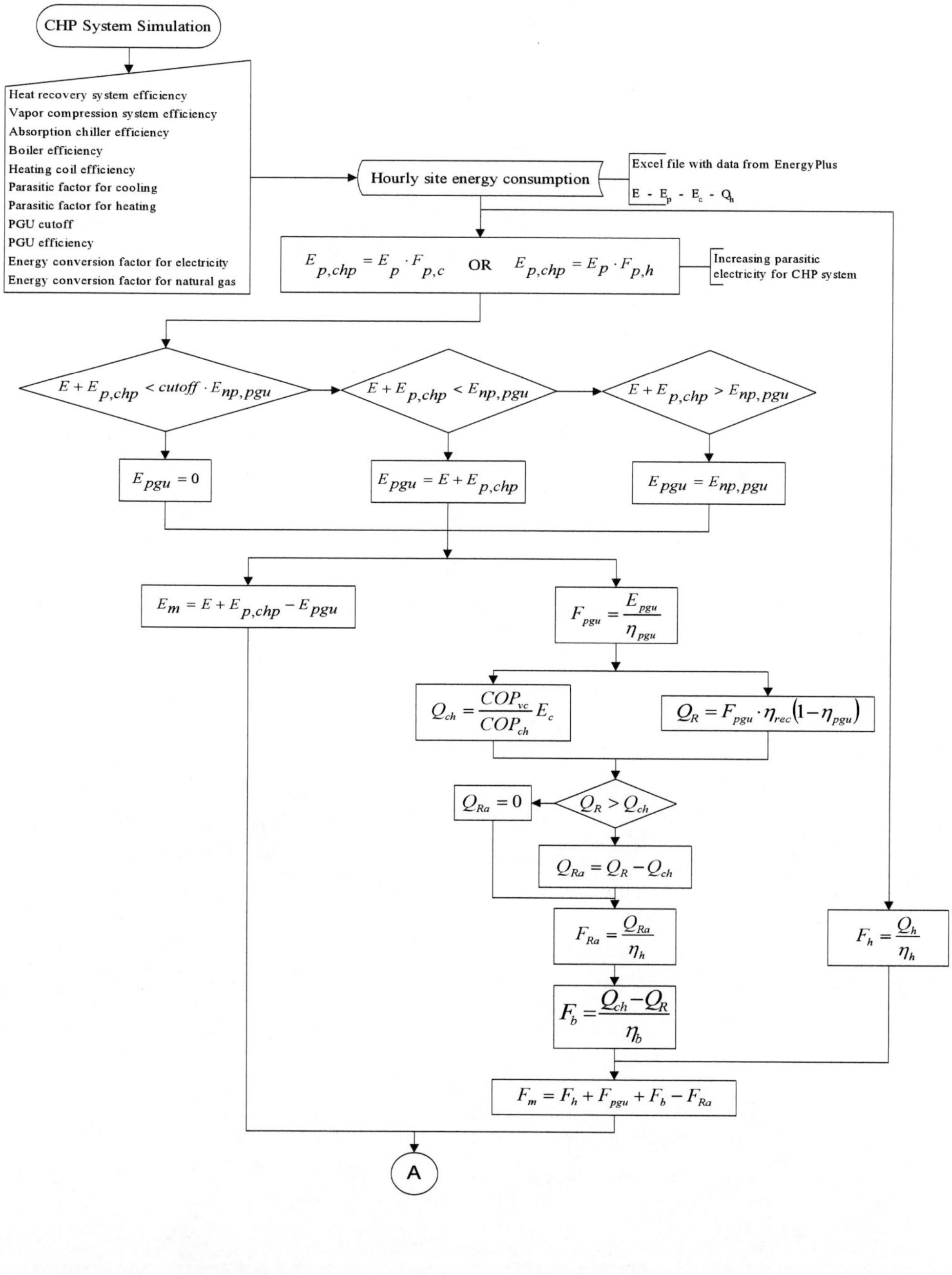

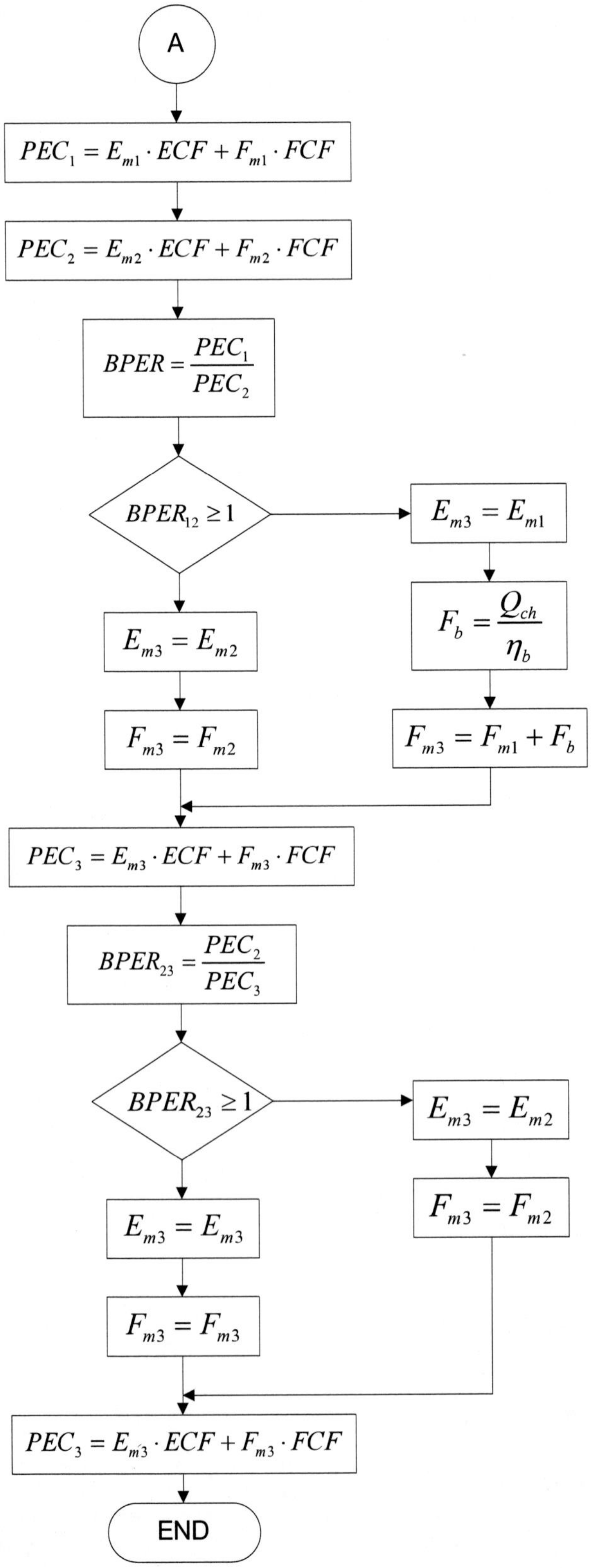
A
$PEC_1 = E_{m1} \cdot ECF + F_{m1} \cdot FCF$
$PEC_2 = E_{m2} \cdot ECF + F_{m2} \cdot FCF$
$BPER = \frac{PEC_1}{PEC_2}$
$BPER_{12} \geq 1$
$E_{m3} = E_{m1}$
$F_b = \frac{Q_{ch}}{\eta_b}$
$F_{m3} = F_{m1} + F_b$
$E_{m3} = E_{m2}$
$F_{m3} = F_{m2}$
$PEC_3 = E_{m3} \cdot ECF + F_{m3} \cdot FCF$
$BPER_{23} = \frac{PEC_2}{PEC_3}$
$BPER_{23} \geq 1$
$E_{m3} = E_{m2}$
$F_{m3} = F_{m2}$
$E_{m3} = E_{m3}$
$F_{m3} = F_{m3}$
$PEC_3 = E_{m3} \cdot ECF + F_{m3} \cdot FCF$
END

Appendix B. Site Energy Consumption by the Type of Source

		Building			CHP			CHP-BPER		
		Electricity	Natural Gas		Electricity	Natural Gas		Electricity	Natural Gas	
City (Zip Code)	PGU Efficiency	kWh	kWh	MBTU	kWh	kWh	MBTU	kWh	kWh	MBTU
Denver (80210)	0.25	205061	306705	1046	132974	487155	1662	141707	453075	1546
	0.30	205061	306705	1046	38191	753259	2570	60843	677972	2313
	0.35	205061	306705	1046	22591	741338	2529	22591	741338	2529
Chicago (60610)	0.25	208373	378750	1292	136059	549355	1874	142529	523997	1788
	0.30	208373	378750	1292	39864	804307	2744	57063	747154	2549
	0.35	208373	378750	1292	23030	798147	2723	23030	798147	2723
Sterling (20165)	0.25	217396	276821	945	141152	467988	1597	148783	438203	1495
	0.30	217396	276821	945	42542	746283	2546	62609	679664	2319
	0.35	217396	276821	945	23570	744392	2540	23570	744392	2540
San Francisco (94110)	0.25	180490	110212	376	134601	235738	804	146061	190191	649
	0.30	180490	110212	376	66593	433006	1477	100494	320143	1092
	0.35	180490	110212	376	17752	530414	1810	17752	530414	1810
Tampa (33610)	0.25	280066	23410	80	182025	315584	1077	187423	294448	1005
	0.30	280066	23410	80	143205	401629	1370	150163	378428	1291
	0.35	280066	23410	80	27474	711191	2427	27474	711191	2427

Appendix C. Region Fuel Mix Comparison

The figure shows the fuel mix comparison for the grid regions corresponding to each of the cities considered in this study. This figure was developed based on the information provided by Power Profile. The fuel mix will define the amount of pollutants when electricity is consumed, and consequently will define the impact of CHP systems on the reduction of emission of pollutants. For the cities of Denver, Chicago, and Sterling, more than 50% of the electric power comes from coal which is highly pollutant. While for the cities of San Francisco and Tampa, more than 70% of the electric power comes from sources other than fossil fuels. Then, for the same energy consumption among the evaluated cities, more pollutants will be generated in the cities of Denver, Chicago, and Sterling, than in the cities of San Francisco and Tampa.

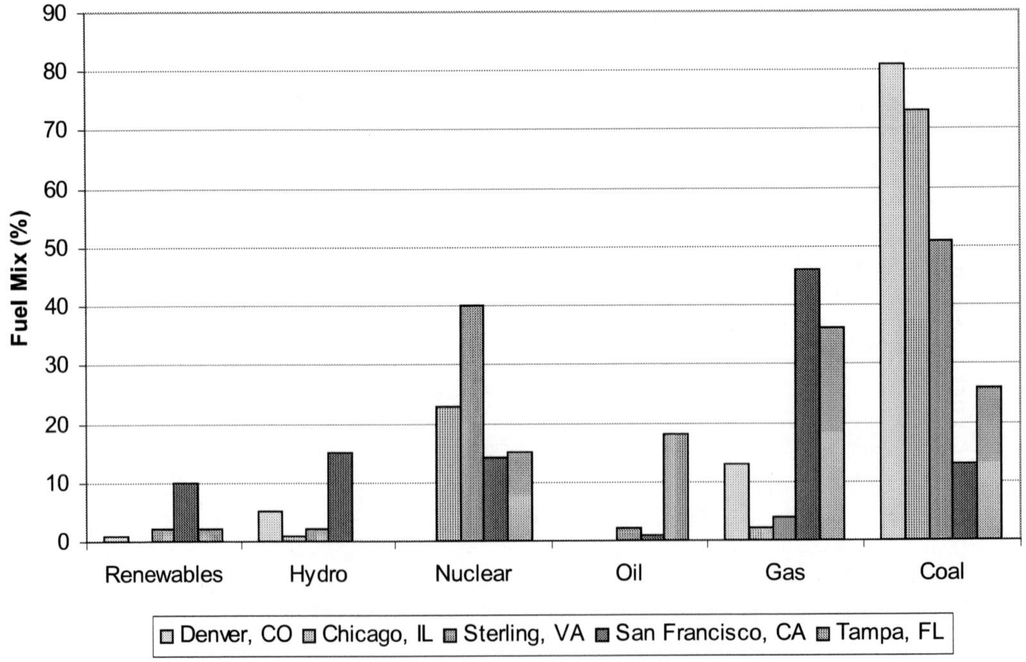

References

[1] U.S. Department of Energy, Office of Energy Efficiency and Renewable Energy, Federal Energy Management Program, *DER/CHP*. http://www1.eere.energy.gov/femp/ *der/index.html. Feb. 1, 2008.*

[2] Sayane S. and Shokrollahi S., "Selection and sizing of prime movers in combined heat and power systems," *Proceedings of ASME Turbo Expo: Power for Land, Sea and Air Conference*, June 14-17, 2004, Vienna, Austria.

[3] Carbon Trust, Action Energy Program, Good Practice Guide (GPG388), "Combined heat and power for buildings," April 2004. http://www.carbontrust.co.uk/publications/ *publicationdetail.htm?productid=GPG388andmetaNoCache=1*

[4] Zogg, R., Roth, K., and Brodrick, J., "Using CHP Systems In Commercial Buildings," ASHRAE Journal, September 2005, Vol. 47 Issue 9, pp. 33-36.

[5] Arthur D. Little, Inc., "Cooling, Heating, and Power (CHP) for Commercial Buildings Benefits Analysis," *Report for the U.S. Department of Energy, Office of Energy Efficiency and Renewable Energy, Distributed Energy Program,* April 2002.

[6] Siddiqui A. S., Marnay C., Bailey O., Hamachi K., "Optimal selection of on-site generation with combined heat and power applications," *International Journal of Distributed Energy Resources*, November 2004, Vol. 1, No. 1, pp. 33-62.

[7] Fischer S. K., and Glazer J., "CHP self analysis," Proceedings of the ASME IMECE Conference, New Orleans, LA, November 2002, pp. 523-527.

[8] Turner, Wayne C. 1997. *Energy Management Handbook.* 3er Edition, The Fairmont Press, Inc.

[9] Cardona E. and Piacentino A., "A methodology for sizing a trigeneration plant in mediterranean areas," *Applied Thermal Engineering*, September 2003, Vol. 23 Issue 13, pp. 1665-1680.

[10] Sun Z. G., Wang R. Z., and Sun W. Z., "Energetic efficiency of a gas-engine-driven cooling and heating system," *Applied Thermal Engineering*, April 2004, Vol. 24 Issues 5-6, pp. 941-947.

[11] Cardona, E., and Piacentino, A., "A Validation Methodology for a Combined Heating Cooling and Power (CHCP) Pilot Plant, *Journal of Energy Resources Technology*, December 2004, Vol. 126 Issue 4, pp. 285-292.

[12] Li, H., Fu, L., Geng, K., and Jiang, Y., "Energy utilization evaluation of CCHP systems," *Energy and Buildings*, March 2006, Vol. 38 Issue 3, pp. 253-257.

[13] U.S. Department of Energy, Energy Information Administration, *Glossary*. http://www. *eia.doe.gov. Jan. 10, 2007.*

[14] Kowalski, G. J., and Zenouzi, M., "Selection of Distributed Power-Generating Systems Based on Electric, Heating, and Cooling Loads," *Journal of Energy Resources Technology*, September 2006, Vol. 128 Issue 3, pp. 168-178.

[15] Rizy, D. T., Zaltash, A., Labinov, S. D., Petrov, A. Y., Vineyard, E. A., Linkous, R. L., "CHP integration (or IES): maximizing the efficiency of distributed generation with waste heat recovery," *Proceedings of the Power Systems 2003 Conference*, March 2003, Clemson, South Caroline.

[16] Pilavachi, P. A., Roumpeas, C. P., Minett, S., Afgan, N.H., Pilavachi, P. A., Roumpeas, C. P., Minett, S., Afgan, N. H., "Multi-criteria evaluation for CHP system options," *Energy Conversion and Management*, December 2006, Vol. 47 Issue 20, pp. 3519-3529.

[17] Fischer, S., and Glazer, J., "CHP self analysis," *Proceedings of IMECE*, November 17-22, 2002, New Orleans, Louisiana.

[18] U.S. Department of Energy, Office of Energy Efficiency and Renewable Energy, Building Technology Program, *EnergyPlus Energy Simulation Software*. http://www. *eere.energy.gov/buildings/energyplus/*

[19] U.S. Department of Energy, Environmental Protection Agency, *Energy Star. http://www. energystar.gov*

[20] U.S. Department of Energy, Environmental Protection Agency, Energy Star Program, *Target Finder. http://energystar.gov/index.cfm?c=new_bldg_design.bus_target_finder*

[21] Newborough, M., "Assessing the benefits of implementing micro-CHP systems in the UK," *Proceedings of the Institution of Mechanical Engineers - Part A: J. Power and Energy*, June 2004, Vol. 218 Issue 4, pp. 203-218.

[22] Keppo, I., and Savola, T., "Economic appraisal of small biofuel fired CHP plants," *Energy Conversion and Management*, April 2007, Vol. 48 Issue 4, pp. 1212-1221.

[23] Jablko, R., Saniter, C., Hanitsch, R., and Holler, S., "Technical and economical comparison of micro CHP systems," *Future Power Systems, 2005 International Conference*, November 16-18, Amsterdam, Netherlands.

[24] Tucker, C. T., "Rethinking the benefits of CHP," *Proceedings of the ASME Power Conference* 2005, April 5-7, 2005, Chicago, Ilinois.

[25] De Paepe, M., D'Herdt, P., and Mertens, D., "Micro-CHP systems for residential applications," *Energy Conversion and Management*, November 2006, Vol. 47 Issue 18/19, pp. 3435-3446.

[26] Zoog, R. A., "Cooling, heating, and power (CHP) for commercial buildings benefits analysis," *Cogeneration and Distributed Generation Journal*, Fall 2004, Vol. 19 Issue 4, pp. 14-44.

[27] Mago, P.J., Chamra, L.M., and Moran, A., "Modeling of micro-cooling, heating, and power (Micro-CHP) for residential or small commercial applications," *ASME International Mechanical Engineering Congress and Exposition (IMECE2006)*, November 5-10, 2006, Chicago, Illinois.

In: Progress in Management Engineering
Editors: L.P. Gragg and J.M. Cassell, pp. 237-279
ISBN: 978-1-60741-310-3

Chapter 9

RHEOLOGICAL INVESTIGATIONS IN SOIL MICRO MECHANICS: MEASURING STIFFNESS DEGRADATION AND STRUCTURAL STABILITY ON A PARTICLE SCALE

Wibke Markgraf **and Rainer Horn**
Christian-Albrechts-University zu Kiel,
Institute for Plant Nutrition and Soil Science, Kiel, Germany

Abstract

Rheology is regarded as the science of flow behavior, where, based on isothermic equations, the deformation of fluids and plastic bodies subjected to external stresses may be described. Hooke's law of elasticity, Newton's law for ideal fluids (viscosity), Mohr-Coulomb's equation, and finally, Bingham's yielding are well known relationships and parameters in the field of rheology.

Rheometry is a well established measurement technique to determine the specific rheological properties of fluid and plastic bodies. In order to explain point contact processes and strength an extrapolation of such findings to data of triaxial, direct shear or oedometer tests is still missing.

A parallel-plate-rheometer MCR 300 (Modular Compact Rheometer, Paar Physica, Ostfildern, Germany) has been used to conduct oscillatory tests. From the stress-strain relationship parameters and specific characteristics as storage modulus G' and loss modulus G", loss factor tan δ (= G''/G'), viscosity η, yield stress τ_y and the linear viscoelastic deformation (LVE) range including a limiting value γ_L were determined and calculated, respectively.

Thus, this work aims to introduce rheometry as a suitable method to determine the mechanical behavior of soils, as viscoelastic material, and mineral suspensions when subjected to external stresses. To do this a Na-bentonite, Ibeco Seal-80, has been used for preliminary tests; the suspensions were equilibrated with NaCl solutions in different

[*] E-mails address: w.markgraf@soils.uni-kiel.de. Olshausenstr. 40, D-24098 Kiel, Germany.

concentrations in order to determine the ionic strength effects on interparticle strength and changes in mechanical properties. Furthermore, a Dystric Planosol, a Calcaric Gleysol from North Germany and loess material from Israel, saturated with NaCl and/or $CaCl_2$ in several concentrations were analyzed.

In order to demonstrate clay mineralogical and/or textural effects as well as of leaching of organic matter and iron oxides, the degree of stiffness and structural stability of clay rich substrates from Brazil, - a smectitic Vertisol and a kaolinitic Ferralsols - were quantified. In addition, scanning electron microscopy was applied for visualizing structural characteristics. Due to the modification of microstructural analysis by such visual investigations, structural changes and consequences for upscaling considerations become evident as well as the need of research in soil mechanical processes on the particle-particle scale. It is shown that rheometry is an applicable method to detect microstructural changes by using a rotational rheometer.

Introduction

In this chapter, rheology will be introduced as a method for investigating microstructural behavior of soil on a micro scale (particle-particle).

For a basic understanding of soil (micro)mechanical behavior the particulate nature of soils needs to be recognized: physicochemical properties, the interaction between single particles, which are defined by their size, shape, mineralogy etc., the interparticle arrangement and interconnected porosity, inherently non-linear non-elastic contact phenomena, and particle forces based on principle laws of Newton (gravity and viscosity), Hooke (elasticity) and Coulomb as introduced in Terzaghi and Jelinek (1954), Kézdi (1974), Fredlund and Rahardjo (1993), Hartge and Horn (1999), or Mitchell and Soga (2005).

The rheological investigation of micromechanical behavior and microstructural changes in soils needs to be intertwined to such fundamental knowledge of soil mechanics. In a sensitive intergranular system, parameters such as physicochemical properties, single grain characteristics as well as hydraulic properties (water content) are of great importance for shear behavior and deformation in soils on the microscale. Rheology as soil mechanical topic is limited to phenomenons of yielding or creeping. In early works of Atterberg (1914) definitions of consistence limits and sensitivity (Terzaghi, 1936) of clay rich or pure clay materials are given.

Table 1. Rheology is a discipline of both solid and fluid mechanics, in detail of plasticity and non-Newtonian fluids

Continuum mechanics	Solid mechanics or strength of materials	Elasticity	
		Plasticity	Rheology
	Fluid mechanics	Non-Newtonian fluids	
		Newtonian fluids	

A new aspect of rheometry, either from the classical rheological or soil scientific point of view, lies in the linkage of both, soil micromechanical behavior with respect to soil or single particle properties and rheological characteristics. States of soil cannot be divided strictly into two classes of elastic or plastic (viscous). Moreover, soil obtains several transitional states between elasticity and plasticity, depending on applied stress – steady stress, oscillatory stress - and strain as well as intensity, time and frequency (under oscillatory condition). Small stress-strain considerations are basically focused in rheological measurements.

Figure 1. a) modular compact rheometer MCR 300; 1 pneumatic ball bearing, and instrument lift; 2 manual control; 3 rotating bob (25 mm); 4 measuring plate with Peltier unit; b) 5 control display; c) 6 profiled parallel plate measuring system (25PP MS) in detail (setting of zero-gap); d) 7 sheared surface of a sample after an applied amplitude sweep test (AST) under oscillating (OSC) conditions with controlled shear deformation (CSD); duration of one test: 15-18 minutes (f = 0.5 Hz).

In Markgraf et al. (2006) rheology is introduced as a new methodological approach to soil mechanics with special regard to scale considerations. In order to investigate micromechanical shear behavior, rheological methods – in this case amplitude sweep tests – are adapted and combined with common methods and fundamental knowledge derived from soil physics, clay mineralogy, and rheology itself. Whenever physicochemical or mechanical interactions at the contact points of single soil particles, aggregates or soil as a whole specimen need to be investigated, diverse measuring methods are applied. Common methods and characteristics in soil physics as the compression behavior and derivable parameters of precompression and shear resistance - the angle of inner friction, cohesion, shear stress, deformation and the yield point - are mainly used for process analyses in soils that are defined as three phase system. However, in rheology these parameters are used for the quantification of shear behavior and deformation of fluids or plastic (viscous) materials. A fundamental introduction to the research area of rheology is described in works of Keedwell (1984), Vyalov (1986), Barnes et al. (1989), Whorlow (1992), Lagaly et al. (1997), Collyer and Clegg (1998), Schulz (1998), Schramm (2002), and Mezger (2006). Rheology has important applications in engineering, geophysics and physiology. In geotechnical research, where the application of ring shear apparatus' (Suklje, 1969; Sonderegger, 1985) or simple rigidity tests of rocks are rather common, solid earth materials that exhibit viscous flow over long time scales are known as *rheids*. In engineering, rheology has had its predominant application in the development and use of polymeric materials. Rheological investigations of suspensions, gels and other (in)organic materials on a smaller scale (meso and microscale) are well known in inorganic chemistry, polymer sciences, and applied material sciences (Brandenburg, 1990; Güven and Pollastro, 1992; Cristescu and Gioda, 1994; Kosmulski et al., 1999; Neaman and Singer, 2000; Akroyd and Nguyen, 2003). Rheology is principally concerned with extending disciplines of elasticity and (Newtonian) fluid mechanics to materials, whose mechanical behavior cannot be described by classical theories (*Table 1*). Due to its application predictions for mechanical behavior (on the continuum mechanical scale) based on the micro- or nanostructure of the material are established. It furthermore unites the seemingly unrelated fields of plasticity and non-Newtonian fluids by recognizing that both of these types of materials are unable to support a shear stress in static equilibrium. In this sense, a plastic solid is a fluid and shows a creeping or yielding character. Granular rheology refers to the continuum mechanical description of granular materials.

One of the subjects of rheology is to establish empirically the relationships between deformation and stress, respectively their derivatives by adequate measurements. These experimental techniques are known as rheometry. Such relationships are then amendable to mathematical treatment by the established methods of continuum mechanics. Microstructural investigations of soils as non-Newtonian, partly linear viscoelastic material are rather uncommon in the wide spectrum of rheology. Linear viscoelasticity had been defined mathematically by Dafarmos and Nohel (1980) using the Volterra-equation. Basically rheological methods are related to dimensions, the degree of homogeneity, and viscosity of the investigated substance. Furthermore, accuracy and controllability of the appropriate measuring system are of great importance. Rheological tests, which are used for quality control, are based on simple relaxation or rotational test (Mezger, 2006). The application of a parallel plate measuring system, commonly in rotation mode, has been limited to substances as oil (Newtonian fluid), dyes or coatings (Meichsner et al., 2003), ceramics or clay suspensions of low viscosity (Kosmulski at al., 1999; Neaman and Singer, 2000; Tarchitzky and Chen, 2002). An approach to rheological measurements was done by

Ghezzehei and Or (2000, 2001), who investigated comparatively steady stress (=internal capillary tensions) and oscillatory stress-strain correlations (=vibration effects, compaction). However, with regard to soil scientific aspects, especially to micromechanical behavior, this approach seems to be insufficient. Soil substrates, which are completely saturated, are in a state of yielding already; structuring processes, among this aggregation, swelling and shrinkage (Tariq and Durnford, 1993) cannot be proved or quantified rheometrically and are intertwined with or transferred to methods as mentioned above. A new approach of rheometry, which can be adapted to soil micromechanical investigations and be related to common shear tests (oedometer, triaxial test), has had to be developed. Due to this, it has to be proved, whether it is possible to detect micromechanical shear behavior, with special regard to contact level considerations, of homogenized soil samples (<2mm; specimen volume approximately $4cm^3$) under saturated or pre-drained conditions. Amplitude sweep tests were conducted with a modular compact rheometer MCR 300 (*Figure 1*); an applicable method, investigated substrates, and test results will be presented in detail. Recent collected data show significant differences of resulting parameters, which are correlated to the water content, and other factors that may affect the matric potential: texture, single particle properties, pore size distribution, influence of (cat)ions (osmotic potential), and other physicochemical properties.

1.2. Fundamentals of Soil Micromechanics

In the following paragraphs an introduction to general considerations of fundamental soil micromechanics will be given, which are of relevance for a rheological approach to microstructural changes.

1.2.1. Particle Associations

Microstructural changes in soils are related to several kinds of particle associations, basically on the association of clay platelets and of single grains. According to van Olphen (1977) clay particles in suspensions can be described as

(1) dispersed with no face-to-face association of particles,
(2) as aggregated in a face-to-face (FF) association with several particles, flocculated in an edge-to-edge (EE) or edge-to-face (EF) state,
(3) and deflocculated under non-associated conditions.

The aggregation of clay particles is correlated to the actual pH value (Jasmund and Lagaly, 1993; Chorom et al., 1994; Lagaly et al., 1997; Dörfler, 2002), and ion concentrations. Cardhouse structures can be formed by edge-to-face and edge-to-edge associations. Under applied stress (compression) these voluminous structures collapse. Higher structural stability is obtained by larger and thicker face-to-face associations (Rosenqvist, 1959, 1962; Kézdi, 1974). As stated by Smith and Reitsma (2002), a different mechanical behavior is evident, if kaolinite (1:1 layer clay mineral) and montmorillonite (2:1) are compared (Markgraf et al., 2006). In addition, kaolinite is defined as low active clay (LAC), and montmorillonite as high active clay (HAC) with a high swelling and shrinkage capacity.

Montmorillonite has a more distinctive diffuse double. According to the Derjaguin-Landau-Verwey-Overbeek (DLVO) diffuse double layer theory (Derjaguin and Landau, 1941; Verwey and Overbeek, 1948) no physical contact between clay particles is necessary for taking up charges or the adsorption of cations. Hence, the osmotic pressure is the actual effective stress between clay platelets in a Na-montmorillonitic soil. Thus, the residual shear strength of kaolinite remains high (20-25°) even under high stress, in opposite to montmorillonite (Smith and Reitsma, 2002). Here, the residual shear strength decreases rapidly and remains stable at 5°. This is reflected in sliding shear behavior versus turbulent shear behavior in kaolinite or kaolinitic soils.

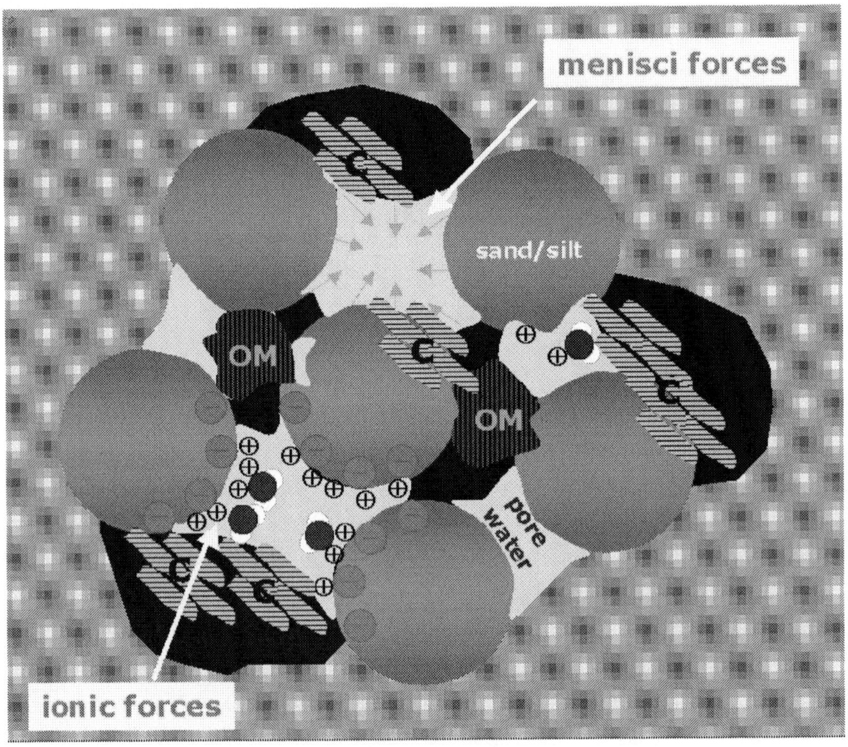

Figure 2. Sand/silt grains and clay platelets are formed to an aggregate; the three phases of a soil - gaseous, liquid, and solid - are combined by forces as generated due to several bonding mechanisms: ionic (Na^{+}), covalent (Ca^{2+}, Mg^{2+}), hydration mechanisms, capillary forces (matric potential, influenced by osmotic potential), and associated menisci forces. Other bonds may be built upon organic matter (OM) compounds and are influenced by mineralogical factors as well as parameters, which define pore water characteristics (e.g. salt concentration).

Particle association of compacted clays, single grains in sediments, and finally all soils show a variety of aggregation. In comparison to the particle association in clay suspensions, aggregation is generally related to combinations of the configurations of clay particles, but show differences in water content and density of the soil mass. Three groupings of fabric elements can be identified: elementary particle associations as e.g. individual grains, particle assemblages as organized unit of grains with physical limits and mechanical function, and water or gas filled pores (*Figure 2*). Furthermore, the stability of aggregates depends on contents and distribution of texture, organic matter, clay minerals and (hydr)oxides, and water (Emerson, 1967, 1983; Tisdall and Oades, 1982; Oades and Waters, 1991) as well as salt concentrations (Sharpley, 1990; Rengasamy and Olssen, 1991; Shainberg et al., 1981; Shainberg and Levy, 1992; Shani and Dundley, 2001) and other physicochemical properties such as pH, CEC etc. Results of investigated kaolinitic Fe-(hydr)oxide rich Ferralsols (Typic Hapludox, USDA 2003) and a smectitic Vertisol (Typic Calciudert, USDA 2003) from Brazil show significant differences in microstructural stability. Theses differences are related to stabilizing effects due to variations in clay mineralogy, organic matter content, goethite, and hematite. Successive leaching of Na-dithionite soluble iron and organic matter by H_2O_2 in combination with visual analyses (scanning electron microscopy and polarized light microscopy) lead to distinctive conclusions.

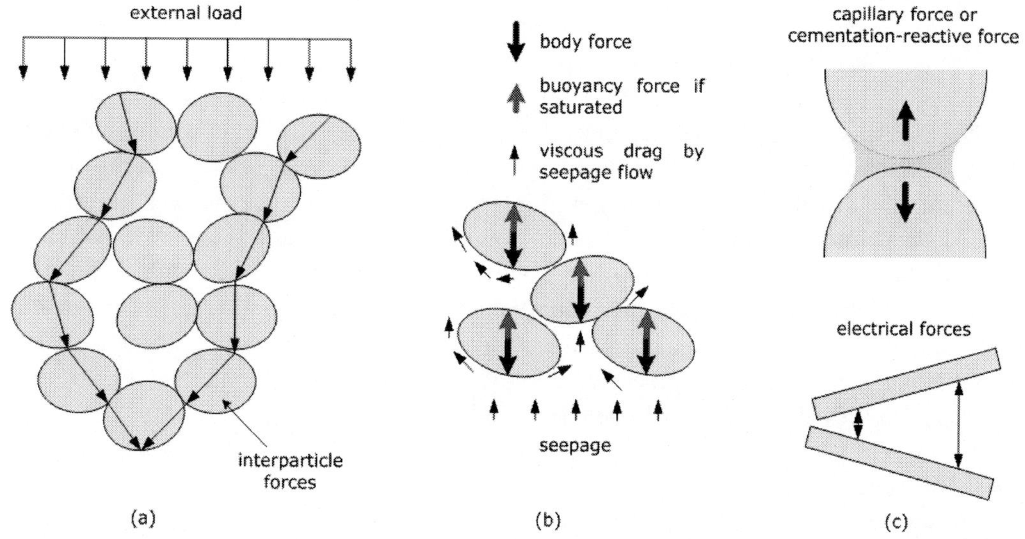

After Santamarina, 2003.

Figure 3. Interparticle forces at the particle level: (a) skeletal forces by external loading, (b) particle level forces, and (c) contact level forces.

Similar approaches were made by Bohor and Hughes (1971), Davey (1979), FitzPatrick (1993), Terribile and FitzPatrick (1995), Boivin et al. (2004), and Reed (2005), who applied transmission and scanning electron microscopy for the identification of soil mineralogy and structure using thin sections. Le Bissonnais and Arrouays (1996, 1997) adapted this method to microstructural analyses with focus on soil aggregate stability, which is affected by organic matter compounds. Beside the identification of soil compounds and their association, the quantification of particle forces is of great importance for microstructural analyses, including considerations of interparticle forces and intergranular stresses.

1.2.2. Particle Forces

Particle forces at the microscale can be separated into several categories according to Santamarina (2001, 2003), based on earlier works of Ingles (1962) (*Figure 3*):

(1) Forces due to applied boundary stresses: forces are transmitted along granular chains that form within the soil skeleton. Capillary effects at high degree of saturation prior to air entry fall under this category.
(2) Particle-level forces: particle weight, buoyancy and hydrodynamic forces are included. A particle can experience these forces even in the absence of a soil skeleton.
(3) Contact-level forces: capillary forces at a low degree of saturation, electrical forces, and the cementation-reactive force belong to this category. The first two mentioned forces can cause strains in the soil mass even at constant boundary loads. Conversely, the cementation-reactive force opposes skeletal deformation.

The investigation and quantification of particle forces is still on an establishing state. Numerical micromechanics were pioneered by Cundall and Strack (1979), who provided an insight into the distribution and evolution of interparticle skeletal forces in soils. Information about skeletal force distribution was gained from photoelastic studies conducted by Valdes (2002). Micromechanical analyses and mathematical approximations with simulations demonstrate the significance of particle coordination, rotational frustration (after Santamarina 2001) and the buckling of chains on the ability of a soil to mobilize internal strength (Santamarina, 2001). The relevance of rheological methods in micromechanics and their applicability are established and will be introduced.

1.2.3. Interparticle Forces

Interparticle forces and intergranular stresses occur in association with particle forces or any external applied stress that lead to deformation, viscous or even plastic behavior. With regard to clay particles, interparticle repulsive forces such as electrostatic forces, surface and ion hydration need to be considered. This mechanism occurs favorable in clay minerals of high swelling potential as i.e. sodium montmorillonites. In this case, hydration beside valency effects predominate soil structural processes. Cations (i.e. K^+, Na^+, Ca^{2+}, Mg^{2+}, Al^{3+}) dissolved in e.g. capillary water or under laboratory conditions in form of chlorides in distilled water, and in a certain concentration may either lead to a microstructural stabilization or destabilization (Emerson and Bakker, 1973). When focused on single grains or microaggregates, several attraction and repulsive forces of both, chemical and physical origin are effective: electrostatic and/or electromagnetic attraction, primary valence bonding, cementation, and capillary stresses (Emerson, 1962; Yimsiri and Soga, 1999, 2000, 2002; Mitchell and Soga, 2005).

Furthermore, soil adhesion can be generally classified into normal adhesion and tangential adhesion. On any scale, the system of soil adhesion consists of soil, solid surface and their interface. Furthermore, the structure of and properties of the interface layer dominate soil adhesion. This aspect can be transferred to the application of a parallel plate rheometer with a specific surface character, especially with regard to soil-tool interface considerations and micromechanical shear behavior. In Tong et al. (1994), Santamarina (2001, 2003) and Jia (2004) particle forces at the soil-tool interface are modeled and transferred to practical application. According to Santamarina (2001) the normal adhesion forces (N_A) at the interface come from forces caused by intermolecular attraction of bare soil particles (N_{As}), attraction of water meniscus (N_{Am}), attraction of water film due to viscosity (N_{Av}), which is influenced by chemical characteristics of (cat)ions that may change the viscosity and surface tension of a liquid (water), and capillary negative adsorption (N_{ca}).

1.2.4. Effective and Intergranular Stress

In soil physics mechanical analyses of shear parameters, compression behavior and resulting hydraulic and mechanical characteristics are conducted on undisturbed structured soil samples and larger single aggregates with respect to further influencing factors. However, soil physical methodology does not include standardized analyses of forces on the scale of contact points (Hartge and Horn, 1999). In general, intergranular stress σ_i' (*Figure 4*) is used synonymously with effective stress σ' [hPa]. In context with the soil stability discussion and effective tensions the concept of effect stress according to Bishop (1960), and Bishop and Bjerrum (1960) needs to be mentioned. The following equation describes occurring forces between single particles:

$$\sigma' = \sigma - u_a + \chi\,(u_a - u_w) \text{ with } \sigma'_i = \sigma' \qquad (1)$$

σ' = effective stress [hPa]
σ = total stress [hPa]
σ'_i = intergranular stress
χ = χ-factor [-] in dependency on the degree of saturation of a soil;
χ = 0 (unsaturated) χ = 1 (saturated)
u_a = pore air pressure [Pa]
u_w = pore water pressure [Pa]

The intergranular stress σ_i' [hPa] is given by

$$\sigma_i' = \sigma + A - u \qquad (2)$$

where u is the hydrostatic pressure [Pa] between particles or $u = h_m \gamma_w$

$$\sigma_i' = (\sigma - u_0) + (A - R) \qquad (3)$$

where $(\sigma - u_0)$ is the skeletal force, and (A – R) electrochemical forces;

A [N] are long range forces and R [Pa] the long range pressure,

with

$$R = h_s \gamma_w$$

where h_s is the osmotic head pressure [kPa], and γ_w the unit weight of water [g].

However, forces between single grains cannot be quantified as well as effects of chemical and physical interaction between particles, aggregates or loose assemblages in natural soils depending on the water content due to a parameterization of the osmotic potential. The defined χ-factor of this equation has been related to water filled pores and used as parameter for a fundamental understanding of the rigidity of a soil with respect to menisci forces. Terzaghi (1936), Bishop (1960), and Skempton (1960) established and modified this equation, which has been mentioned in numerous fundamental works of Kézdi (1974), Drescher et al. (1988), Fredlund and Rahardjo (1993), and Mitchell and Soga (2005).

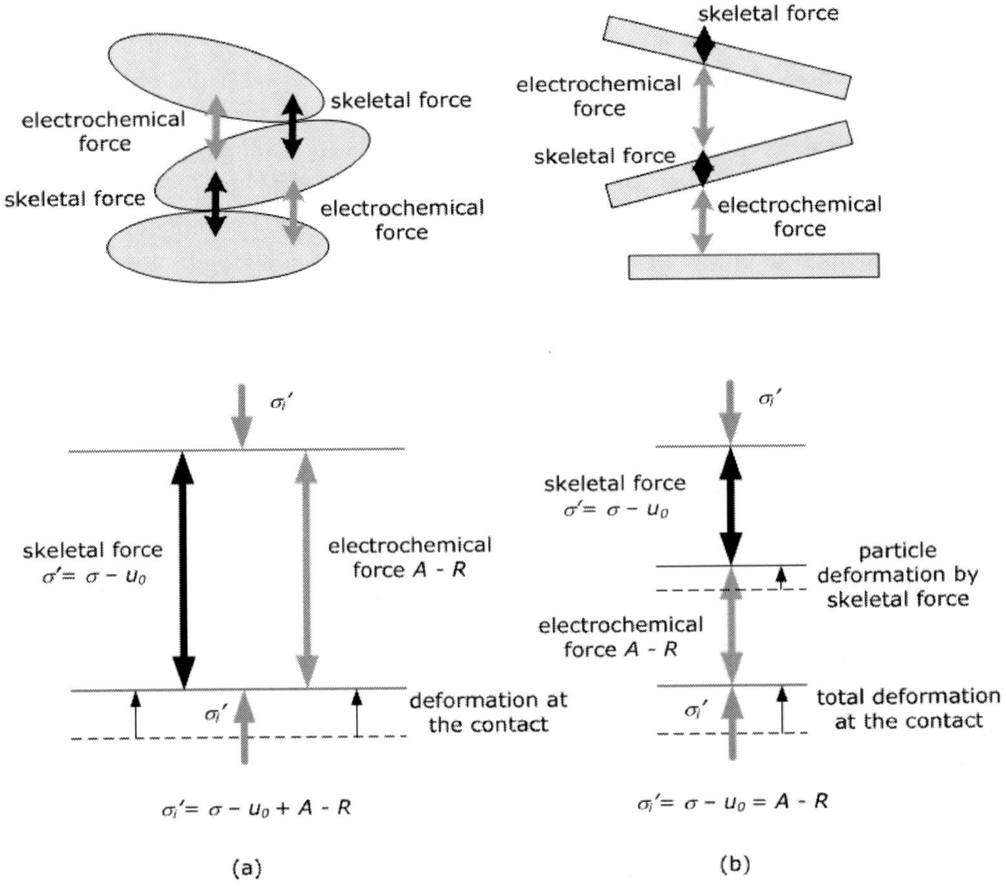

Adapted from Mitchell and Soga, 2005.

Figure 4. Contribution of skeletal force ($\sigma - u_o$) and electrochemical forces (A - R) to intergranular force σ_i: (a) parallel model and (b) series model.

Furthermore, Diamond (1970, 1971), Emerson et al. (1978), Emerson (1983), Fiés and Bruand (1998), Grant et al. (2002), Hartge and Horn (1999, 2002) concluded in their investigations of bonding mechanisms between particles and aggregates that there are different contact points between liquid and solid phase in a soil-tool interface system. Jia (2004) and Shi-qiao et al. (2005) give a mathematical approximation to this research topic. In correlation to changes in water tension and water content due to chemical or mechanical influence, shape and size of particles etc. different shear behavior occurs (Warkentin and Yong, 1962; Yong and Warkentin, 1966; Kézdi, 1974; Berilli et al., 2002; Smith and Reitsma, 2002).

As intergranular stress is influenced by electrochemical forces (A - R) cation constituents and their effects on the χ-factor have to be taken into account (*equation 3*). Bresler et al. (1982) quantify structural stability indirectly due to salt depending changes in hydraulic conductivity, but they do not mention any interaction of particle characteristics and ion concentration. Warkentin and Schofield (1962), Yong and Warkentin (1966), Mitchell and Soga (2005) give a detailed definition of potential interactions with the solid phase, although it is not possible to make any conclusions with regard to soil strength.

Based on rheological investigations of suspensions (water contents >90%) (Jasmund and Lagaly, 1993; Schulz, 1998), thixotropy effects could be proved, which are correlated to clay mineralogy characteristics and (cat)ion properties. In the area of soil micromechanics Ghezzehei and Or (2001), Tuller and Or (2003) showed that osmotic effects are of interest for soil stabilization in clays, whereas they did not consider other indicators but swelling behavior, without taking interparticle forces or shear strength into account. Peng et al. (2005) demonstrated structural changes by the influence of salts as they occur in irrigation areas, especially with respect to shrinkage and swelling behavior on the meso scale.

Homogenized suspensions, pastes, as well as natural soil material show on any scale very different hydraulic properties, if compared with undisturbed samples. In this case, additional variations of chemical and physical kind may occur in dependency on the scale (single grain, microaggregate, and undisturbed cylinder sample), changes in texture, bulk density, degree of aggregation, pore tortuosity, water content, and salt effects etc., which altogether affect hydraulic parameters. These differences need to be obtained rheologically and measured with appropriate techniques.

1.3. Deformation Characteristics

Kézdi (1974), and Mitchell and Soga (2005) give a general introduction to deformation characteristics. They state that strains consist of elastic and plastic parts. Transferred to rheological understanding, combinations of elastic and plastic or plastic and viscous compounds are possible. Soil can be defined as visco-elastic material (Markgraf et al., 2006). In soil mechanics, it is assumed that plastic strains develop only when the stress state satisfies some failure criterion.

According to this fundamental definition, the yield point differentiates the state of soil between elastic and plastic (viscous). From a soil mechanical point of view there is no distinct transition from elastic to plastic behavior, which is related to cyclic loading. In solid mechanics, the small strain shear modulus G [Pa] (or Young's modulus E) is defined as

$$G = \tau_c/\gamma_c \qquad (4)$$

τ_c is the applied shear stress [Pa] and γ_c is the corresponding shear strain [%].

Under small stresses, strains of solid materials are more or less proportional to the applied stress. The constant of proportionality (elasticity), is given by the inverse of Young's modulus of elasticity, which is a parameter of stiffness. Thus, elasticity is typically modeled by using the linear relationship between stress and strain. The classic theoretical example of linear elasticity is the perfect spring, whose behavior is described by Hooke's law. However, linear elasticity is an approximation, whereas natural, real materials exhibit some degree of non-linear behavior.

Viscoelastic material models are frequently used to describe the behavior of plastics, polymers or, as presented in this work, of soil. Commonly applied viscoelastic models are the Kelvin-Vogt and the Maxwell model. Each model can be represented by springs and dashpots sets in combinations of series and parallel elements (Kézdi, 1974; Mezger, 2006).

Fundamental knowledge of soil stiffness in the linear elastic region is important for evaluating soil response under dynamic loadings such as mechanical or vehicle vibrations (Garciano et al., 2001). It also provides indirect information regarding the state and structure of natural soil. Therefore, stiffness values can be used to assess the quality of soil samples. The linear elastic stiffness of soils is evaluated from measurements of elastic wave velocities or use of displacement transducers. Theoretical analyses of elastic waves are described in detail in Santamarina (2001). For a characterization of deformation it is useful to differentiate between four zones and three states (*Figure 6*) (Jardine, 1992):

Zone 1: true elastic region
Zone 2: nonlinear elastic region
Zone 3: preyield plastic region
Zone 4: full plastic region

In amplitude sweep tests (oscillatory conditions), a transgression from elastic to plastic (viscous) state is evident. Here, three phases of deformation or, according to Jardine (1992), Jardine et al. (2004), of stiffness degradation can be pointed out:

Phase 1: an initial state of full elasticity, including a linear viscoelastic deformation range;
Phase 2: a transition state, in which stiffness degradation and plastic (viscous) strain development occur;
Phase 3: a final state of yielding or creeping.

Above a certain stress known as the elastic limit or the yield strength of an elastic material, the relationship between stress and strain is breaking down. Beyond this limit, the solid may deform irreversibly, exhibiting plasticity. This phenomenon is often observed using stress-strain curves.

Furthermore, not only solids exhibit elasticity. Some non-Newtonian fluids, such as viscoelastic fluids, will also exhibit elasticity in certain conditions. In response to a small, rapidly applied, and removed strain, these fluids may deform and then return to their original shape. Under larger strains, or strains applied for longer periods of time, these fluids may start to flow, exhibiting viscosity.

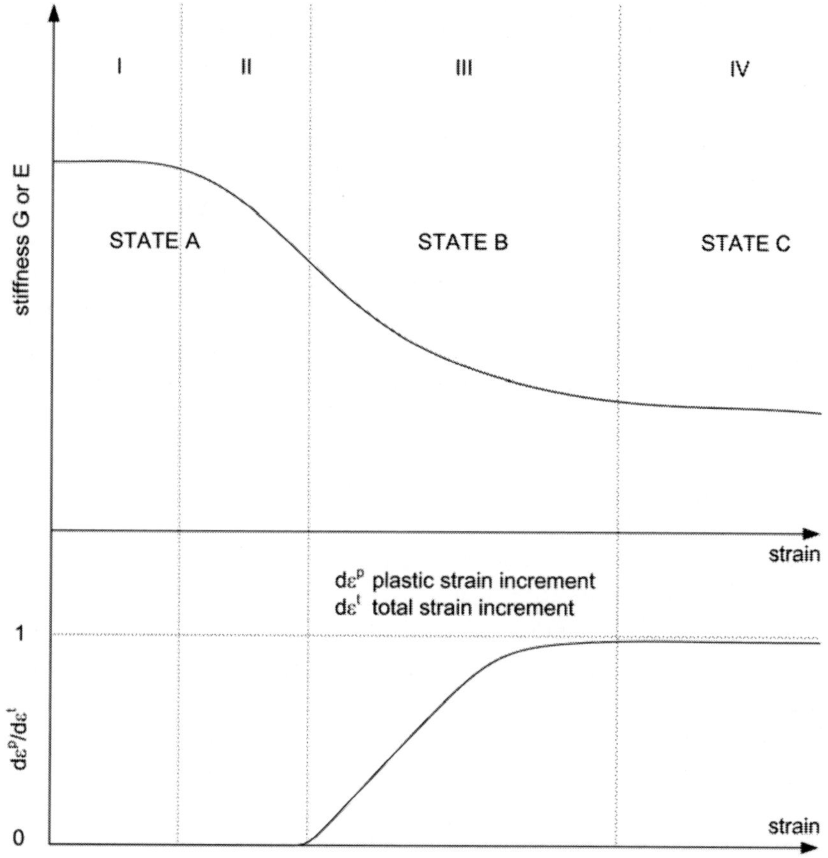

After Jardine, 1992.

Figure 5. Four zones of deformation characterization: stiffness degradation and plastic strain development; an increasing stiffness degradation is associated by an increase of plastic strain. The part of plastic strain ($d\varepsilon^P$) is 0 in state A, the region of true and non-linear elasticity, is approximating in state B, the region of preyielding, and equals total strain ($d\varepsilon^t$) in state C, the full plastic region.

1.4. Soil Mechanics at the Microscale and Rheometry

With respect to mechanical shear behavior, the particle form needs to be taken into account. Interactions of single particles – single grains, tactiles, clay platelets or micro aggregates (<250μm according to Tisdall and Oades, 1982; Oades and Waters, 1991; Sumner and Naidu, 1998) – may occur with great variability, depending on physical properties and on scale considerations. Well known mechanisms like slaking or simple disperging effects have been already described in several works of Emerson (1954, 1983, 1994). Furthermore, in fundamental works of soil mechanics a general introduction to interparticular effects is given (Kèzdi, 1974; Fredlund and Rahardjo, 1993; Mitchell and Soga, 2005) with special regard to capillary forces (menisci forces).

Li (2003) defined a stress-like quantity called the quasi effective stress tensor for unsaturated soils. Suction-induced shear is affected by the pore fluid fabric. He concluded: "as the degree of saturation changes and/or soils deform, this fabric anisotropy could change significantly, […] (and) for any phenomenological relations proposed for unsaturated soils, the impact of this fabric tensor and its evolution must be carefully considered". Tuller et al. (1999), Tuller and Or (2001; 2002, 2005) describe the influence of particle distribution, resulting flow paths (pore characteristics) and pore water film continuity, which are relevant for mechanical stability. According to Cho et al. (2006) the size and shape of soil particles reflect the formation of a single grain. Chemical processes determine the size and shape of clay and silt particles, whereas mechanical processes are predominantly responsible for surface properties in case of sand and coarser particles. Increasing particle irregularities lead to a decrease in structural stability. An additional loss of elasticity is obtained depending on the state of stress (pressure). This instance can be also transferred to oscillatory shear. Based on earlier works of Wadell (1932), Krumbein (1941), Powers (1953), Krumbein and Sloss (1963), Barrett (1980), and Cho et al. (2006) there are three scales in particle shape:

(1) sphericity that is conferred to eccentricity or platiness,
(2) roundness conferred to angularity, and
(3) smoothness, which is conferred to roughness.

Grant et al. (1990) pointed out that the roughness of soil fracture surfaces is an important measure of soil microstructure; thus, micromechanical shear behavior is affected by this parameter. Several particle-level mechanisms that are associated with increasing irregularities, and include a decrease in sphericity and/or roundness, are responsible for a certain macroscale response: hindered rotation, slippage and ability for particle rearrangement, lower interparticle coordination, increased particle level dilatation, lower contact stiffness, and higher proneness to contact deformation (Cho et al., 2006).

1.5. Research Objectives

Open questions remained with respect to interparticular processes on the microscale. The following research objectives derive from the general introduction of rheological considerations to soil micromechanics:

In general, the application of amplitude sweep tests with a rotational rheometer MCR 300 for (oscillatory) shear stress investigations in soil mechanics on a particle-particle scale (contact point scale) has to be established and proved.

In this context, relevant parameters in rheology will be introduced in detail that are transferable to soil micromechanical characteristics: shear modulus G, storage and loss moduli G' and G", the linear viscoelastic (LVE) deformation range, and the deformation limit γ. Hooke's law of ideal elastic bodies, Newton's law of ideal fluids (viscosity), viscoelasticity, shear behavior and micromechanical mechanisms will be defined and brought into context of fundamental soil mechanical considerations.

An approach to rheometry was made by Markgraf et al. (2006), who presented results of conducted amplitude sweep tests with sodium bentonite, clayey and silty soils.

It is unclear, how far the osmotic potential is responsible for the accumulation of fine particles in the contact point region and for the stabilization of assemblages or particle packages in the three-phase system soil. It has had to be proved that rheometry is adaptable for the investigation of microstructural changes, which are affected by salts and/or carbonates.

Hence, rheological strength analyses of under laboratory conditions treated samples with NaCl and $CaCl_2$-solutions, as well as Al^{3+}-rich and $CaCO_3$ dominated soils may show a certain trend of stiffness development (Markgraf and Horn, 2006). This trend may be confirmed by investigated microstructural changes in a Calcaric Gleysol and a Dystric Planosol.

An additional aspect lies in the influence of Fe-(hyrd)oxides and soil organic matter; results from conducted amplitude sweep tests with South-Brazilian soils lead to interesting findings that are presented in Markgraf and Horn (2007).

Textural effects in correlation to particle properties and water content may be identified by resulting single parameters (e.g. the deformation limit γ_L) of amplitude sweep tests, curve characteristics of the storage modulus G' and the loss modulus G", and loss factor tan δ (=G''/G').

Particle associations can be visualized by SEM micrographs; in addition, energy dispersive scan (EDS) analyses support laboratory findings (x-ray diffraction) with regard to microstructural compounds. As shear behavior is dependent on particle size, shape, and surface roughness, results, which derive from amplitude sweep tests may be explained, and supported by such visual findings. Hence, in Markgraf and Horn (2007) an interaction between SEM/EDS-analyses and rheological investigations of South-Brazilian soils will be presented to prove that rheometry (amplitude sweep tests) and scanning electron microscopy (SEM) are a reasonable amendment.

2. Method and Material

2.1. Rheological Techniques in Soil Mechanics

An approach of rheometry in soil mechanics has already been introduced by Markgraf et al. (2006), based on works of Keedwell (1984), Whorlow (1992), Macosko (1994), Collyer and Clegg (1998), Schulz (1998), Ghezzehei and Or (2001), Schramm (2002) and Mezger (2006). Natural substances i.e. soils consist of elastic and viscous compounds. The proportion of those depend on physical and chemical properties, i.e. texture, porosity, water content, and ionic composition of the liquid phase. Elasticity and viscosity of substrates and liquids can be described mathematically by the fundamental laws of Hooke and Newton.

Generally, deformation and flow behavior depend on the type, degree and duration of applied stress, either oscillating or steady stress. The shear stress τ is defined as

$$\tau = F/A \quad \tau = \text{shear strength [Pa]} \tag{5}$$

F = force [N]
A = area [m²]

Hooke's law: Elastic Flow Behavior and Shear Modulus G

Hooke's law defines an ideal elastic substance, which can be expressed in the mechanical analogue of a spring (equations [7] and [8]).

The energy, which was invested in deforming the body is completely stored; as soon as the stress, that caused the deformation, has been removed, the original shape is restored and the energy is recovered. The linear relation between the extent of elastic strain and the applied stress is:

Derived from the Young's Modulus E [Pa]

$$E = \sigma/\varepsilon \quad \text{where } \sigma = \text{tensile stress [Pa]} \tag{6}$$

ε = tensile strain [ ] or [%]

the shear modulus is

$$G = \tau/\gamma \quad \text{where } G = \text{shear modulus [Pa]} \tag{7}$$

τ = shear stress [Pa]
γ = shear deformation [ ] or [%]

with the shear deformation γ

$$\gamma = s/h \quad s = \text{deflection [mm]} \tag{8}$$

$s/h = \tan\varphi$ h = distance between the plates [mm]
φ = deflection angle [°]
if s = h or $\varphi = 45°$, then $\gamma = 1$ which is equivalent to 100 %.

The shear deformation γ [ ] or [%] depends on the plate distance h [mm] and the diameter d [mm] of the rotating measuring device (Mezger, 2006).

A representative result of an amplitude sweep test is shown in *Figure 6*. The plots of storage and loss modulus (G' and G") are generated automatically during a test. Three phases of elasticity loss can be identified:

Phase 1: initial or plateau phase, G'>G"; an elastic behavior is observed, represented by a spring for ideal elastic substances according to Hooke's law. A linear viscoelastic (LVE) range and the included deformation limit γ_L are parameters needed to quantify 'stored elasticity' of any viscoelastic substrate, e.g., soils.

Phase 2: stage of transgression or the intersection of G' and G".

Phase 3: final stage of structural collapse, G'<G"; a viscous character predominates; substances are creeping or running; this behavior can be represented by a dashpot (a device for damping shock or vibration), an analogue for ideal fluids according to Newton's law.

More parameters need to be identified for curve progressions:

(1) the duration of each phase of elasticity dissipation,

(2) characteristics of the plateau phase; distances between graphs of G' and G"; slope progression in phase 2; intersection of G' and G"; and
(3) whether the state G'<G" is reached or not in phase 3; this is the stage of microstructural collapse, where the final level of G' and G" is considered.

Samples are repacked in 100 cm^3 cylinders at a bulk density d_B of 1.4 g/cm^3 and saturated with distilled water. Half of the prepared samples were drained at -60 hPa (pF 1.8), three replicates of each were prepared.

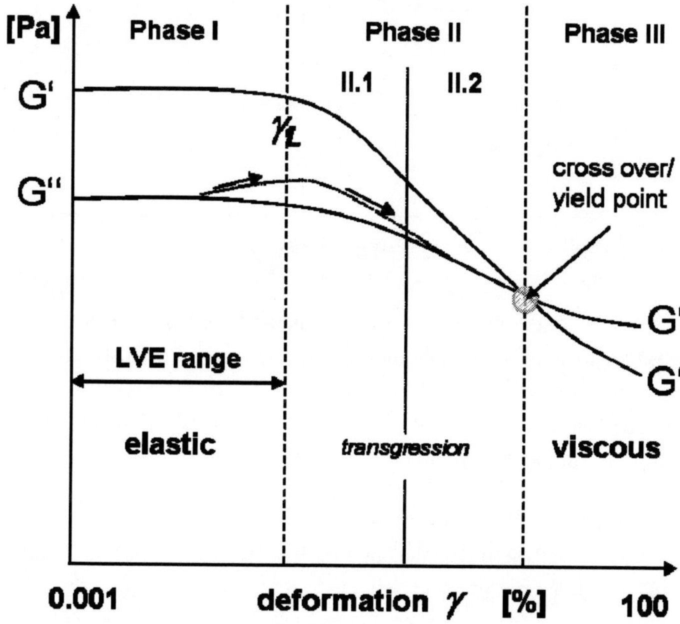

Figure 6. Idealized generated plots of storage modulus G' [Pa] and loss modulus G" [Pa]. In general, three stages of elasticity loss can be defined, showing a gradual transition of an elastic (G'>G") to a viscous (G'<G") character. In phase 1 an elastic behavior predominates, and a parallel run of G' and G'' is given. At this stage of oscillation with controlled shear deformation a complete or partially recovery of soil structure can be assumed. By reaching deformation limit γ_L [%] a yielding character is already given in phase 2; an intersection of G' and G" marks an absolute yield point. In phase 3 a microstructural collapse occurs and a viscous character is obtained.

Table 2. General pre-settings of an amplitude sweep under oscillatory conditions, with controlled shear deformation (CSD)

Parameter	
plate distance	d = 4 mm (>2 μm)
plate radius	R = 25 mm (profiled)
shear deformation	γ = 0.0001... 100% (>2 μm)
angular frequency (frequency)	ω = π 1/s (f = 0.5 Hz)
measuring points	30 pts.
duration	appr. 15 min.

Adapted from Markgraf et al., 2006.

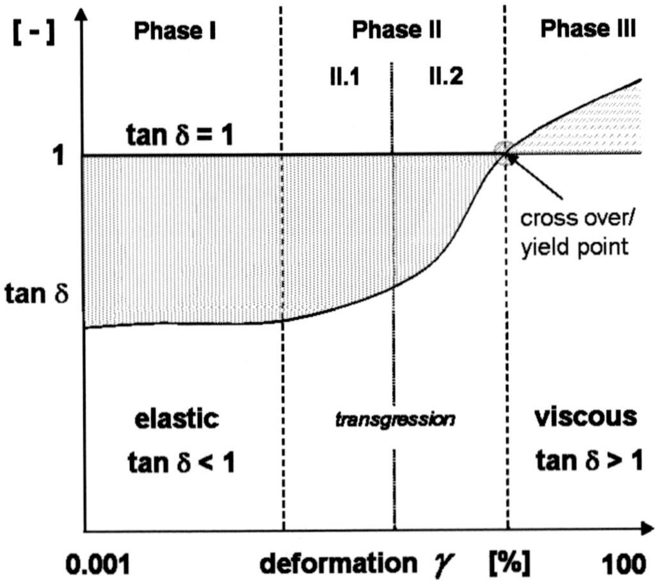

Figure 7. Deriving from one data set, results can be also plotted in a loss factor vs. deformation coordinate system. Loss factor tan δ [ ] equals the ratio of loss modulus to storage modulus (=G''/G'), and may function as analogue expression of elastic (tan δ<1), viscoelastic (tan δ≤1) or viscous (tan δ>1) behavior. For further comparison, the integral of tan δ (γ) lim γ=0.00%1 to lim γ="cross over point", with tan δ=1 as defined limit on the y-axis can be calculated.

An amount of approximately 4 cm^3 of the prepared samples was taken out of the cylinders by using a small spatula and was placed carefully between the plates.

Table 2 informs about parameters, which were chosen for the amplitude sweep tests (AST).

The plate distance varies with the texture: basically, for substrates with a grain size ≤ 2 μm a gap of 1 to 2 mm should be preset, for substrates of coarser fraction (silt, loam, fine sand) a distance of 4 mm is recommended. The plate geometry of the rheometer with a parallel plate measuring system (PP MS) is determined by the plate radius R, according to DIN 53018. The rotating bob is 25 mm in diameter and has a profiled surface. During all tests a constant temperature of 20°C is given, regulated by a Peltier unit. The generated normal force averages between +0 to +12 N and was not exceeded during tests. The operational maximum limit is +/-50. A resting period of 30 seconds in the beginning was inserted in the test software (US200) to ensure a reorientation of the particles and an undisturbed measurement. Both cases might be induced by the application of the material on the measuring plate. The deformation γ is preset as logarithmic range from 0.0001 to 100%, the frequency at a constant value of 0.5 Hz, or π 1/s in SI unit. One run, including 30 measuring points, lasts approximately 15 minutes. The correlation between frequency and duration is inversely proportional, e.g. 1 Hz leads to test duration of 7 minutes.

For calculations of the linear viscoelastic (LVE) deformation range (*Figure 6*), deformation-limit γ_L and yield-stress τ_y (that is, the "yield stress II" in US200) analyses were executed after each completed test run (Ghezzehei and Or, 2001; Mezger, 2006; Markgraf et al. 2006; Markgraf and Horn 2006, 2007). To do this under oscillatory conditions a tangent is fit to the G' curve which is based on the minimum γ value and is limited by a decline of G' that has a deviation of >5% in relation to this calculated tangent.

Natural viscoelastic substances react with a temporal delay. This is represented by the phase shift angle δ, whereas tan δ equals the relation of the loss modulus G" [Pa] to the storage modulus G' [Pa], defining the relation of imaginary ("lost") to stored elasticity. If tan δ<1, G' prevails G", a gel character is given. Viscous behavior is defined in case of tan δ>1 and G" predominates G' (*Figure 7*). Furthermore, a correlation between in *Figure 6* and *7* presented stages (Phases I-III) of stiffness degradation become obvious. Due to a decrease of G', the ration of G''/G' (= tan δ) increases; if tan δ=1 is reached, elastic and viscous parts are equivalent, and an absolute yield point (=cross-over) is given at a defined deformation [%]. If tan δ>1, a viscous character predominates, and a structural collapse occurs; at this stage deformation is irreversible. For further comparison, the integral of tan δ (γ) lim γ=0.00%1 to lim γ="cross over point", with tan δ=1 as defined limit on the y-axis can be calculated. This method may allow an even more precise definition of elasticity, rigidity, or stiffness of a soil at the particle-particle scale.

2.2. Scanning Electron Microscopy

Scanning electron microscopy (SEM) was done with a CamScan CS 44 (E.O. Electron-Optik Service GmbH, Dortmund, Germany), which also can be used to for energy dispersive scan investigations (EDS) (Markgraf and Horn, 2007). Oven-dried (at 40 °C) samples and aluminum holders are connected with a self-adhesive carbon die. For SEM, required conductivity was achieved by applying a gold-palladium coating under high vacuum conditions (sputtering). SEM micrographs were obtained at 15 KeV at a working distance of 15 mm. Monochrome photographs were taken with a small picture reflex camera, which is integrated into the CamScan CS 44 working station as an external unit. A detailed description of SEM and microanalysis (and their applications) is given by Henning and Störr (1986; see their tables) and Schmidt (1994). Visual images of polarizing microscopy were prepared according to the method of van Reeuwijk (2002).

Table 3. Physicochemical properties of the investigated, mostly fine-grained, clayey or silty substrates

	Sand	Silt	Clay	Na	Mg	Ca	Al	pH	C_t	$CaCO_3$	Fe_2O_3	Fe_d
		[%]			[mmol$_c$/kg]			*$CaCl_2$*		[%]		[‰]
Na-Bentonite Ibeco Seal 80	2	30	72		443*		n.v.[§]	6.5	n.v.	n.v.	n.v.	n.v.
Avdat Loess	17	64	19	10	35	98	n.v.	7.6	1.8	52.4	n.v.	n.v.
Dystric Planosol	20	15	65	7.0	28.3	115.0	58.3	4.5	0.5	n.v.	n.v.	n.v.
Calcaric Gleysol	7	58	32	0.9	12.0	181.0	n.v.	7.6	1.4	25.2	n.v.	n.v.
Smectitic Vertisol (Livramento)	3	32	65	2.7	157	396	n.v.	5.5	3.4	n.v.	n.v.	n.v.
Sandy Ferralsol (Cruz Alta)	75	7	18	0.4	2	5	n.v.	4.1	0.6	0.04	2.0	14
Clayey Ferralsol (Cruz Alta)	46	9	45	0.3	17	28	n.v.	5.5	1.0	0.02	2.6	59
Clayey Ferralsol F[†] (S. Ângelo)	6	25	69	1.5	17	41	n.v.	4.4	6.6	0.03	3.9	99
Clayey Ferralsol NT[‡] (S. Ângelo)	5	19	75	1.1	21	43	n.v.	4.9	1.1	0.03	4.1	102

*CEC_{eff} in total [§] n.v. no value [†] F natural forest [‡] NT no tillage.

2.3. Physicochemical Analyses

Analyses were conducted according to standard methods as described in Schlichting et al. (1995), and van Reeuwijk (2002). Sieved (<2 mm) and homogenized samples were taken to measure exchangeable cations, which were extracted by 1 M ammonium acetate. Concentrations of Ca^{2+} and Mg^{2+} were measured by an atomic absorption spectrometer, whereas Na^{+} was measured by flame emission. Iron oxides were extracted by Na-dithionite according to Mehra and Jackson (1960). Soil organic matter was removed with H_2O_2.

2.4. Investigated Material

In Table 3 properties of the investigated substrates are summarized. Sodium Bentonite "Ibeco Seal 80", an industrial material with a high swelling potential due to a high montmorillonite content, Avdat Loess (Negev, Israel), as very silt-rich and carbonatic substrate, a Dystric aluminum-rich Planosol, and a Calcaric Gleysol, both regional soil types from Schleswig-Holstein, and five soil substrates from South Brazil, as smectitic Vertisol (WRB; Typic Calciudert, USDA, 2003) and four kaolinitic Ferralsols (WRB; Typic Hapludox, USDA, 2003).

Na-Bentonite (Ibeco Seal-80)

Ibeco Seal-80, a Na-bentonite, can be assigned to the group of activated bentonites. These are bentonites with smectites whose initial composition of alkaline-earth-cations has been replaced with Na^{+}-ions in a technical process named alkali-activation. In natural Na-bentonites (e.g. Wyoming, Greek or Southern German Bentonites) smectites are predominantly occupied with Na^{2+}-ions in the intermediate layers. In Na-bentonites, Ca^{2+} or Mg^{2+}- ions also occur frequently and in varying concentrations. Commonly, IS-80 is applied in construction and civil engineering e.g. as landfill sealing, for Caisson constructions as well as for Geosynthetic Clay Liners (GCL, bentonite mats) and for use in deep-mining (Markgraf et al., 2006).

Avdat Loess (Negev, Israel)

Avdat Loess, which origins from the Negev Desert in Israel, has a silt content of 64% and a natural high calcium carbonate content of 52%. X-ray diffractometry shows a high fraction of strongly weathered and swellable smectites and vermiculites. Montmorillonites and magnesium rich forms of smectites besides kaolinite and illite exist in low amounts only (Markgraf et al., 2006).

Exchangeable cations were extracted by 1M ammonium acetate. According to standardized methods (Schlichting et al., 1995), Na^{+} was measured by flame emission, concentrations of Mg^{2+} and Ca^{2+} were detected by atomic absorption spectrophotometer.

Dystric Planosol and Calcaric Gleysol (Schleswig-Holstein, Germany)

Homogenized (ground and sieved < 2 mm), soil substrates have been prepared according to Markgraf and Horn (2006) deriving from a (Calcic) Calcaric Gleysol (Ritzerau, E Schleswig-Holstein, "Holsteinton"), taken from a G(c)o in 40 cm depths (Richter, 2005), and a Dystric Planosol, from a IISd in 60 cm depths, which originates from Wacken (SW Schleswig-Holstein, Germany).

Table 1 summarize physical and chemical properties of the tested material. The Calcaric Gleysol (Soil 03) is characterized by a silt rich texture (58%) accompanied by a high clay content of 32% (moderate silty clay), as well as a basic pH value of 7.5 in average due to $CaCO_3$ and low content of organic matter. High pH values are typical for Calcaric (calcaric, calcic or gypsic) Gleysols. In opposite to this, a low pH value of about 4.5 (strongly acid) in average, influenced by Al^{3+} (58 $mmol_c$/kg) with exception of the $CaCl_2$ 1.0M treated sample, and a clay rich texture (65%, weekly sandy clay) are given in case of the Dystric Planosol (Soil 04). The electrical conductivity (EC) was measured in 1:5 w/w extract solutions. Exchangeable cations were extracted by 1M ammonium acetate. According to standardized methods, K^+ and Na^+ were measured by flame emission; concentrations of Mg^{2+} and Ca^{2+} were detected by atomic absorption spectrophotometer. For scanning electron microscopy (SEM) samples (all treatments) were oven-dried, prepared on a specimen stage and Au-metalized.

Ferralsols (Typic Hapludox) and Vertisol (Typic Calciudert) – South Brazil

Samples were taken from three locations: ≥40 cm depth under no tillage (NT) and campo-natural conditions (CN meadow, F natural forest) in Santo Ângelo (two clayey Ferralsols, kaolinitic); in Cruz Alta (sandy and clayey Ferralsol, kaolinitic); and in Santana do Livramento under pasture (Typic Calciudert, smectitic) (Markgraf and Horn, 2007). Homogenized air-dried samples were sieved to <2 mm and were repacked in 45 cm^3 cylinders (n=3; d_B=1.4 g/cm^3). Thereafter they were completely saturated with distilled water; parallel samples (n=3) were prepared and drained at -60 hPa. Altogether, 78 samples were prepared: 30 of untreated, natural soil material from Santana do Livramento, Cruz Alta, and Santo Ângelo, 24 of H_2O_2- treated, and 24 of Fe_d-leached samples (Cruz Alta and Santo Ângelo).

2.4. Calculations and Statistics

Approximately two hundred and fifty samples were tested, with three repetitions (n=3) each amplitude sweep. Seven to eight samples with different treatments - either saturated untreated, NaCl or $CaCl_2$-treated, Fe_d- and SOM-leached, or pre-drained at -60 hPa - were measured per day, which resulted in twenty-one to twenty-four measurements per day, with 14 to 18 minutes for each test. From data of the deformation limit γ_L, including an analogue yielding point τ_y, and a absolute yielding point ("cross over"), arithmetic mean values were calculated based on a pre-set range (in US200) of tolerance with ±5% deviation included in "LVE range/yield stress II" analyses. Hence, a high level of accuracy can be assumed. This is also shown by the fact that a minimum deformation of 0.0001% equals a deflection of 1μm under oscillatory conditions.

A clearer illustration and comparison of elastic compounds and an absolute yielding point is possible by plotting test results using loss factor tan δ vs. deformation γ, and calculating the integral γ=0.001…intersection with tan δ=1; an idealized graph is presented in *Figure 7*, whereas a representative result of a conducted amplitude sweep test with is plotted in *Figure 8*.

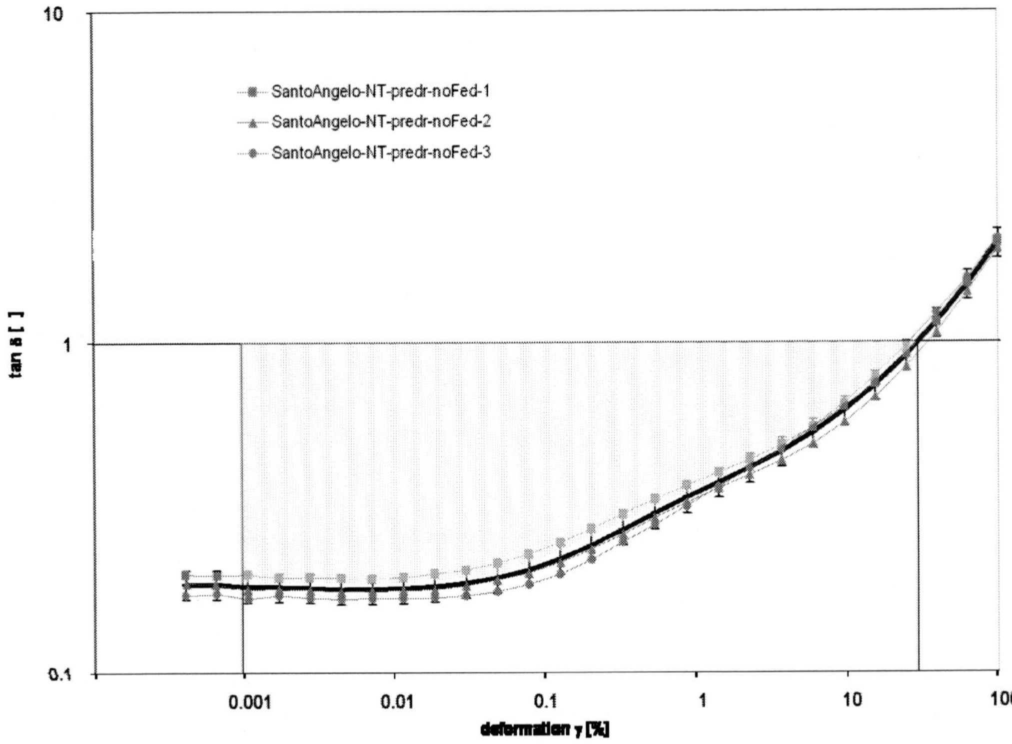

Figure 8. Representative result of a conducted amplitude sweep test (CSD) is shown. Loss factor tan δ [ ] vs. deformation γ [%] is plotted. Pre-drained soil samples (Ferralsol according to WRB, Typic Hapludox, according to U.S. Soil Taxonomy) from Santo Ângelo, South-Brazil were tested (n=3). Black box and whisker plot outlines arithmetic mean with standard deviation.

3. Results

3.1. Normal Force

Normal force F_N [N] shows a certain development during an amplitude sweep test. Mainly in dependency on water content - saturated or pre-drained conditions - and texture, F_N ranges from a minimum of 1-2N to a maximum of >30N in the beginning, which is analogue to phase 1, the true elastic region of G' and G". With increasing deformation and progressing stiffness degradation, normal force is decreasing; this instant can be transferred to phase 2, a state of transgression. According to phase 3, a phase of structural collapse, in which a viscous character is obtained, normal force tends to 0N. In general, in phase 1 a normal force of 12-15N should not be exceeded in case of sandy and silt-rich substrates. Pre-drained, clay-rich or coarser substrates may lead to F_N values >25 to 30N, which is the exception. Values of F_N can be used as orientation for structural disturbances that may occur due to an unsensitive application of the testing material. If a normal force of >15N is already reached in the beginning i.e. in case of the silty Avdat Loess, a repetition is recommended, as mechanical disturbance was generated in advance; deriving parameters cannot be taken as correct. In case of clay-rich, and very rigid or pre-drained (-150hPa) material F_N should not exceed 30N.

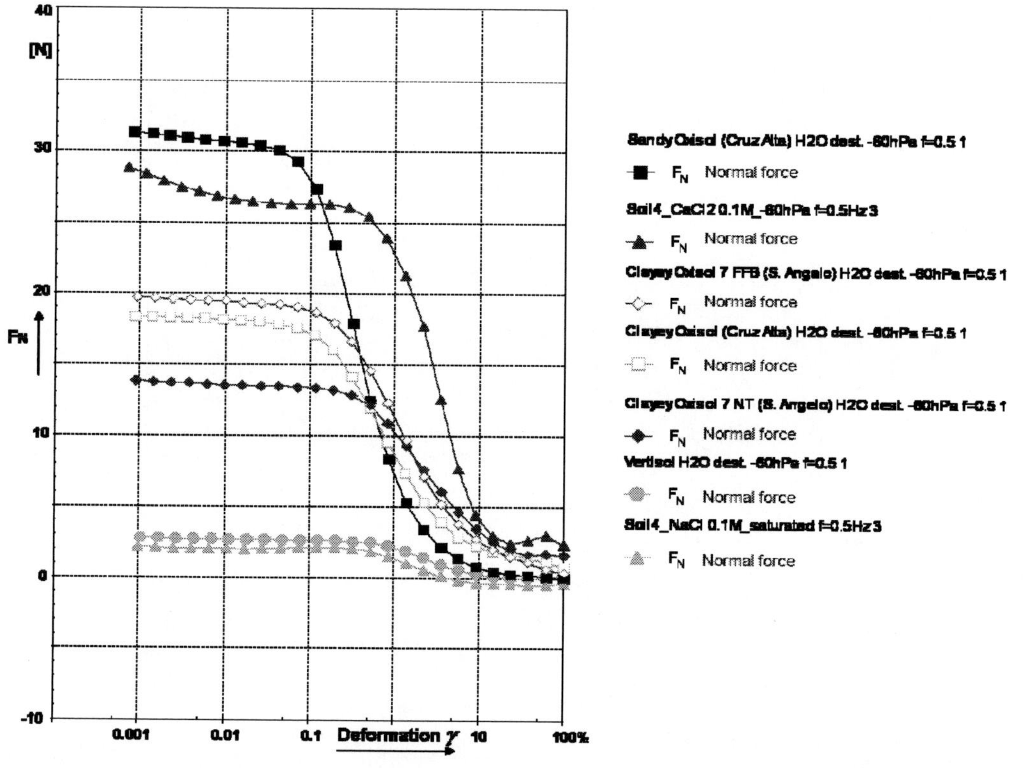

Figure 9. Normal force F_N [N] is recorded during each amplitude sweep test with controlled shear deformation (CSD). Generated work is force-deflection-controlled, hence, G' and G", as well as tan δ derive from deflection angle φ [°] and plate distance d [mm] (equation 8). Level of plotted graphs of F_N (γ) is dependent on water content (saturated, pre-drained), texture, and other diverse physicochemical properties (Fe-oxides, clay minerals, organic matter).

3.1. Na-Bentonite "Ibeco Seal 80" and Avdat Loess

For preliminary tests a very clay-rich and a silty, carbonatic substrate were chosen. First, pastes of sodium-bentonite "Ibeco Seal 80" were prepared and tested. Deriving from these results as presented in Markgraf et al. (2006) a set-up for subsequent amplitude sweep tests had been developed as described in 2.1.

Figures 10 and *11* show results from amplitude sweep tests conducted on sodium-bentonite "Ibeco Seal 80" and Avdat Loess with controlled shear deformation (CSD). Generally G' prevails G", a gel character is given at a resting state in phase 1.

In *Figure 10* graphs from a conducted amplitude sweep test with Ibeco Seal 80 are illustrated. In comparison to silt-rich substrates, Ibeco Seal has a more distinctive plateau within the LVE range and a well defined, clear deformation limit γ_L. When γ_L IS-80 is exceeded, G" is increasing until a deformation of about 12% is reached. The state of elastic behavior passes into viscous behavior, the yield point has been exceeded.

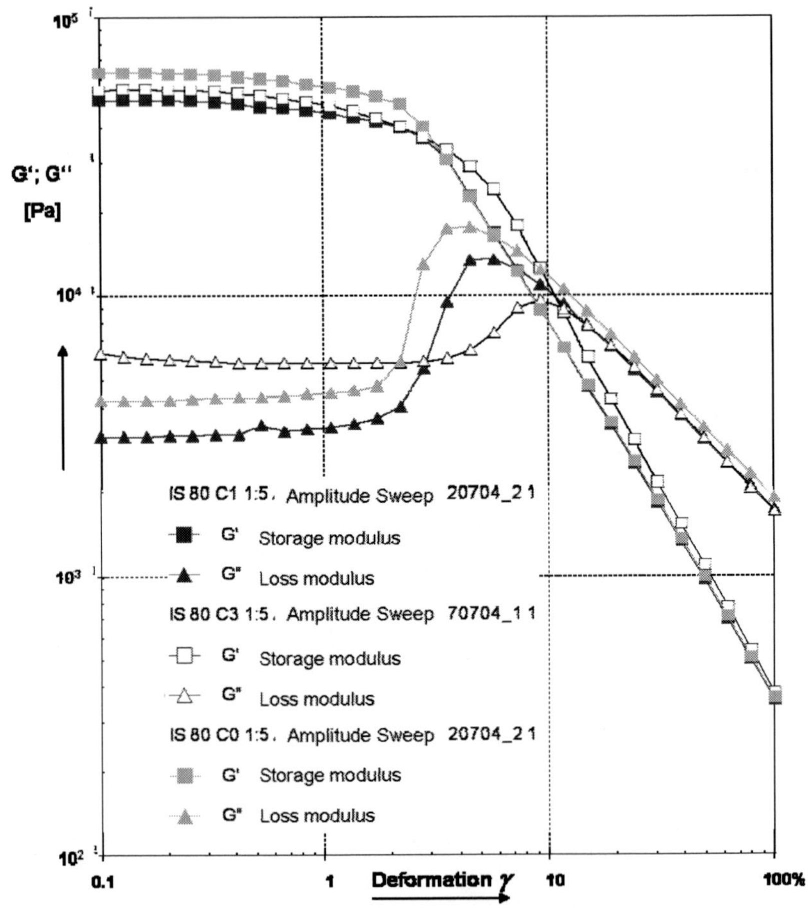

Figure 10. Plotted results of an amplitude sweep test with controlled shear deformation, conducted with untreated, and NaCl-saturated samples (C0: deionized H_2O; C1: 0.01M/L; C3: 0.1M/L) of 'Ibeco Seal-80'. Rectangular symbols represent the storage modulus G' [Pa], triangles represent loss modulus G'' [Pa]. At a deformation of γ= 2…10% an increase of G'' is significant; structural heat, which is set free at this stage, is the reason for this 'knee' formation.

In *Figure 11* $G'_{\text{(distilled water, predrained)}} >> G'_{\text{(NaCl 0.01 M)}} < G'_{\text{(NaCl 0.1 M)}}$; the input deformation has a similar effect to the elastic behavior in silty samples, which are saturated with distilled water or NaCl solutions of low concentrations, here 0.01M. At higher salt concentrations differences in elastic components become apparent, the primary level of G' is $10^{5.5}$ Pa. The response to low deformations is congruent in all cases, the graphs decrease gradually. When γ = 0.01%, which equals a deflection of 0.001 mm, is reached, a transition can be observed. The elastic part decreases more distinctively with increasing deformation, at a constant (angular) frequency.

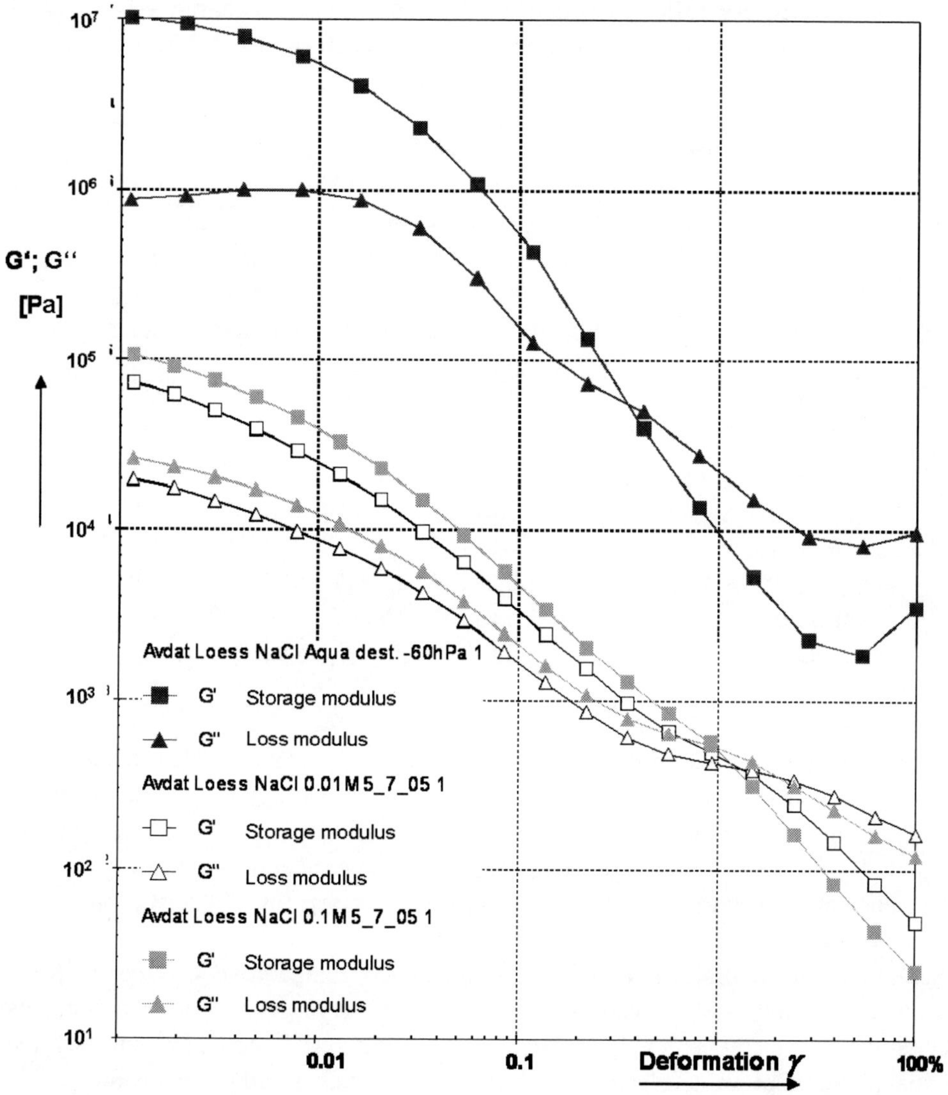

Figure 11. resulting graphs of conducted amplitude sweep tests (with CSD) with Avdat Loess: pre-drained at -60hPa, saturated with NaCl 0.01M, and NaCl 0.1M. Due to formed menisci forces, a strengthening effect occurs; an increase of G' and G'' is recorded, and curve characteristics are changed, e.g. cross-over of G' and G''.

3.2. Dystric Planosol and Calcaric Gleysol: Aggregation vs. Cementation

In *Table 4* summarizes mean values (n=3) of the deformation limit γ_L and the yield stress τ_y of both, the Al^{3+}-dominated Dystric Planosol (Soil 04) and the Calcaric Gleysol (Soil 03), treated with NaCl and $CaCl_2$ salt solutions.

Table 4. Summarized collected data of amplitude sweep tests conducted on Soil 03, Calcaric Gleysol, and Soil 04, Dystric Planosol. For each treatment three passes were absolved (n = 3)(courtesy of Catena)

treatment	γ_L [%]	τ_y [Pa]	γ_L [%]	τ_y [Pa]
Calcaric Gleysol (Soil 03)	saturated		-60 hPa	
DW	0.00912	42.5	0.03487	200.3
NaCl 0.1	0.00833	127.7	0.00845	161.0
NaCl 1.0	0.00808	40.8	0.00836	185.0
CaCl2 0.1	0.00813	83.8	0.00844	619.3
CaCl2 1.0	0.00798	38.8	0.00768	91.9
Dystric Planosol (Soil 04)	saturated		-60 hPa	
DW	0.01457	1568.0	0.01280	2006.7
NaCl 0.1	0.01167	603.7	0.01169	2176.7
NaCl 1.0	0.01183	1106.7	0.01063	859.5
CaCl2 0.1	0.01107	1058.7	0.01200	1455.0
CaCl2 1.0	0.01046	640.7	0.00966	1033.7

The three phases of elasticity loss are well defined (*Figure 12 and 13*): starting with the initial or plateau phase, including the LVE deformation range, which is separated by the deformation limit γ_L from stage two, the phase of transgression, passing into the intersection of G' and G", resulting in stage three, an irreversible structural collapse. These characteristics are less distinctive regarding the Dystric Planosol, i.e. the point of intersection, and phase three.

Saturated and pre-drained conditions are illustrated comparatively. Either the effect of water content as well as of ion concentrations can be observed. At lower water contents, the structural stability increases; the levels of G' and G" is higher in case of drained samples. In the initial phase G' prevails G", an elastic behavior is given, a small plateau is formed clearly. Running through a stage of transgression and an intersection of G' and G", a structural breakdown occurs in the end. Levels of G' and G" of the samples, which were treated with NaCl 0.1M are higher under saturated conditions than those ones of NaCl 1.0M treated substrates. In addition, the deformation limit γ_L as well as the corresponding yield stress τ_y show conformable values. Lower NaCl concentrations have almost no sensitivity regarding changes in water content in opposite to higher concentrations. The deformation limit γ_L and the yield stress τ_y show a congruent curve progression of investigated NaCl 0.1M and 1.0M treated samples when drained at -60hPa. A similar development occurs in case of saturated, $CaCl_2$ treated samples, but become more distinctive under drained conditions. A $CaCl_2$ concentration of 0.1M leads to a noticeable rise of τ_y. Whereas the levels of $G'_{NaCl\ 0.1\ and\ 1.0M}$ and $G''_{NaCl\ 0.1\ and\ 1.0M}$ cover a relatively small range, the difference of $G'_{CaCl2\ 0.1M}$ and $G''_{CaCl2\ 0.1M}$ is one order of magnitude higher. Furthermore, the phase of transgression (phase 2) is characterized by a steeper slope, hence, a shift of the phase durations derives from this instance. A direct comparison of the Calcaric Gleysol and the Dystric Planosol makes clear that $G'_{soil\ 04}$ and $G''_{soil\ 04} >> G'_{soil\ 03}$ and $G''_{soil\ 03}$.

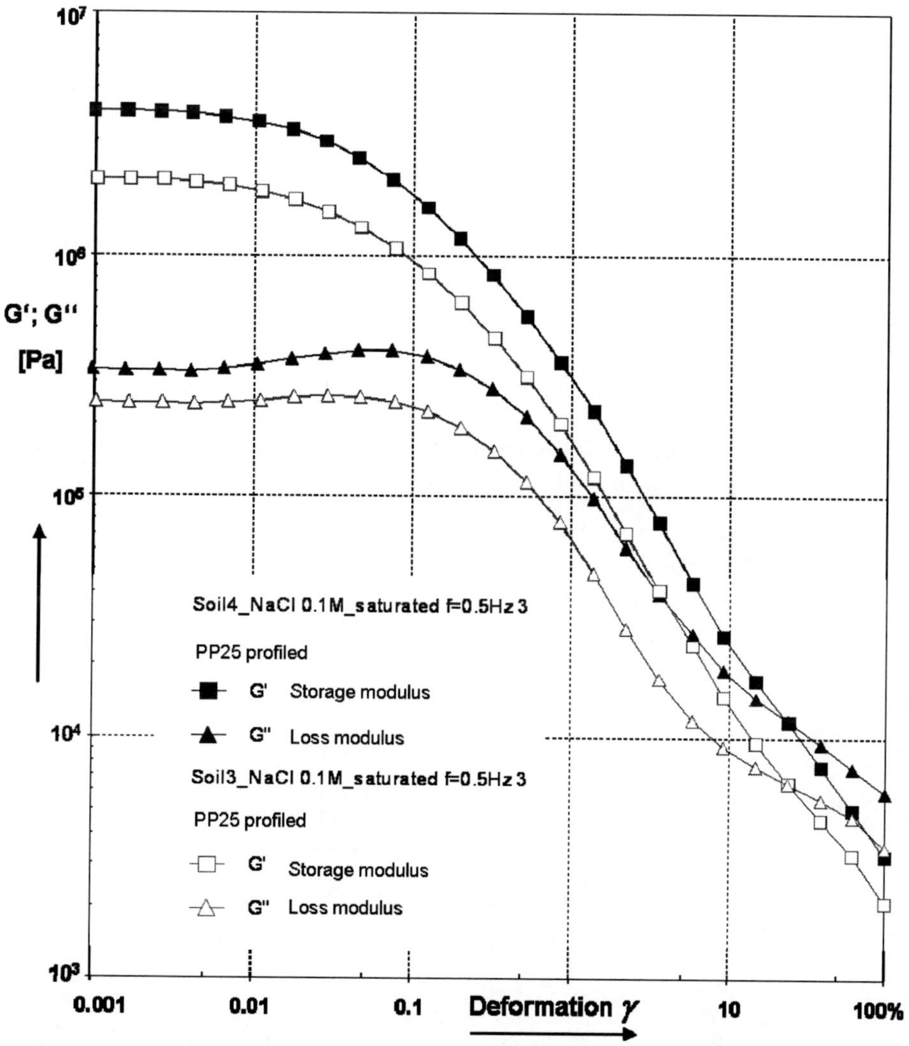

Figure 12. Curves of G' and G'' deriving from amplitude sweep tests conducted with NaCl 0.1M-saturated samples of Calcaric Gleysol (Soil 03) and Dystric Planosol (Soil 04).

SEM micrographs of Soil 04 (*Figure 14*) illustrate exemplarily diverse effects of NaCl and $CaCl_2$ on the surface structure of illite (identification according to Henning and Störr, 1986), in a textural assemblage (Markgraf and Horn, 2006). In general, NaCl 0.1M retains the original structural state of illite and leads to an expansion of the border area at higher concentrations. In opposite to this, $CaCl_2$ has a smoothening influence. Single grains, either sand or silt, are encased by a more or less uniform, even clayey cover. A previously ragged flaky structure cannot be defined in this case.

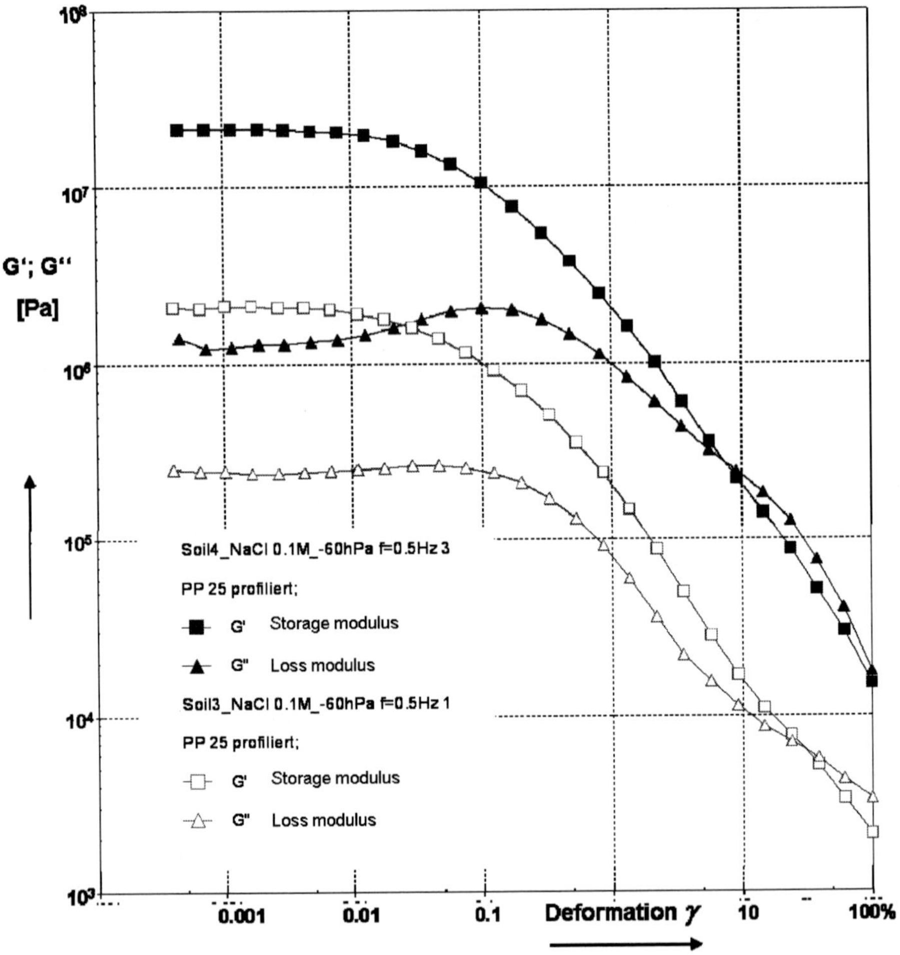

Figure 13. Curves of G' and G'' deriving from amplitude sweep tests conducted with pre-drained, NaCl 0.1M-saturated samples of Calcaric Gleysol (Soil 03) and Dystric Planosol (Soil 04).

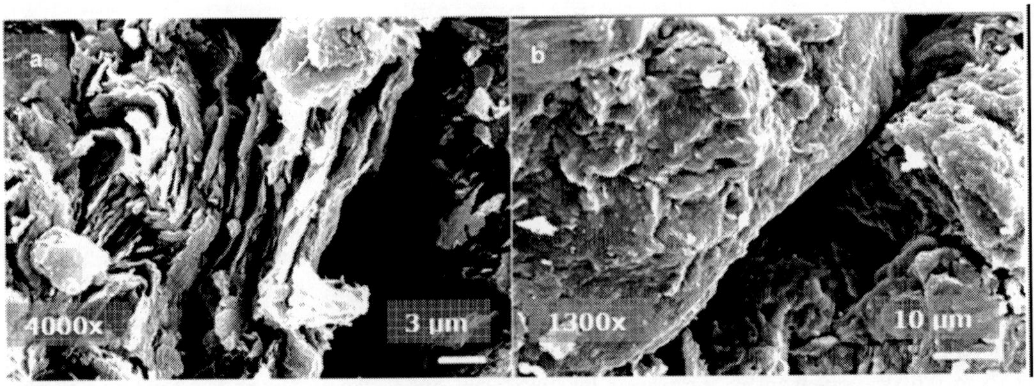

Figure 14. SEM micrographs of Soil 04 (Dystric Planosol), NaCl 0.1M (a) and $CaCl_2$ 0.1M (b) treated and oven dried, showing a typical illitic structure (ragged flakes), which is even intensified in case of NaCl salt solution treatment, due to disperging effects. (Courtesy of Catena).

As a generalized result, Figure 15 outlines stiffness degradation of the Dystric Planosol and Calcaric Gleysol under pre-drained conditions. It can be stated that a clay rich texture and Al^{3+} has a greater stabilizing effect than a silty, carbonatic structure. Hence, aggregation leads to a higher degree of stiffness or structural stability than cementation. This instance can also be proved, if elastic compounds of the Dystric Planosol and the Calcaric Gleysol are compared directly due to calculating the integral (=z in US200) as illustrated in Figure 7: $z_{\text{Dystric Planosol}}=7.352$; $z_{\text{Calcaric Gleysol}}=3.494$. In addition, points of intersection of the graphs with tan $\delta=1$ indicate the same trend: Viscous parts above this line are smaller in case of the Dystric Planosol, if compared to the Calcaric Gleysol; or, in other words: a structural collapse occurs in the Calcaric Gleysol at smaller deformation values than in the Dystric Planosol.

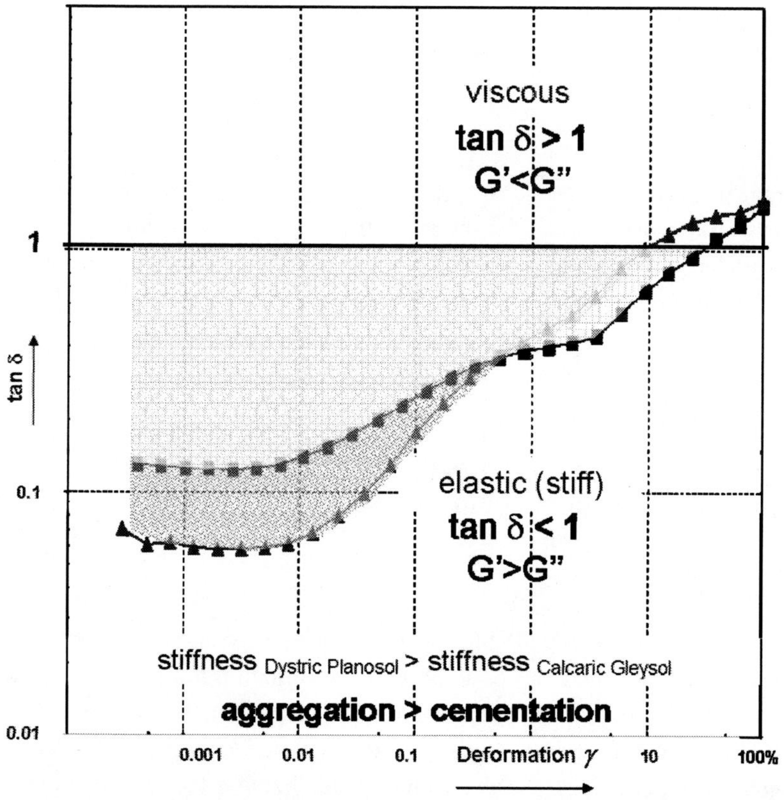

Figure 15. generalized graphs of tan δ (γ), in which the stiffness degradation of the more stable, and elastic, Al^{3+}-rich Dystric Plansol (filled square) and less stable Calcaric Gleysol (filled triangle) is demonstrated. Aggregation of the Dystric Planosol leads to a higher degree of stiffness (microstructural stablitiy) than cementation of the Calcaric Gleysol.

3.3. Ferralsol and Vertisol: Clay Mineralogy and Fe-Oxides

Microstructural effects of SOM and Fe_d, which depended on the water content w/w [%], were obvious (Markgraf and Horn, 2007). Values of γ_L and τ_y showed significant differences (*Table 5*).

Table 5. Summarized results from amplitude sweep tests (CSD). Values of γ_L and τ_y are arithmetic means, n=3. Generally, under pre-drained conditions γ_L and τ_y increase, except untreated clayey Typic Hapludox samples from Santo Ângelo. Furthermore, a decrease of γ_L and τ_y becomes obvious, if untreated, SOM and Fe_d leached samples are compared

Untreated	γ_L [%]	τ_y [Pa]	w/w [%] saturated	γ_L [%]	τ_y [Pa]	w/w [%] -60hPa
Vertisol (Santana do Livramento)	0.02630	255	44	0.02835	532	42
Sandy Ferralsol (Cruz Alta)	0.00682	437	32	0.00661	941	18
Clayey Ferralsol (Cruz Alta)	0.01037	499	37	0.01022	897	19
Clayey Ferralsol F (Santo Ângelo)	0.01677	1105	41	0.00917	*804*	22
Clayey Ferralsol NT (Santo Ângelo)	0.01277	1067	36	0.01277	*919*	18

SOM leached	γ_L [%]	τ_y [Pa]	w/w [%] saturated	γ_L [%]	τ_y [Pa]	w/w [%] -60hPa
Sandy Ferralsol (Cruz Alta)	0.00287	52	31	0.00560	288	25
Clayey Ferralsol (Cruz Alta)	0.00584	101	46	0.00610	293	39
Clayey Ferralsol F (Santo Ângelo)	0.00679	149	43	0.00771	350	34
Clayey Ferralsol NT (Santo Ângelo)	0.00768	215	49	0.00856	502	43

Fe_d leached	γ_L [%]	τ_y [Pa]	w/w [%] saturated	γ_L [%]	τ_y [Pa]	w/w [%] -60hPa
Sandy Ferralsol (Cruz Alta)	0.00295	9.6	25	0.01487	140	16
Clayey Ferralsol (Cruz Alta)	0.01413	49.7	37	0.02443	194	19
Clayey Ferralsol F (Santo Ângelo)	0.01903	40.4	26	0.00585	114	23
Clayey Ferralsol NT (Santo Ângelo)	0.02570	42.7	44	0.03423	208	17

Adapted from Markgraf and Horn, 2007.

Untreated samples, including all natural compounds, were more stable compared to SOM or Fe_d-leached samples. In the latter case, both the deformation limit γ_L and the yield stress τ_y decreased noticeably, which can be summed up as follows: untreated > SOM leached > Fe_d leached. In addition, in almost every case, γ_L and τ_y increased under pre-drained conditions (-60 hPa, *Figure 16a-d*, and *Table 5*). A secondary stabilizing effect was demonstrated: levels of G'_{-60hPa} and G''_{-60hPa} plots were higher than under saturated conditions. Increases of G' and G" showed a higher rigidity, assuming that G' is higher than G" in the first phase.

An ideal curve is shown by the untreated saturated and pre-drained Vertisol samples (*Figure 16a*). A sliding shear behavior predominates, which results from a smectitic clay mineralogy. The plateau phase, which is characterized by a parallel run of G' and G", is well defined, followed by a typical progression in phase 2. It ends with the intersection of G' and G" - at a lower deformation input under pre-drained conditions - and finally leads into phase 3, the stage of viscous behavior.

If patterns of the variously treated Ferralsol samples are compared with the untreated Vertisol samples, characteristics regarding textural compounds, properties of micro cracks, cavities, and pores, and clay mineralogy (HAC, LAC) are evident. Single grains, kaolinite piles, and small roots are embedded as an assemblage with a Fe-oxide coat containing fissures. The latter are preferentially at connection points under untreated conditions. Organic matter is completely removed in H_2O_2-treated samples.

It can be assumed that binding mechanisms, which have been affected by root exudates, are reduced or disabled. After the Na-dithionite treatment, bare surfaces of single grains result. Kaolinite piles may function as single grains with regard to shear behavior, if one assumes stable structural conditions of partially sharp-edged grains. In this case, a direct surface-to-surface or edge-to-edge contact can be assumed during AST, which leads to a higher angle of friction. This is indicated by a lower level of G' and G" and a decelerating set in phase 3. Based on such visual findings, a link to rheological parameters can be done, assuming that single grains show a different mechanical behavior compared to more coherent structures such as micro aggregates. When combined with SEM/ EDS findings, characteristics of G' and G" lead to an understanding of friction processes on a particle-to-particle scale.

Phase 1 is less pronounced with substrates that have much kaolinite or are of coarse texture (Figure 16c). Results with these substrates show a more or less rapid loss of elasticity.

Increases of G" in phase 2 of all untreated Ferralsol samples (blank squares) either saturated or pre-drained, indicate frictional heat (=phase II.1 in *Figure 6*, 'knee'). It results from the reorientation of particles, e.g., kaolinite platelets, packages, and/or single grains. Consequently deformation γ increases, because frictional heat needs to be generated to cause a structural breakdown in phase 3. In general, viscoelastic and cohesive substrates also show this micromechanical behavior. It typically reacts with a temporal delay (Mezger, 2006).

Intersections of G' and G" of Ferralsol samples occur within a range of 50 to 80% deformation under saturated conditions and 70 to >90% under pre-drained conditions (Figure 16b-e), respectively. In Figure 16c and e intersections of G' and G" of untreated clay-rich Ferralsol samples are absent, and G' is higher than G", which indicates a very rigid character of the substrates, despite a relatively rapid decline of G' and G" in the beginning.

Leaching effects occur in almost every case, depending on differences in texture, organic matter, clay, or iron oxide contents at different water contents. With regard to the sandy Ferralsol soil, SOM-leached samples show few deviations in comparison to untreated samples. The effect of frictional heat decreases gradually from untreated > SOM > Fe_d-leached substrates. This effect also can be derived from the curve behavior, which changes from a flat s-character (untreated) to a straightened graph (leaching), followed by a decreasing elastic behavior, a lowering of G' and G" levels, and a reduced distance between graphs of G' and G" in the plateau phase.

Comparing calculated integrated values of each graph, the following results confirm that structural stability decreases due to leaching: $z_{untreated}=29.05$; $z_{H_2O_2\text{-leached}}=10.6$; $z_{Na\text{-dithionite-leached}}=8.537$. Furthermore, curve characteristics reflect the influence of organic matter and iron-oxides on structural stability. Plots of untreated and H_2O_2-leached samples show a similar behavior: an increase of tan δ between $\gamma=0.01\ldots1\%$, and a subsequent slight decrease at $\gamma=5\ldots10\%$ are obvious. If both, organic matter and Fe-oxides are removed; either z becomes small and the curve character changes: tan δ increases constantly.

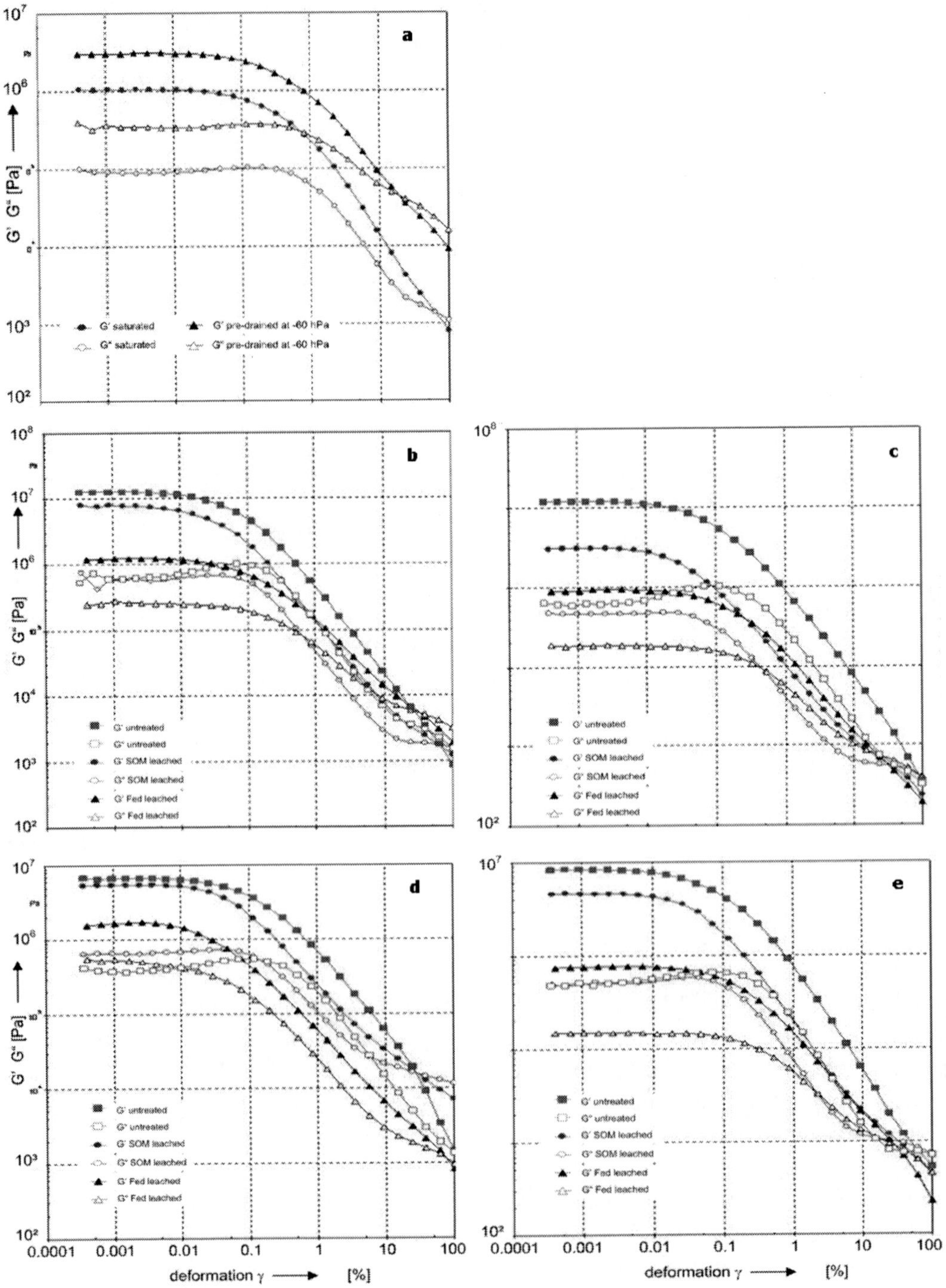

Adapted from Markgraf and Horn, 2007.

Figure 16 Results of amplitude sweep tests with saturated – both conditions are displayed in case of a Vertisol in a) only - and pre-drained (at –60hPa) samples of b) Sandy and c) Clayey Ferralsol from Cruz Alta, d) Clayey Ferralsol, Santo Ângelo, under natural forest (F), and e) Clayey Ferralsol, Santo Ângelo, under no tillage (NT) conditions.

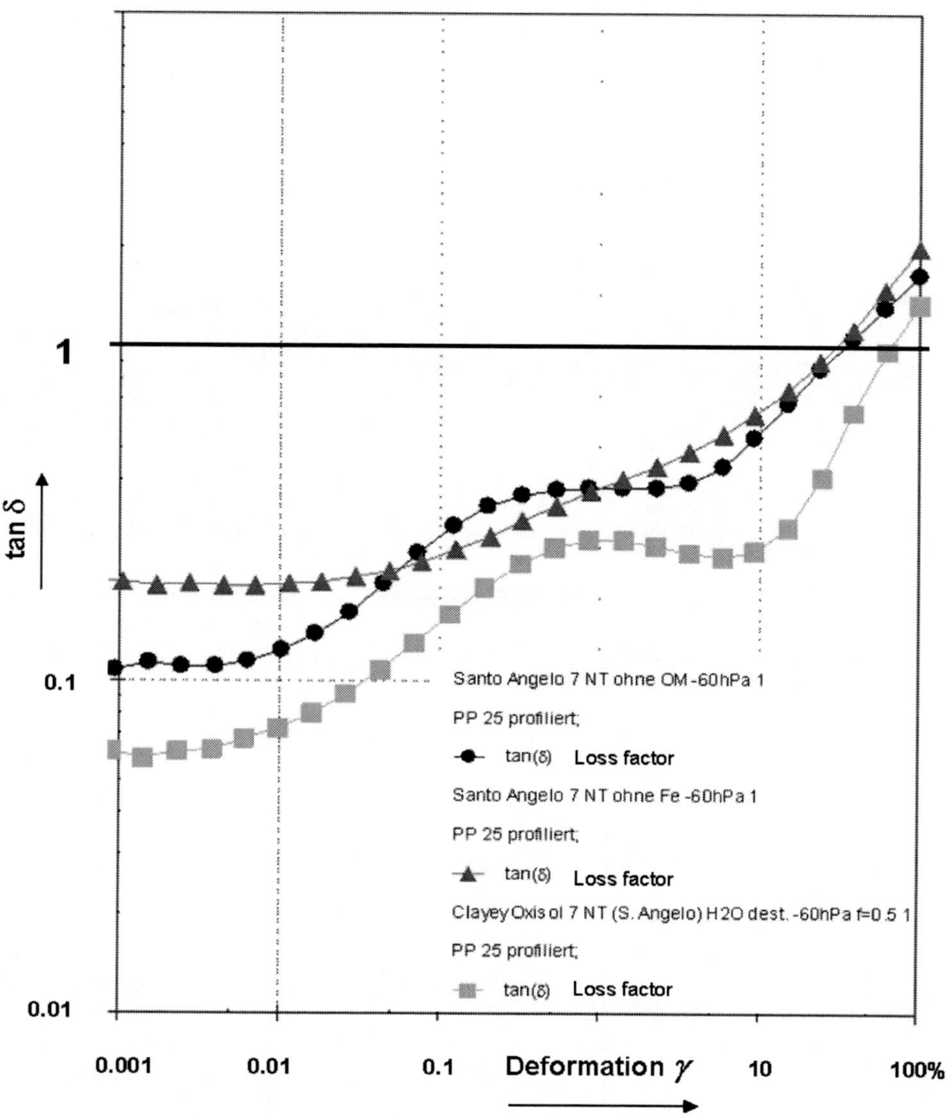

Figure 17. Resulting plots of conducted amplitude sweep tests with pre-drained clayey Ferralsol NT (S. Ângelo). Filled squares: untreated; filled circles: without organic matter (H_2O_2-leached);filled triangles: without Fe_d, Na-dithionite-leached. Due to successively leaching the structural stability is decreased. Organic matter has a significant influence on the stability; the highest degree of stiffness can be reached, if Fe-oxides are considered additionally.

To explain the rheological findings, SEM (scanning electronic microscopy) and EDS (energy dispersive scan) analyses were carried out. Based on visual findings of SEM micrographs, a typical smectitic structure is given for the Vertisol soil (Santana do Livramento). X-ray diffractometry (XRD) graphs confirm these findings. High feldspar compounds, as well as kaolinite, occur with smaller contents of Fe-(hydr)oxides.

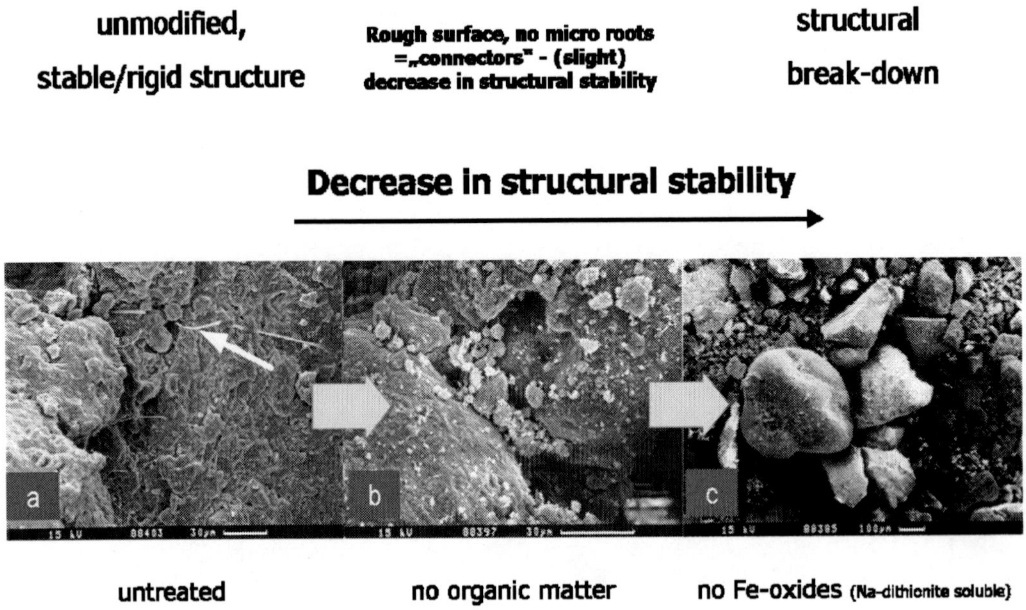

Figure 18. Scanning electron micrographs of a) untreated, b) H_2O_2-treated, and c) Na-dithionite-treated Ferralsol samples, Santo Ângelo, RS, South Brazil.

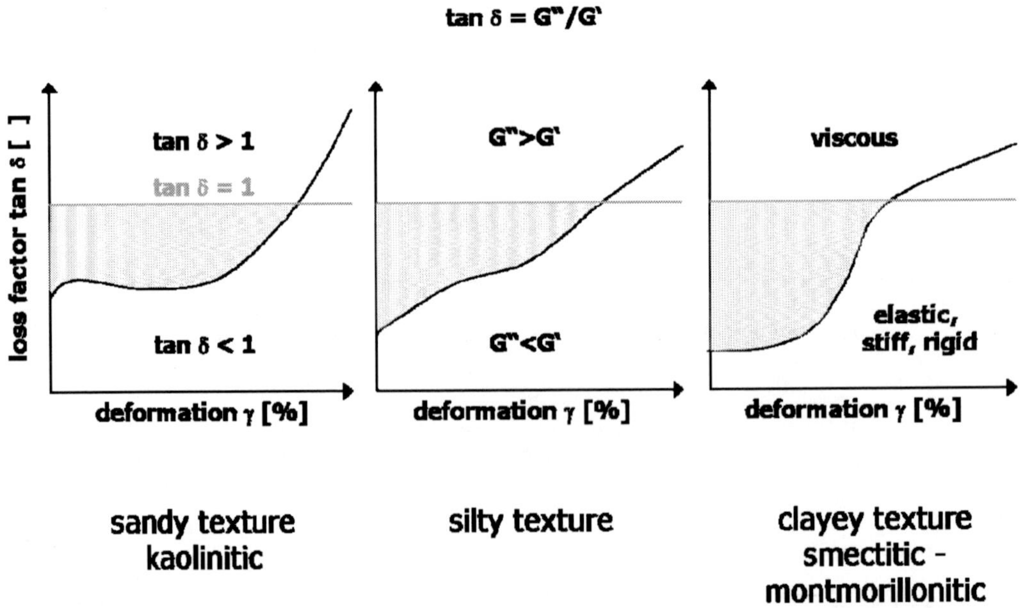

Figure 19. Classification of viscoelastic behavior and stiffness degradation of soil in dependency of texture, clay mineralogy, and water content.

Untreated kaolinitic samples of the Ferralsol soils from Cruz Alta and Santo Ângelo show the phenomenon of *pseudosand* aggregation (Figure 18a). Surfaces are covered with kaolinite and Fe-(hydr)oxides, whereas samples from Cruz Alta are dominated by hematite. This is in contrast to a greater number of goethite compounds in Ferralsol samples from Santo

Ângelo. Additionally, single basalt grains (Figure 18c, after Na-dithionite treatment) are stuck together by fine roots, which merge into micro cracks and cavities on the surface and kaolinite piles. High contents of elementary iron were detected with EDS, which support the existence of magnetite as well as various amounts of silicon in the surrounding structures, i.e., kaolinite.

3.4. Classification of Stiffness

In Figure 19 a synopsis of rheological findings regarding stiffness degradation is illustrated. By classifying results of conducted amplitude sweep tests, which are plotted in tan δ (γ), the transgression from an elastic to a viscous character can be described. If a sandy and/or kaolinitic texture is compared with a silty or a clayey, smectitic substrate, the intersection with the tan δ=1 line occurs with a delay. This is, expressed in G' and G'' the cross-over point and transition to phase 3. Furthermore, if integrals of tan δ (γ) are calculated, limited by tan (δ)=1 on the y-axis, and γ=0.001%... γ at the point of intersection with tan δ=1 on the x-axis, and compared, differences in an elastic, a rigid or stiff character become obvious. Hence, although an intersection with tan δ=1 is obtained at lower deformation, clayey substrates have either elastic and viscous compounds in more or less equal parts, and calculated integrals are greater than those of either sandy or silty material. In addition, curve progressions allow a detailed interpretation of diverse parameters, which may have an influence on structural stability as also presented in Figure 17. Comparable to phase 2.1 (='knee') of G'', an increase of tan δ (γ) is given in case of sandy or silty substrates; if pure sand is investigated, or, as presented, if substrates are leached, and single grains are 'ground', an increase is better defined. The decrease of stiffness in silty substrates occurs proportionally: this is analogue to a proportional increase of G'' (=viscosity) to a decrease of G' (=elasticity, rigidity). This behavior is illustrated as almost straight line, the intersection with the tan δ=1 line occurs earlier, if compared with sandy substrates, but later or at a lower deformation, if compared with clayey material. In the latter case, a s-curve character is given. Structural stability is decreasing at a lower deformation level, showing the most significant elastic behavior in phases 1 and 2 (see also Figure 7). This instant may also be referred to different particle shapes, surface characteristics, and deriving shear behavior in dependency of water content and effective menisci forces, which additionally may increase interparticle forces.

6. Discussion and Conclusion

Amplitude tests were modified and conducted with a modular compact rheometer MCR 300. Knowledge of soil micromechanics, including single particle considerations was intertwined with rheological principles. In conducted amplitude sweep tests, small stress-strain relationships have been considered. Stiffness degradation as introduced by i.e. Jardine (1992) was transferred to results deriving from collected data and plotted graphs of storage and loss modulus G' and G"; those show significant curve characteristics. Further parameters as the shear modulus G, the complex shear modulus G*, tan δ, the ratio of G"/G', deformation γ, strain ε, the shear stress τ, and the shear rate $\dot{\gamma}$ are of relevance, if correlated with general aspects of shear behavior and strength analyses not only in particular in soil physics, but also in geotechnical research. With respect to scale considerations, terms such as the cohesion of soil, the angle of inner friction, residual friction, stiffness, particle forces, and intergranular stress are of interest in the practice of geoengineering in general. Besides a general adaptation of a parallel-plate rheometer to soil micromechanics, the applicability was proved by testing a variety of soil materials. One of the main features that were considered is related to textural effects.

Regarding shear behavior considerations, textural effects have been worked out. It was observed that coarser materials with average grain sizes between 630µm and 2mm tend to turbulent shear behavior in opposite to fine or very fine substrates (≤630µm). Data were presented of a high variety of substrates with loamy sand, loamy silt, loess, clayey to very clayey texture (kaolinitic, illitic, and smectitic) and of pure bentonites. It could be demonstrated that microstructural stability of smectitic substrates is lower than of kaolinitic or illitic ones. However, stiffness degradation occurs gradually in smectitic soils, which leads to a higher structural stability in the end, if compared to a more rapid dissipation of stiffness in kaolinitic or illitic soils. Furthermore, kaolinitic soils show, similar to sandy or silty soils, turbulent shear behavior, whereas sliding shear behavior is evident for smectitic, montmorillonitic soils. Reasons for these different shear behaviors can be found in both, particle properties and particle forces. These findings are in consensus with investigations of Smith and Reitsma (2002), who refer to aspects of different friction angles, as well as of Kézdi (1974), Fredlund and Rahardjo (1993), Mitchell and Soga (2005). Based on works that are more specified to rheological aspects as presented e.g. in Ghezzehei and Or (2000, 2001) a link between soil mechanics in general and rheology could be made.

In proving the significance of water content lies a second important aspect of this soil micromechanical approach to rheometry. Although technical properties are limited to specimens that need to have a minimum water and/or certain clay contents, and have to be a more or less homogeneous paste of a maximum volume of 4-5 cm³, - a flow or creeping character has to be obtained - a methodological advance was achieved by using pre-drained samples (-60hPa). In this case, it could be assumed that particle forces - adhesion, changes of the surface tension of capillary water due to cation effects - occur. Increasing values of G', G", the deformation limit γ_L and significant differences in curve characteristics and the linear viscoelastic deformation range confirm this hypothesis. With respect to Bishop's equation for effective stress, Li (2003) concluded that a fabric factor needs to be examined carefully and involved into any effective stress consideration, in particular with respect to microstructural analysis in unsaturated soils. Rheology can be adapted with appropriate modification of measuring systems, which allow i.e. an online measurement of the matric potential (including the osmotic potential), to effective stress considerations (χ-factor in Bishop's equation for effective stress).

With regard to salt affected microstructural changes that were introduced in Rimmer and Greenland (1976), Shani and Dudley (2001) or Peng et al. (2005), valence effects could be proved. By adding NaCl or $CaCl_2$ solution in different concentrations to pure, natural or already salt affected substrates, it was shown that strengthening effects, induced by ionic forces (monovalent) have a smaller influence than covalent bonds.

Data of conducted amplitude sweep tests with Na-bentonite (Ibeco Seal-80) were introduced, confirm that the microstructural stability was increased due to higher NaCl concentrations, especially when regarded to $CaCO_3$ rich (Avdat Loess) substrates (Markgraf et al., 2006). NaCl and $CaCl_2$ solutions were added to a $CaCO_3$ rich Calcaric Gleysol and an Al^{3+} rich Dystric Planosol (Markgraf and Horn, 2006). Microstructural changes were caused by different concentrations of Na^+ and Ca^{2+} dissolved in distilled water, high $CaCO_3$ and Al^{3+} contents, clay mineralogical - with geological background - and textural constituents and water content (saturated, pre-drained).

Moreover, effects of cultivation systems and associated fertilising concepts to the microstructure could be demonstrated for samples that originate from Brazil (Rio Grande do Sul). In the latter case, the influence of Fe-(hydr)oxides, variations of clay contents and characteristics, as well as differences in organic matter compounds were considered, and led to significant conclusions. Findings corroborate the hypothesis that Fe(hydr)oxides are in first instance responsible for microstructural stability, followed by soil organic matter. In addition, as soil organic matter content is also correlated to cultivation systems - conventional tillage, no tillage, natural conditions under forest or grassland - a first approach to up-scaling could be made (Markgraf and Horn, 2007).

A link contribution to research works of Emerson (1967, 1983), Tisdall and Oades (1982), Rengasamy and Olssen (1991), Oades and Waters (1991), Ohtsubo et al. (1991), Schwertmann et al. (2005) is given. The different significances of soil compounds for soil aggregation were elaborated, taking scale considerations into account. The association of particles to microaggregates in comparison to aggregation on a mesoscale, were induced by the same factors: clay mineral compounds, pH values, soil organic matter, Fe-(hydr)oxides, cations (salts) etc.

With respect to these relevant factors of aggregation and particle association, visualizing methods were compared. In dependency on particle shape, size, roughness, sphericity, and degree of aggregation curve characteristics of G' and G", and decline in stiffness were parametrized and related to information that derived from scanning electron microscopy. Based on fundamental findings of i.e. Santamarina (2001) and Cho et al. (2006), who showed a significant influence of the particle shape to the shear behavior with respect to soil-tool interface models, the rheological data and SEM micrographs could be intertwined to these considerations.

Due to a combination of this modified amplitude sweep test with methods that are commonly used in clay mineralogy such as x-ray diffractometry (XRD), scanning electron microscopy (SEM), and energy dispersive scans (EDS), the applicability in other research fields could be demonstrated. Hence, for instance SEM cannot only be used for the identification of specimen compounds (clay minerals), but also for microstructural characteristics: particle associations, particle shape, roughness and/or size, connectors as e.g. micro roots, clay mineral features, physical changes as affected by NaCl or $CaCl_2$ solution treatments, micro fractures and fissures etc. By taking SEM micrographs as an auxiliary information source for the interpretation of microstructural changes, (de-)stabilizing effects in a Dystric Planosol and a Calcaric Gleysol could be detected. In addition, an even closer interrelationship of rheology and SEM/EDS analyses was pointed out. Structural compounds, in particular Fe-(hydr)oxides, single grains, clay minerals, particles in assemblages, microaggregates and the aggregation phenomenon of pseudosand could be identified in four South Brazilian kaolinitic Ferralsol (Hapludox) and a smectitic, montmorillonitic Vertisol

(Calciudert). Further microstructural features as micro roots, micro cracks and micro pores are of importance for aspects that are regarded to swelling and shrinkage, the interaction at the plant-root-soil interface, and to structural processes as aggregate formation, including pores. Micro mechanical behavior under oscillatory conditions is on the one hand parameterized by rheological data and on the other hand complemented and visualized by SEM/EDS findings.

Due to improving the applicability of amplitude sweep tests in soil mechanical testing, soil-tool interface aspects that were introduced by Tong (1994), Santamarina (2001, 2003), and Cho et al. (2006), intertwined with meso- or even macroscale considerations, as farm implements have a defined surface with occurring forces between soil (particles) and the tool. Conclusion regarding the influence of particle shape to shear behavior were made, which derived from observations in amplitude sweep tests. Increases of the loss modulus G" in the end of phase 1 and the beginning of the transgression state indicated clearly frictional heat: The contact points of coarser particles were increased according to this phenomenon. Comparable aspects could be transferred to clay mineralogical aspects; former existing card house structures collapsed under oscillatory shear conditions, but show a significant resistance against strain, which was also indicated by this rise of G" values.

The detected normal force F_N showed significant differences that depend on substrate specifications. Higher values, partly >25N, were measured in very stiff or coarse and pre-drained material, in which turbulent shear behavior predominated. In the initial state, in which the linear viscoelastic (LVE) deformation range is defined, a parallel run of G' and G" is not well pronounced, whereas lower F_N values (0-12N) were evident in case of fine textured, clay rich and saturated substrates. A state of sliding shear behavior is either given from the beginning (smectitic clay mineralogy) or reached at the end of phase 1. Due to oscillatory shear a reorientation and alignment of particles are caused. In a final state of stiffness degradation a complete structural collapse was reached. If compared to turbulent shear behavior, the state of creeping is reached earlier (=intersection of G' and G") in fine textured and/or saturated material.

A rotational rheometer can be integrated to the soil-tool interface theory. The parallel-plate measuring device exposes two surfaces, which are in contact with a soil testing specimen. In addition, vibration effects can be simulated that may be produced by any instrument; frequencies up to 10Hz are common for farm implements (Garciano et al., 2001). In the present studies the frequency was kept constantly at 0.5 Hz throughout conducted amplitude sweep tests; however, preliminary test results, which included variations of frequencies, show a trend of vibration effects. With increasing frequency not only the tests duration was shortened, but levels of G' and G" decrease, or in other words, stiffness degradation occurs. In this context, it has also to be considered that homogenized samples (<2mm) were prepared and tested. Thus, if transferred to natural soil structures, a higher degree of stiffness can be assumed considering a higher degree of aggregation, which also lead to different hydraulic properties. At present primarily a methodological approximation of rheological measurements to soil mechanics was done starting with paste-like substrates. The water content had been reduced stepwise, in order to simulate in-field conditions. Due to technical limitations, measurements with undisturbed cylinder samples and water contents lower than 20 w/w [%] were not applied. Problems occurred due to fringe effects in cylinder samples, the limited specimen volume of 4-5cm³ and a minimum water content that is needed for any rheological investigations; under (slightly wet to) dry conditions particle movement is inhibited (Santamarina, 2001), and a high degree of friction is generated by the contact of bare particles. This finally leads to incorrect results. Nevertheless, it can be assumed that the functionality of microaggregates and single particles was kept. Intact microaggregates were observed in SEM micrographs. Hence, the focus of upcoming research is to be extended from microscale considerations (rheometer-soil interface) to the macroscale (in field).

Acknowledgement

The first author would especially like to thank the reviewers for their endeavor and support. Furthermore, a special thank is addressed to our co-operation partners from USFM, Rio Grande do Sul, Brazil, the technical staff of the Institute for Plant Nutrition and Soil Science of the CAU zu Kiel for physicochemical analyses, the technical staff of the Institute of Geoscience (CAU) for SEM and EDX analyses, as well as to national and international colleagues for their advice and vivid, fruitful discussions.

References

Akroyd, T.J.; Q.D Nguyen. *Min. Eng.* 2003, 16:731-738.

Atterberg, A. *Kolloidchem. Beihefte.* 1914, 6:55-89.

Barrett, P.J. *Sedimentology*1980, 27:291-303.

Barnes, H.A.; J.F. Hutton;K. Walters. *An Introduction to Rheology.* Elsevier: Amsterdam, Oxford, New York, Tokyo, 1989; 199 pp.

Berilli, M.; T.A. Ghezzehei; D. Or. In *Environmental Geomechanics - Monte Verità 2002.* L. Vulliet; Laloui, L.; Schrefler, B., Eds. EPFL Press: Lausanne, Monte Verità. 2002.

Bishop, A.W.; L. Bjerrum. *Proc. ASCE Research Conference on Shear Strength of Cohesive Soils*, ASCE: Boulder (CO), 1960.

Bishop, A.W. *Publikasjon Norges Geotekniske Institutt.* 1960, 23:1-5.

Bohor, B.F.;R.E. Hughes. *Clays Clay Miner.* 1971,19:49-54.

Boivin, P.; P. Garnier;D. Tessier. *Soil Sci. Soc. Am. J.* 2004, 68:1145-1153.

Brandenburg, U. *Fließverhalten von Bentonit-Dispersionen.* PhD thesis. Christian-Albrechts-Universität zu Kiel: Kiel, Germany. 1990. 268 pp.

Bresler, E.; B.L. McNeal; D.L. Carter. *Saline and Sodic Soils: Principles-Dynamics-Modeling.* Springer: Berlin, Germany. 1982. 236 pp.

Cho, G.-C.; J. Dodds; J.C. Santamarina. *J. Geotech. Geoenv. Eng.* 2006, 132:591-602.

Chorom, M.; P. Rengasamy; R.S. Murray. *Austr. J. Soil Res.* 1994, 32:1243-1252.

Collyer, A.A.; D.W. Clegg. *Rheological Measurement.* 2nd edition. Chapman and Hall: London (UK). 1998. 779 pp.

Cristescu, N.D.; G. Gioda *Visco-Plastic Behaviour of Geomaterials.* International Centre for Mechanical Sciences, Courses and Lectures No. 350. Springer: New York (NY). 1994. 289 pp.

Cundall, P.A.; O.D.L. Strack. *Géotechnique* 1979, 29:47-65.

Dafarmos, C.M.; J.A. Nohel. *A nonlinear hyperbolic Volterra equation in viscoelasticity.* Mathematics Research Center: Madison (WI). 1980.

Davey, B.G. In *Modification of Soil Structure.* W. W. Emerson; R.D. Bond; A.R. Dexter, Eds. John Wiley and Sons: New York (NY). 1979. pp. 97-102
Derjaguin, B.V.; L.D. Landau. *Acta Physicochim. URSS* 1941,14:633-662.
Diamond, S. *Clays Clay Miner.* 1970, 18:7–23.
Diamond, S. *Clays Clay Miner.* 1971, 19:239-250.
Dörfler, H.-D. *Grenzflächen und kolloid-disperse Systeme: Physik und Chemie.* 989 pp. Springer: Berlin. 2002.
Drescher, J.; R. Horn;M. de Boodt, Eds. *Impact of Water and External Forces on Soil Structure.* Catena Supplement 11. Catena: Cremlingen, Germany. 1988. 171 pp.
Emerson, W.W. *J. Soil. Sci.* 1954, 5:235-250.
Emerson, W.W. *J. Soil. Sci.* 1962, 13:31-45.
Emerson, W.W. *Austr. J. Soil Res.* 1967, 5:47-57.
Emerson, W.W.; A.C. Bakker. *Austr. J. Soil Res.* 1973, 11:151-157.
Emerson, W.W.; R.D. Bond; A.R. Dexter, Eds. *Modification of Soil Structure.* John Wiley and Sons: New York (NY). 1978. 438 pp.
Emerson, W.W. In *Soils: An Australian Viewpoint.* CSIRO, Ed. Academic Press: London (UK). 1983. p. 928
Emerson, W.W. *Austr. J. Soil Res.* 1994, 32:173-184.
FAO-Unesco. *Soil Map of the World, South America, 1:5,000,000.* FAO Unesco (printed by Tipolitografia F. Failli, Rome): Paris. 1971. 193 pp.
Fiés, J.C.; A. Bruand. *Eur. J. Soil Sci.* 1998, 49:557-567.
FitzPatrick, E.A. *Soil Microscopy and Micromorphology.* Wiley: Aberdeen. 1993. 304 pp.
Fredlund, D.G.; H. Rahardjo. *Soil Mechanics for Unsaturated Soils.* John Wiley and Sons, Inc.: New York (NY). 1993. 517 pp.
Garciano, L.; R. Torisu; J. Takeda; J. Yoshida. *J. Jap. Soc. Agricul. Mach.* 2001, 63:45-50.
Ghezzehei, T.A.; D. Or. *Water Resour. Res.* 2000, 36:367-379.
Ghezzehei, T.A.; D. Or. *Soil Sci. Soc. Amer. J.* 2001, 65:624-637.
Grant, C.D.; A.R. Dexter;C. Huang. *Soil Sci.* 1990, 41:95-110.
Grant, C.D.; P.H. Groenevelt; G.H. Bolt. In *Environmental Mechanics - Water, Mass and Energy Transfer in the Biosphere.* Raats, P.A.C.; D. Smiles; A.E. Warrick, Eds. The Philip Volume. American Geophysical Union: Washington (D.C.). 2002. 345 p.
Güven, N.; R.M. Pollastro, Eds. *Clay-Water Interface and its Rheological Implications.* T.C.M. Society: CMS Workshop Lectures. The Clay Mineral Society: Boulder (CO). 1992. 244 pp.
Hartge, K.H.; R. Horn. Einführung in die Bodenphysik. (Introduction to Soil Physics – in German) Enke: Stuttgart, Germany. 1999. 304 pp.
Hartge, K.H.; R. Horn. *Wasser and Boden* 2002, 54:34-38.
Henning, K.-H.; Störr, M. *Electron micrographs (TEM, SEM) of clays and clay minerals.* Akademie-Verlag: Berlin, Germany. 1986.
Ingles, O.G. *Bonding Forces in Soils - Part 3 (1).* Pages in Proceedings of the First Conference of the Australian Road Research Board. 1962:1025-1047
Jardine, R.J. *Soils and Foundations* 1992, 32:11-124.
Jardine, R.J.; D.M. Potts; K.G. Higgins, Eds. *Advances in Geotechnical Engineering: The Skempton Conference.* Proceedings of the Skempton Memorial Conference on Advances in Geotechnical Engineering, held at the Royal Geographical Society. American Society of Civil Engineers (Thomas Telford Ltd.): London, UK. 2004. 1400 pp.

Jasmund, K.; G. Lagaly, Eds. *Tonminerale und Tone: Struktur, Eigenschaften, Anwendungen und Einsatz in Industrie und Umwelt.* Steinkopff-Verlag: Darmstadt, Germany. 1993. 490 pp.

Jia, X. *Biosystems Eng.* 2004, 87:489-493.

Keedwell, M.J., Ed. *International Conference on Rheology and Soil Mechanics. International Conference on Rheology and Soil Mechanics.* Elsevier Applied Science: London, New York (NY). 1984. 371 pp.

Kézdi, Á. *Handbook of soil mechanics - Volume 1: Soil Physics.* Akadémiai Kiadó: Budapest, Hungary. 1974. 294 pp.

Kosmulski, M.; J. Gustafsson; J.B. Rodenholm. *Colloid Polym. Sci.* 1999, 277:550-556.

Krumbein, W.C. *J. Sediment. Petrol.* 1941, 11:64-72.

Krumbein, W.C.; L.L. Sloss. *Stratigraphy and Sedimentation.* Freeman and Company: San Fransisco (CA). 1963. 497 pp.

Lagaly, G.; O. Schulz; R. Zimehl. *Dispersionen und Emulsionen - Eine Einführung in die Kolloidik feinverteilter Stoffe einschließlich der Tonminerale.* Steinkopff: Darmstadt, Germany. 1997. 560 pp.

Le Bissonnais, Y. ; D. Arrouays. *Eur. J. Soil Sci.* 1996, 47:425-437.

Le Bissonnais, Y. ; D. Arrouays. *Eur. J. Soil Sci.* 1997, 48:39-48.

Li, X.S. *Géotechnique* 2003, 53:273-278.

Macosko, C.W. *Rheology - Principles, Measurements, and applications.* VCH Publishers: New York (NY). 1994. 550 pp.

Markgraf, W.; R. Horn; S. Peth. *Soil Till. Res.* 2006, 91:1-14.

Markgraf, W.; R. Horn. In *Sustainability - Its Impact on Soil Management and Environment.* R. Horn; H. Fleige; S. Peth; Xh. Peng, Eds. Catena: Reiskirchen, Germany. 2006. pp. 47-58

Markgraf, W.; R. Horn. *J. Soil Sci. Soc. Am.* 2007, 71:851-859.

Mehra, O.P.; M.L. Jackson. In *7th National Conference on Clays and Clay Minerals.* Swineford, A., Ed.; Adlard and Son Ltd.: Washington (D.C.). 1960. pp. 317-327.

Meichsner, G.; T. Mezger; J. Schröder. *Lackeigenschaften messen und steuern: Rheologie - Grenzflächen – Kolloide.* Vincentz Verlag: Hanover, Germany. 2003. 236 pp.

Mezger, T. *The Rheology-Handbook - For users of rotational and oscillatory rheometers.* 2nd edition. Vincentz Verlag: Hanover, Germany. 2006. 252 pp.

Mitchell, J.K.; K. Soga. *Fundamentals of soil behavior.* 3rd edition. John Wiley and Sons: Hoboken (NJ). 2005. 577 pp.

Neaman, A.; A. Singer. *Soil Sci. Soc. Amer. J.* 2000, 64:427-436.

Oades, J.M.; A.G. Waters. *Austr. J. Soil Res.* 1991, 29:815-828.

Ohtsubo, Y.A.; A. Yoshimura; S.I. Wada; R.N. Yong. *Clays Clay Min.* 1991, 39:347-354.

Peng, X.; R. Horn; D. Deery; M.B. Kirkham; J. Blackwell. *Austr. J. Soil Res.* 2005, 43:555-563.

Powers, M.C. *J. Sediment. Petrol.* 1953, 23:117-119.

Reed, S.J.B. *Electron Microprobe Analysis and Scanning Electron Microscopy in Geology.* 2nd edition. Cambridge University Press, Cambridge. 2005. 192 pp.

Rengasamy, P.; K.A. Olssen. *Austr. J. Soil Res.* 1991, 29:935-952.

Richter, F. *Vergesellschaftung und Eigenschaften von Böden unterschiedlicher geomorpher Einheiten einer Jungmoränenlandschaft des Ostholsteinischen Hügelandes.* Christian-Albrechts-University zu Kiel, Kiel. PhD Thesis. 2005. 132 pp.

Rimmer, D.L.; Greenland, D.J. *Soil Sci.* 1976, 27:129-139.

Rosenqvist, I.T. *J. Soil Mech. Found. Div., Proc. Am. Soc. Civil Engrs.* 1959, 85:285-312.

Rosenqvist, I.T. *Clays Clay Miner.* 1962, 9:12-27.

Santamarina, J.C. *Proc. Symp. Soil Behavior and Soft Ground Construction, in honor of Charles C. Ladd.*, MIT (MI). 2001. pp. 1-32

Santamarina, J.C. In *Soil Behavior and Soft Ground Construction.* Germaine, J.T.; T.C. Sheahan; R.V. Whitman, Eds. ASCE Geotechnical Special Publications No. 119, ASCE: Reston (VA). 2003. pp. 25-56.

Schlichting, E.; H.-P. Blume; K. Stahr. Bodenkundliches Praktikum - Eine Einführung in pedologisches Arbeiten für Ökologen, insbesondere Land- und Forstwirte, und für Geowissenschaftler. 2nd edition. Blackwell: Berlin, Germany. 1995. 295 pp.

Schmidt, P.F., Ed. *Praxis der Rasterelektronenmikroskopie und Mikrobereichsanalyse.* Kontakt and Studium 444. Expert Verlag: Esslingen, Germany. 1994. 810 pp.

Schramm, G. *Einführung in Rheologie und Rheometrie.* 2nd edition. Haake GmbH: Karlsruhe, Germany. 2002. 360 pp.

Schulz, O. *Berichte aus der Chemie - Strukturell-rheologische Eigenschaften kolloidaler Tonmineraldispersionen.* Christian-Albrechts-University zu Kiel. PhD thesis. 1998. 501 pp.

Schwertmann, U.; F. Wagner; H. Knicker. *Soil Sci. Soc. Am. J.* 2005, 69:1009-1015.

Shainberg, I.; J.D. Rhoades; R.J. Prather. *Soil Sci. Soc. Am. J.* 1981 24:273-277.

Shainberg, I.; G.J. Levy. *Physico-chemical effects of salts upon infiltration and water movement in soils. Interacting processes in soil science.* In R. J. Wagenet; P. Baveye; B.A. Stewart, Eds. Advances in Soil Science. Lewis Publications: Boca Raton (FL). 1992. pp. 37-94

Shani, A.; L.M. Dudley. *Soil Sci. Soc. Amer. J.* 2001, 65:1522-1528.

Sharpley, A.N. *Soil Sci.* 1990, 149:44-51.

Shi-qiao, D.; R. Lu-quan; L. Yan; H. Zhi-Wu. *J. Bionics Eng.* 2005, 2:33-46.

Skempton, A. In *Pore Pressure and Suction in Soils: Conference organised by the British National Society of the International Society of Soil Mechanics and Foundation Engineering at the Institution of Civil Engineers held on March 30th and 31st, 1960.* ISSMFE, Ed. Butterworths: London (UK). 1960. pp. 4-16

Smith, D.W.; M.G. Reitsma. In *Environmental Geomechanics - Monte Veritá 2002.* Vulliet, L.; L. Laloui; B. Schrefler, Eds. EPFL Press: Monte Veritá, Switzerland. 2002. pp. 27-44

Sonderegger, U.C. *Das Scherverhalten von Kaolinit, Illit und Montmorillonit.* PhD thesis. ETH Zurich, Switzerland. 1985. 165 pp.

Suklje, L. *Rheological aspects of soil mechanics.* Wiley Interscience: London (UK). 1969. 571 pp.

Sumner, M.E.; R. Naidu. *Sodic Soils: distribution, properties, management and environmental consequences.* Oxford University Press: New York (NY). 1998. 207 pp.

Tarchitzky, J.; Y. Chen. *Soil Sci. Soc. Amer. J.* 2002, 66:406-412.

Tariq, A.U.R.; D.S. Durnford. *Soil Sci. Soc. Amer. J.* 1993, 57:1183–1187.

Terribile, F.; E.A. Fitzpatrick. *Eur. J. Soil Sci.* 1995, 46:29-46.

Terzaghi, K. *The shearing resistance of saturated soils.* Proceedings of the 1st International Conference on Soil Mechanics, Cambridge (MA). 1936. pp. 54-56

Terzaghi, K.; R. Jelinek. *Theoretische Bodenmechanik.* Springer: Berlin, Germany. 1954. 505 pp.

Tisdall, J.M.; J.M. Oades. *Soil Sci.* 1982, 33:141-163.
Tong, J.; L. Ren; B. Chen; A.R. Qaisrani. *J. Terramech.* 1994, 31:93-105.
Tuller, M.; D. Or; L.M. Dudley. *Water Resour. Res.* 1999, 35:1949-1964.
Tuller, M.; D. Or. *Water Resour. Res.* 2001, 37:1257-1276.
Tuller, M.; D. Or. *Vadose Zone J.* 2002, 1:14-37.
Tuller, M.; D. Or. *J. Hydrol.* 2003, 272: 50-71.
Tuller, M.; D. Or. *Water Resour. Res.* 2005, 41:W09403 (09401-09406).
US200, RHEOPLUS/32 V3.21. Anton Paar Germany. 2007.
USDA. *Keys to Soil Taxonomy.* 9th edition. NRCS: Washington (D.C.). 2003.
Valdes, J.R. *Fines Migration and Formation Damage.* PhD thesis, Georgia Institute of Technology (GA). 2002. 178 pp.
van Olphen, H. *An Introduction to Clay Colloid Chemistry.* 2nd edition. John Wiley and Sons: New York (NY). 1977. 318 pp.
van Reeuwijk, L.P. *Procedures for soil analysis.* 6th edition. International Soil Reference and Information Centre (ISRIC): Wageningen, The Netherlands. 2002.
Verwey, E.J.W.; J.T.G. Overbeek. *Theory of the Stability of Lyophobic Colloids: the Interaction of Soil Particles having an Electric Double Layer.* Elsevier: New York (NY). 1948.
Vyalov, S.S. *Rheological fundamentals of soil mechanics.* Elsevier: Amsterdam, The Netherlands. 1986. 564 pp.
Wadell, H. *J. Geol.* 1932, 40:443-451.
Warkentin, B.P.; R.N. Yong. *Clays Clay Miner.* 1962, 9:210-218.
Warkentin, B.P.; R.K. Schofield. *Soil Sci.* 1962, 13:98-105.
Whorlow, R.W. *Rheological Techniques.* 2nd edition. Ellis Horwood Ltd.: Chichester (UK). 1992. 460 pp.
Yimsiri, S.; K. Soga. *Effect of surface roughness on small-strain modulus: Micromechanics view.* In Pre-failure Deformation Characteristics of Geomaterials. Balkema: Rotterdam. 1999. pp. 597-602
Yimsiri, S.; K. Soga. *Géotechnique: Intern. J. Soil Mech.* 2000, 50:559-572.
Yimsiri, S.; K. Soga. *Soils Found.* 2002, 42:15-26.
Yong, R.N.; B.P. Warkentin. *Introduction to soil behavior.* In G. Norbby, Ed. Macmillan series in civil engineering. Macmillan: New York. 1966. pp. 281-349

Reviewed by:
Dr. Paul Hallett
SCRI, Invergowrie, Dundee, DD2 5DA,
Scotland, United Kingdom.
Email: Paul.Hallett@scri.ac.uk

Prof. Dr. Mary Beth Kirkham
Department of Agronomy, 2004 Throckmorton Hall
Kansas State University, Manhattan, Kansas 66506-5501, USA
E-mail: mbk@ksu.edu

In: Grid Technology and Applications...
Editors: G.A. Gravvanis et al, pp. 281-305
ISBN 978-1-60692-768-7

Chapter 10

On Heuristic Methods for the Project Scheduling Problem*

Dallas B.M.M. Fontes*†*and Portio L.A. Liana-Ignes
Faculdade de Economia da Universidade do Porto
Rua Dr. Roberto Frias, 4200-464 Portio, PORTUGAL

Abstract

Project Management (PM) has emerged from different fields of application and it entails planning, organizing, and managing resources to bring about the successful completion of specific project goals and objectives, while controlling the resources (time and money) and the quality. The operational research contribution for PM has mainly been done through providing tools (model, methods, and algorithms) to solve project scheduling problems. The project scheduling problem involves the scheduling of project activities subject to precedence constraints and resource constraints. Although this problem has been the subject of extensive research since the late fifties, there have been some publications reporting extreme budget over runs and/or extreme time delays, thus proving that there is still the need for further research. this chapter intends to be a guided tour through the most important recent developments in algorithmic methods to solve the project scheduling problem. Since these problems are NP-hard, our main focus is on heuristic methods, particularly on meta-heuristics. The paper concludes with an examination of areas that in the opinion of the author would particularly benefit from further research.

Key words: Project Management, Heuristic Methods, and Review.

1. Introduction

Project Management (PM) has emerged from different fields of application and it entails planning, organizing, and managing resources to bring about the successful completion of specific project goals and objectives, while controlling the resources (time and money) and

*The financial support of FCT, POCI 2010 and FEDER, through project POCI/EGE/61823/2004, is gratefully acknowledged.
†E-mail address: fontes@fep.up.pt. Tel: 351-225 571 100, fax: 351-225 505 050.

the quality. The operational research contribution for PM has mainly been done through providing tools (models, methods and algorithms) to solve project scheduling problems. Therefore, in this chapter we look to the Project Scheduling Problem. This problem has been the subject of much research due to the enormous number of practical applications in various industries and organizations and also because it can be treated as a generalized case of many scheduling problems, such as job-shop, flow-shop, and assembly line balancing.

The basic Project Scheduling Problem (PSP) involves the scheduling of a set of project activities (usually termed jobs or tasks) under precedence constraints with the objective of minimizing a given performance measure. Typically, this measure is given by the total project throughput duration also known as makespan. Research has commonly focused on the Resource Constrained PSP (RCPSP), given its realist setting. In the RCPSP resource constraints are also considered, since it involves assigning jobs or tasks to a resource or set of resources with limited capacity in order to meet some predefined objective. The precedence constraints usually correspond to the temporal constraints, whereas the resource constraints usually correspond to the renewable resources (e.g., manpower, material, and machines) which are required by the project. Although many different objectives are possible, depending on the goals of the decision maker, the most common of these is to find the minimum makespan. This version of the problem is the most well known and studied. Nevertheless, many researchers have studied other versions. For example, RCPSP with discounted cash flows where, given the considered cash flows, the project managers aim to maximize the net present values instead of minimizing project durations, see [24]. The Preemptive RCPSP includes the flexibility to interrupt the activities since no extra cost exist if they are interrupted and restarted later. For a very recent discussion on such problem see [5]. Other issues have also been addressed, for example [50, 30] studied the Multi-mode RCPSP, [44] studied Stochastic RCPSP, [76] studied the PSP with discounted cash flows, among others.

There exist several algorithms which are able to optimally solve small-sized[1] RCPSPs in reasonable time, however, no optimal solution algorithms exist to solve large-size problem instances[1]. This is not surprisingly since the RCPSP is one of the most intractable problems in the operations research area [32, 78]. Its hardness can be emphasized by observing that the graph theory node coloring problem can be formulated as RCPSP [59, 7]. Thus, RCPSP is strongly NP-hard and hence, it is also difficult to approximate. Therefore, we choose to address heuristic methods for this problem.

In this paper, we deal with the classical RCPSP. The contribution of the paper is twofold. Firstly, we introduce some recent meta-heuristic methods that have been used to addressed the classical RCPSP. Secondly, based on the results the different methods have obtained, we point out advantages and disadvantages of the approaches developed and also the areas where there is still the need for further research.

[1]Problem sets commonly used to compare methods' performance. The problems data, the best known solution, and the CPM lower bound are available for download in the PSP library, see [38].

2. The Resource Constrained Project Scheduling Problem (RCPSP)

The resource constrained project scheduling problem (RCPSP) can be described as a set J of activities where each activity $j \in J$ has to be processed in order to complete the project. The activities are interrelated by two kinds of constraints, namely precedence constraints, that force some activity j not to be started before all its immediate predecessor activities have been finished, and resource constraints, since performing the activities requires resources with limited availability. An activity is therefore, characterized by resource requirements, say r_{jk} units of resource type k, a non-preemptable processing time duration p_j and a set of precedence constraints. Resources are characterized by their availability, say R_k at any point in time. All parameters are assumed to be nonnegative and deterministic. The objective of the RCPSP is to find the activity schedule that satisfies all precedence and resource constrains, while minimizing the overall completion time, i.e., the project makespan.

Recent years have witnessed a tremendous increase in research for the RCPSP, both in terms of heuristic and optimal procedures. For previous developments we refer to the surveys provided in [29, 52, 26, 10, 40, 16]. More recently, [41, 22] have given overviews of heuristic methods for the RCPSP. The heuristic methodologies developed may be classified into several categories, such as priority-rule-based X-pass methods [38, 36, 61], local-search-oriented approaches [9, 73, 18, 69], and population-based approaches [27, 21, 49, 71, 13, 15].

Over the years many heuristics have been developed, but more recently the research effort has been mainly on evolutionary heuristics, since these have been the ones with better performances; see for example the comparisons made in [22]. In this chapter, we describe heuristics developed for the RCPSP in recent years, since much of the early work may already be summarized elsewhere, although some references may be made to previous works, as needed. Most of the work surveyed here is on meta-heuristics.

3. Local Search Heuristics

The first heuristic methods developed for the RCPSP were priority-rule based scheduling methods [33], which had the objective of generating precedence and resource feasible schedules. A multitude of priority rules were proposed and tested experimentally, see for example [11, 8]. Priority-based heuristics have the advantage of being intuitive and easy to implement, also they do not require much computational effort. However, the solutions they provide do not have such a good quality when measured, for instance, as the average deviation from the optimal objective function value. Hence, research interests shifted to more elaborate heuristics. Many meta-heuristic strategies have been applied to solve hard combinatorial optimization problems. Here we review the recent developments on meta-heuristics for the RCPSP. Our focus is on population base heuristics, however, we also describe other heuristics since many of the most recent developments include hybrid approaches.

Here we review neighborhood search heuristics, including simulated annealing and tabu search[2]

Fleszar and Hindi [18] apply a Variable Neighborhood Search (VNS), a meta-heuristic strategy introduced by Mladenovic and Hansen [51], to the RCPSP. They employ the activity list representation, the serial SGS, and an enhanced shift move which allows to shift activities together with their predecessors or successors. During run-time, their approach adds precedence relations on the basis of lower bound calculations. Palpant et al. [53] combine the forward-backward scheduling, the serial SGS, and constraint-based optimization of partial schedules into a local search procedure. An initial solution is generated by applying forward-backward scheduling. Afterwards, a so-called block of activities, activities which are processed in parallel or contiguously, is selected randomly and constraint propagation is employed to determine the selected activities a minimum makespan schedule under the constraints imposed by the non-selected activities. The entire method iterates between the selection of activities, optimization of partial schedules, and forward-backward scheduling until a stopping criterion is met.

Pesek et al. [56] propose a local search heuristic that uses a complex neighborhood based on constructive heuristics. In this work a schedule is represented as a list of activities and the initial solution is obtained by, at each step, adding one feasible activity at the end of the list, serial SGS. The authors propose two neighborhoods. One is sampled by removing a set of activities and then reinsert them, one at the time, in the best position available. Another neighborhood is obtained by a multi-shift, which involves both trying to bring activities forward and trying to send activities backwards. The former neighborhood is used within a local search method, while the latter is used within a TS method. From the results reported it has been concluded that local search heuristic leads to better results. These results are amongst the best in current literature, however, results are not reported for the largest problems in literature.

Local search approaches seem to be unable to obtain good results on their own. However, they have shown that their incorporation in other methods may lead to substantial improvements.

3.1. Simulated Annealing

Simulated annealing (SA) is a generic probabilistic meta-algorithm for the global optimization problem, namely locating a good approximation to the global minimum of a given function in a large search space. It is often used when the search space is discrete. The name and inspiration come from annealing in metallurgy, a technique involving heating and controlled cooling of a material to increase the size of its crystals and reduce their defects. The heat causes the atoms to become unstuck from their initial positions (a local minimum of the internal energy) and wander randomly through states of higher energy; the slow cooling gives them more chances of finding configurations with lower internal energy than the initial one. The SA idea has been introduce by Kirkpatrick et al. [35] to the field

[2] In this chapter we consider simulated annealing and tabu search as local search methods, since they perform neighborhood search. In the literature, however, sometimes these methods are classified as meta-heuristics. Although we believe this not to be the best classification, it may be justifiable due to the fact that these methods, unlike traditional local search methods, probabilistically allow for moves to worst solutions. Nevertheless, it should also be notice that they work with one single solution, rather than with a population of solutions.

of combinatorial optimization. By analogy with this physical process, at each step of the SA algorithm a new random "nearby" solution is created. If this new solution is better then it replaces the current one. Otherwise, if it does not improve upon the current one, then it might be accepted, with a probability that depends on the magnitude of the function deterioration and on a global parameter T (called the temperature). The temperature, typically starts at a large value which is gradually decreased during the process. The dependency is such that the current solution changes almost randomly when T is large, but increasingly "downhill" as T goes to zero. The allowance for "uphill" moves saves the method from becoming stuck at local minima, which are the curse of greedier methods.

Regarding project scheduling problems, recently only a few works can be found using simulated annealing. Zhang et al [82] propose a hybrid simulated annealing algorithm to navigation co-scheduling system for Three Gorges Dam and the Gezhouba Dam. The solution search starts by trimming the existing timetable based on heuristic rules; the result is then used to compute the ship scheduling by the depth-first-search (DFS) algorithm. The solution thus obtained is finally updated by the Metropolis rule.

In [60] a parallel search fuzzy-based adaptive sample-sort simulated annealing is proposed. This heuristic consists of a sample-sort simulated annealing (SSA), a schedule generation scheme (SGS), and a fuzzy logic controller (FLC). The SSA is the one proposed by Thompson and Bilbro [64], where the sequence of temperature is replaced with an array of samplers operating at static temperatures and the single stochastic sampler is replaced by a set of samplers. As it uses an array of samplers that perform simultaneous interacting searches this algorithm can be considered a parallel SA algorithm. Each sampler operates at a different temperature and exchanges information with neighbor samplers in two ways. On one hand, less-cost solutions are propagated; on the other hand, the probability of accepting a higher cost solution is updated having also in consideration the temperature of neighboring samplers. The authors have concluded it to have a performance similar to a sequential SA, but faster convergence. Based on the work of Lee and Takagi [45] the authors use FLC tool for controlling the parameters, this makes the algorithm adaptive in nature by regulating the swapping rate of an activity's priority during an improved schedule generation process. Initially, feasible schedules are generated by a serial SGS [38], and then improved through the use of SSA. The authors have obtained results that are amongst the three best in current literature.

A hybrid-directional planning SA has been prosed by [75], where the schedules are randomly extended by using a forward, backward or bidirectional scheme, with equal probability. The neighborhood is sampled by random swapping and insertion, with equal probability. The temperature is decreased, at a fixed rate, every time a specified number of iterations has been performed. The results obtained show that using the proposed hybrid paths leads to better results than the forward, backward, and bidirectional paths. However, the average deviation of the best solution known, ranks below 20 amongst current literature. Being a lot worst for larger problems (set $J120$).

3.2. Tabu Search

Tabu search is an iterative search procedure, usually attributed to Fred Glover [19]. The basic idea in tabu search in combinatorial optimization is to let a solution move around in the search-space, picking the next solution from a neighborhood. The algorithm keeps a memory of previously visited solutions, and, in order to avoid cycling, the search is usually not allowed to revisit solutions already in the memory. Often this is implemented by having a tabu list of forbidden moves. Every time a move is made, the inverse move is added to the tabu list, where it stays for a number of moves. When a move is made, usually, the best solution in the neighborhood, not forbidden by memory, is picked. Since the tabu list stores moves and not solutions, it is usually necessary to allow the algorithm to implement moves in the tabu list, if they satisfy an aspiration criterion. This search strategy often leads to the solution rolling from one local optimum to the next in the search-space. Tabu search has proven to be very efficient for a number of combinatorial optimization problems.

Not many recent approaches have been developed for the RCPSP addressed here using tabu search, however, for other versions of this problem a few works can be found, see for example [74, 47, 79, 50]. Pan et al [54] propose a tabu search approach where the initial feasible solution is obtained by the heuristic rule MINSLK. Such a solution is then changed by using either an insert or swap move. Moves are kept in the tabu list for a predefined number of iterations. If a solution in the tabu list has better value than the best of the previous cycle then it becomes the current optimal solution. Thus, the authors use a short-memory structure, only. The tabu list is operated in a FIFO mode and solutions obtained by a tabu move can only be used if no better solution has been found in the neighborhood and is better than the best found so far (aspiration criterion). Perhaps, one of the main advantages of this work is the user friendly interface developed. Computational experiments were performed using Patterson problems ([11]), not used in current research. This is probably the reason why the authors have compared their results with those of methods developed more than 10 years ago, e.g., [46].

Yin et al. [75] developed a hybrid-directional planning TS. As was the case for their SA approach (discussed in the previous section), schedules are randomly obtained by using a forward, backward or bidirectional scheme and the neighborhood is sampled by random swapping and insertion. The tabu list has been implemented as a FIFO list of fixed length. If no better solution can be found within the current neighborhood then they select the least worsening solution or a tabu solution if it is better than the best found so far. The computational results show that the TS approach produces solutions with similar quality of those produced by the SA (also proposed in [75]) for problem sets $J30$, $J60$, and $J90$, and better for $J120$.

4. Evolutionary Heuristics

Evolutionary Algorithms (EAs) are computational models for problem solving, optimization or simulation inspired by the evolution of species in nature and the Darwinian principles. Common for all evolutionary algorithms is that they work on one or more sets of potential solutions to a problem. Because of the biological inspiration, each solution in an EA is often termed an individual, while a set of individuals is called a population. The population is manipulated using a cycle of selection of good individuals followed by the generation of new individuals by variational operators such as mutation and recombination. The basic idea in an evolutionary algorithm is to assign each individual a fitness value, measuring how well it solves the problem at hand. Individuals which are known to be good or promising (having a high fitness score) are then selected for reproduction, and new individuals which are related to them are generated. The new individuals are inserted in the population if they are acceptable, while inferior individuals are discarded. By constantly creating variations of the best solutions known, the algorithm gradually improves the solution quality. Every iteration of this cycle is called a generation. The way new individuals are generated is usually stochastic. Because of this, evolutionary algorithms are stochastic search algorithms, in which previously sampled points are used to guide the selection of future sampling points.

4.1. Applying Evolutionary Algorithms

Evolutionary Algorithms (EA) are inspired by biological evolution: reproduction, mutation, recombination, and selection. Candidate solutions to the optimization problem are perceived as individuals in a population, and the fitness function determines the "quality" of these individuals and therefore their chances of survival. To build such an algorithm many choices have to be made. EAs often perform well approximating solutions to all types of problems because they, ideally, do not make any assumption about the underlying fitness landscape; this generality is shown by it successes in diverse fields and problems.

The choice of genetic representation is usually the first and probably most important single decision, but many other decisions affect the effectiveness of the algorithm. When choosing a set of genetic operators (mutation, crossover) the roles of the different operators should be considered. A mutation operator is used to ensure sufficient genetic diversity in the population, however, care must be taken since this may prevent local search and small improvements in a "hill climbing like manner. Parents, with which the next generation is to be obtained are chosen with a bias towards higher fitness. The parents reproduce by copying with recombination (crossover) and/or mutation. Recombination acts on the two selected parents and results in one or two individuals. Mutation acts on one individual and results in a new individual. These operators create the offspring, that is, the set of new individuals. These new individuals compete with old individuals to form the next generation. The populations size and the number of generations have to be balanced since, generally, the larger these parameters are, the better the algorithm will perform, but at the expense of longer run-times, since more fitness evaluations will be involved. However, for a fixed allowed number of fitness evaluations, the choice of these parameters becomes a tradeoff. A large population size means a better exploration of the search-space, while a large number of generations allows for better exploitation of the promising solutions found.

EAs tend to get trapped into a local optimum and also to have most of its population concentrated on a small part of the search space located around the local optimum, which is usually termed premature convergence. This problem can be obviated by using structured populations or by hybridization. In the former case the population is divided into smaller subparts that tend to breed more amongst themselves, the dispersion of genetic material is slowed down, giving the algorithm more time to settle on the most promising part of the search-space. A discussion on structured populations (island and diffusion models) and on sharing, crowding, and tagging methods can be found in chapter 6 of [4].

Since there is no theory adequately describing how EAs work, some of these decisions are largely up to experience and taste of the person designing the algorithm. Due to the large number of design decisions, the interdependence between these and the lack of mathematical understanding of both the evolutionary algorithms and the search-spaces of the problems they are applied to, it is usually infeasible to find the optimal set of parameters for a given problem. This is particularly so since the optimal set of parameters can also be expected to vary from problem instance to problem instance. Usually the parameters are fixed in a trial and error fashion in which a number of possibilities are tried out, and the one observed to work best is chosen.

4.2. Advantages of Evolutionary Computation

Evolutionary algorithms have been applied to a large number of academic and real world problems over the last twenty years or so. These algorithms have achieved good results, particularly for NP-hard problems. Since, as said before, these heuristics can be used as a blind search method to most problems, it can almost always be applied. In addition, the fact that they can be easily hybridized with other problem specific techniques, in order to improved performance, makes them a very flexible tool. Other, so-called modern search heuristics [57] have these properties as well; Simulated Annealing and Tabu Search are also generally applicable optimization methods, and also have been demonstrated to perform well in many applications.

However, EAs have at least one huge advantage over SA, TS and most other methods. That is, EAs work with a population of solutions, while other methods work on a single solution. This makes them more suitable for dynamic problems as well as for multi-objective problems. In the former case, if changes in the environment make the current best solution unacceptable, chances are there is another solution in the population with a better performance. For the latter case, we are typically interested in obtaining not just one solution, but a number of different non-dominated solutions, since a compromise between the different objectives usually has to be made.

Here, other then Genetic Algorithms (GAs), which are the most popular EAs, we also review, Ant Colony Optimization (ACO), Particle Swarm Optimization (PSO), and Electro-Magnetism (EM). ACO is based on the ideas of ant foraging by pheromone communication to form a path. When one ant finds a good path from the colony to a food source, other ants are more likely to follow that path, and positive feedback eventually leads all the ants following a single path. The idea of the ant colony algorithm is to mimic this behavior with "simulated ants" walking around the graph representing the problem to solve, therefore it can be seen as the problem of finding good paths through graphs. PSO algorithms are also population-based evolutionary algorithms, based on the ideas of animal flocking behavior. It is a kind of swarm intelligence that is based on social-psychological principles and provides insights into social behavior. A communication structure or social network is defined, assigning neighbors for each individual to interact with. A population of individuals, also known as the particles, iteratively evaluates the fitness of the candidate solutions and remember the location where they had their best success. Each particle makes this information available to their neighbors. Movements through the search space are guided by these successes, with the population usually converging, by the end of a trial, to a solution better than that of non-swarm approach using the same methods. ElectroMagnetism (EM) is one of the most recent heuristics applied to combinatorial optimization problems.

This method draws its functionality from analogies with electromagnetic principles. As in other evolutionary search algorithms, a population, or set of solutions, is created. Each of these solutions attracts or repels other solutions, the magnitude of which is proportional to the product of the charges and inversely proportional to the distance between the points (Coulombs Law).

4.3. Genetic Algorithms

Genetic algorithms were first proposed by Holland [28] in the mid-seventies. Holland was inspired by the mechanisms of biological evolution and natural genetics. Recombination, or crossover, is the way in which two or more parents are combined to produce offspring. Usually, this is performed by taking part of the genotype of each parent, and combining the two parts to obtain a new genotype, sharing characteristics of both parents. Recombination assumes that it is possible to take part of one genotype and part of another to produce a child that shares characteristics with both parents. If two different individuals are good, crossover may take the good parts from each of them and combine them to form an even better individual. Of course, since there is no way of knowing which part of each genotype is good, it may happen that the new genotype is worst. Thus, the recombination is done in a random fashion. This assumes the two good genotype parts to be compatible, that is form a feasible genotype. Mutation operators are also used in genetic algorithms, but mostly to make sure genetic material lost early in the search process can be reintroduced later. This way premature convergence may be avoided. Another way, of trying to avoid premature convergence, is to introduce new genetic material in the population through immigration, that is by generating new individuals in each generation.

Regarding the GAs developed and due to the fact that most methods have been developed based on previous ones, we start the review with some earlier works, some of which have already been revised elsewhere.

Alcaraz and Maroto [1] develop a genetic algorithm based on the activity list representation and the serial Schedule Generation Scheme (SGS). An additional gene decides whether forward or backward scheduling is employed when computing a schedule from an activity list. The crossover operator for activity lists is extended such that a child's activity list can be built up either in forward or in backward direction. Alcaraz et al. [2] have extended this work by adding two features from the literature. On one hand, they use the self-adaptation mechanism proposed by Hartmann [21], where an additional gene is used to determine whether the serial or the parallel SGS is to be used. Both serial and parallel schedule generation schemes generate a feasible schedule by extending a partial schedule in a stage-wise fashion. For the serial method this consists of all currently unscheduled activities whose predecessors have already been scheduled, while for the parallel method only the precedence-feasible unscheduled activities that do not violate the resource constraints, are considered. Both methods schedule, in each stage or iteration, exactly one activity. Hence, both perform $|J|$ iterations, where J is the set of project activities. On the other hand, they also employ the Forward-Backward Improvement (FBI) of Tormos and Lova [65]. The FBI is a heuristic procedure which in each iteration selects either the serial or the parallel SGS to be used to generate a schedule, by regret-based sampling with the latest finish time (LFT) priority rule.

Valls et al. [70] introduce several non-standard population-based schemes in a study which focuses on forward-backward improvement, which they have called double justification procedure. The double justification procedure shifts all activities within a schedule to the right and subsequently to the left in order to obtain a better schedule. They have proposed and tested several different algorithms incorporating the aforementioned double justification procedure. One of these algorithms was the GA by Hartmann [20], which has outperformed the other state-of-the-art heuristics for an upper limit of 5000 generated schedules. The good results obtained motivated the authors to develop a Hybrid Genetic Algorithm (HGA) [71] incorporating ideas and strategies from that method and others previously developed in [73, 69]. The HGA proposed in [71] uses the peak crossover operator, which exploits the knowledge of the problem in order to identify and combine the good parts of the solution. The double justification procedure, which is a local improvement operator is then applied to all the solutions obtained using genetic operators. The authors also use the two-phase strategy. In the second phase, new populations are generated by using a random biased sampling of the neighborhood of the best solution found so far [73, 69]. The results obtained rank amongst the best and, in addition, are obtained with modest computational times.

Debels and Vanhoucke [13], utilize a bi-population genetic algorithm. One of the populations is made of left justified schedules formulated from a forward pass on a random activity list. The other population consists of right justified schedules formulated from a backward pass on a random activity list. These two populations are then used to employ a forward-backward iterative local search process similar to one used by both Alcaraz and Maroto [1] and Valls et al. [73]. This algorithm performs slightly better than their hybrid electromagnetism/scatter search algorithm discussed below. Debels and Vanhoucke [15] introduce a decomposition-based GA that iteratively solves subparts of the project. The algorithm selects a subproblem from a feasible solution and applies the GA in order to find a high quality solution for the subproblem, which is then re-incorporated into the original solution. Thus, exploring subparts of the problem, as previously done by, e.g., [73, 53]. The authors use two serial SGS, one left justified and another right justified, and a bi-population GA based on that of Debels and Vanhoucke [13]. The method can be summarized as follows: 1. Construct the sub-problem, i.e sub-schedule $Sb \in S$ where S is the schedule of the full problem; 2. Genetic algorithm – obtains an improved sub-schedule Sb^* from Sb; 3. Merge – the improved sub-schedule Sb^* is reintroduced into the original schedule S to create an improved schedule S^*. The results reported compare favorably with current best heuristics.

Mendes et al. [48] proposed a random key-based genetic algorithm, which considers three schedule types: 1. Semi-active schedules – feasible schedules obtained by sequencing activities as early as possible. (No activity can be started earlier without changing the sequence.); 2. Active schedules – Feasible schedules in which no activity can be delayed without delaying some other activity or breaking a precedence relationship. (Optimal schedules are always members of this set of schedules and active schedules are always members of the set of semi-active schedules.); 3. Non-delay schedules – Feasible schedules in which no resource is allowed to be idle when it could start to process an activity. (Non-delay schedules are also members of the set of active schedules.) The representation proposed uses random keys, as activity priorities, and also their delay times. Although the search space is limited to active schedules, it is still very large and contains many solutions with long project durations, typical of random key representations. Therefore, the authors have limited it further by employing parameterized active schedules with a restriction on the project duration. The problem objective is to minimize the makespan, however, the authors propose a modified makespan function to measure the individuals merit, i.e. to be the fitness function, in order to provide feedback to the GA. The results reported show the method performance to be amongst the best known methods.

In a recent study Kolisch and Hartmann [39] demonstrated the power and general applicability of the forward-backward improvement. They perform several tests by adding forward-backward improvement to various sampling methods and meta-heuristics from the literature, as well as, to new approaches (c.f. [39], Section 2). The best results they report are for the genetic algorithm of Hartmann [20]. A similar approach is that of Ying et al [75], since they have proposed a hybrid-directional planning to improve heuristics solving the RCPSP. The authors discuss, implement and report results for hybrid methods combining the hybrid-directional planning with GAs, SA, and TS approach. The hybrid-directional planning proposed by the authors constructs schedules by mixing the forward, backward and bidirectional planning schemes. The planning direction is chosen randomly according to the uniform distribution with equal probability. The authors obtained better results with the hybrid-directional planning scheme than with the other planning directions. However all results obtained were far behind existing literature. Nevertheless, their experiments seem to indicate that the inclusion of the hybrid-directional planning scheme in heuristic approaches could lead to improved results. Their genetic algorithm did not perform well in comparison with the SA and TS algorithms.

4.4. Ant Colony Optimization

Dorigo et al. [17] first introduced Ant Colony Optimization (ACO). It has since been successfully applied to various complex problems including the well studied traveling salesman problem and also more recently to the RCPSP. Ant colony optimization aims to simulate the collective effort of ant colonies to solve problems. When ants travel between the nest and a food source they deposit a pheromone in the form of a trail. This pheromone attracts other ants to follow that same trail, and the more ants travel the path the more pheromone is deposited, and thus the greater the attraction to other ants. This mechanism of depositing and sensing the pheromone is known as stigmergy.

Although a lot of effort has already been put into the development of ACO algorithms, the research for the RCPSP has remained scarce.

Merkle et al. [49] developed ACO algorithm to the RCPSP employing serial SGS in conjunction with a modified latest finish time (LFT) priority rule. Usually, in an ACO based scheduling a pheromone matrix is used, where pheromone is deposited, by ants, in a matrix element when a good solution is found. In the traditional approach, the matrix dimensions represent the position i and the activity j. In other words, T_{ij} represents the possibility of activity j being in the ith position. In previous works, the ants evaluate the desirability of placing activity j as the ith task based on pheromone level, i.e., T_{ij} value. Merkle et al. [49] proposed an alternative measure, which takes into account the desirability of having activity j at the ith position or less, which is given by $\sum_i T_{ij}$. Thus preventing postponing activities, which should be scheduled early, until much later in the sequence. The algorithm also includes an elitist strategy, 2-Opt local optimization and also a low probability of replacing the best solution to date with the best for the current generation. The latter helps to prevent premature convergence due to the elitist strategy.

Herbots et al. [23] studied the applicability of ACO by testing three different algorithm configurations: 1. SSS with normalized latest start time priority rule; 2. PSS with normalized latest finish time priority rule; 3. SSS with normalized weighted resource utilization and precedence priority rule. To build a solution with the serial SGS, an ant starts with a partial activity list that contains only the start activity and selects the next activity from the set of eligible activities iteratively according to the pheromone and the heuristic information available. The pheromone information encodes a long term memory about the whole ant search process, whereas the heuristic information is problem dependent. The selected activity is scheduled at the earliest possible and resource feasible start time. The ants then perform a backtrack step where activities are remove and subsequently re-inserted, using forward, backward and bidirectional scheduling. The pheromone is updated by using the best solution found so far and also through evaporation, in order to reduce the influence of old pheromone on future searches. The authors also use local search to improve solutions. Computational results are reported showing a good performance, however, only a few problems are generated and solved. No thorough comparisons are provided.

4.5. Particle Swarm Optimization

Particle Swarm Optimization (PSO) algorithms are also population-based evolutionary algorithms, initially developed by Kennedy and Eberhart [34]. This type of evolutionary algorithms simulates the social behavior of bird flocking to a desired place. PSO conducts search using a population (called swarm) of individuals (called particles) that are updated from iteration to iteration. Each particle represents a candidate position (i.e., solution) to the problem at hand, resembling the chromosome of GA. A particle is treated as a point in an n-dimension space, where n is the number of activities to be performed, and the status of a particle is characterized by its position and velocity [34]. PSO algorithms are initialized with a swarm of random particles, which are evolved through particle flying along the trajectory. The trajectory adjustments are based on the best experience or position of the one particle (called local best) and also on the best experience or position ever found by all particles (called global best). PSO shares many common points with the former metaheuristics, especially GA, such as random generation of initial population, searching for optima by updating from iteration to iteration and evaluation of the fitness or objective of possible solutions. Unlike GA, PSO updates a population of particles with the internal velocity and attempts to profit from the discoveries of themselves and previous experiences of all other companions. However, PSO has been seldom applied to solve the RCPSP.

Zhang et al. [81] propose a hybrid particle-updating mechanism for the PSO. Both the priority-based representation and the permutation-based (activity list) representation are analyzed. In addition, a serial generation method is incorporated to achieve transformation from the particle-represented solution to a feasible schedule so that the particle can be evaluated during particle flying in terms of the project duration of the schedule. Computational experiments, with the $J30$ set, were performed and the authors were able to conclude that

PSO with the permutation-base representation achieves better results. Comparisons with other methods show the results to be comparable. In a subsequent work [80], the authors developed a PSO-based approach where particles represent activity priorities (in a similar fashion to random key GAs) which are then transformed into feasible schedules by a parallel generation scheme. In this method initial priorities are generated randomly and then evolved by a PSO mechanism that includes a multiple-pass heuristic [8] and a local search [58]. The multiple-pass heuristic uses different priority rules for each pass, while the local search adjusts some activities starting time, based on current solutions. The method's performance has been evaluated using only one problem. The makespan value is compared with some priority rules and a GA algorithm. The authors also show the makespan evolution with the number of iterations for their PSO approach and the aforementioned GA. The GA used has the same parameters as the PSO proposed. Given the insufficient experimental testing it is not possible to draw any conclusions.

A PSO for the a multi-product batch process in the chemical industry is proposed by Tang and Tang [63]. The authors also propose a batch splitting mechanism to change their problem into a RCPSP. The authors PSO method include three stages: stochastic precedence search scheme, since some limits change with decisions taken; improvement SGS, which additionally considers a filtered set of activities, that forbids scheduling all other activities to a machine where the head of a batch has been done until its tail has finished; and a recycle material scheme, which addresses storage violation after relaxing storage limits. Computational testing has been performed by using 22 problem instances, which are benchmark in the chemical industry, and the comparisons with state of the art heuristics shows that PSO can be used to obtain quickly very good solutions.

4.6. Electromagnetism

ElectroMagnetism (EM) is one of the most recent "evolutionary heuristics and it has been introduced by Birbil and Fang [6] for unconstrained optimization problems. Recently Debels et al. and Debels and Vanhoucke [12, 14] have adapted it to combinatorial optimization and successfully applied to the RCPSP. This optimization method, as its name infers, draws its functionality from analogies with electromagnetism principles. As in other evolutionary search algorithms, a population, or set of solutions, is created. Each of these solutions attracts or repels other solutions, the magnitude of which is proportional to the product of the charges and inversely proportional to the distance between the points (Coulombs Law). Therefore, inferior solution points will prevent a move in their direction by repelling other solution points in the population, and attractive solutions will facilitate moves in their direction.

Debels et al. [12] present a scatter search algorithm for the RCPSP, and seed their algorithm with very basic principles taken from the electromagnetism philosophy. More precisely, they restrict the use of the EM philosophy to the description of the hybrid two-point/electromagnetism crossover operator. However, forces are only calculated based on one other population-element, which is not in line with the basic EM philosophy since a point is supposed to exert a force on all other points. In addition, forces are not related to the distance between solutions. Again this is in contradiction to the EM philosophy in which the magnitude of the force is inversely proportional to the distance between points, in order to follow the law of Coulomb. In this work a random key schedule representation is used since it has the advantage that each solution corresponds to a point in Euclidian space, therefore, allowing for geometric operations on its components. However, since one single schedule may have different representations, the authors use an improved random key representation, standardized random key, for which each solution is uniquely associated with a schedule. The scatter search, a population based heuristic, is used to combine solutions. Its main advantage lies in the fact that solutions are obtained based on generalized path constructions in Euclidean space and by utilizing strategic designs. It also uses diversification and intensification to enhance the search. Some see scatter search as a generalization of GAs, see for instance Taillard et al. [62]. The computational results, on $J30$, $J60$, $J90$, and $J120$ show that the procedure outperforms some of the previous state-of-the-art heuristics in the literature, therefore being competitive to solve RCPSPs.

In a subsequent work Debels and Vanhoucke [14] developed an EM approach that employs fully the EM philosophy. Forces are computed using information of all solutions in the population, as well as, information on the current best solution. After applying the forces, i.e. determining the movement of each solution, a local search procedure is applied. The local search chosen is the iterative FBI. As was the case in [12] the standardize random key representation is used. The authors also incorporate diversification, which is achieved through mutation. Further improvements are obtained by considering an extended neighborhood, which is obtained by generating more schedules out of a schedule, and by allowing forces to be applied to sub-schedules, that is, by updating only part of the random key. The computational results show that this heuristic is amongst the best and also that only hybrid heuristics perform better.

5. Hybrid Heuristics, the New Trend

A noteworthy trend is that researchers are clearly moving away from "pure evolutionary methods and towards hybrid algorithms. Actually, most of the methods reviewed in the previous sections are hybrid methods.

Hybrid algorithms have their main structure given by the evolutionary algorithm, which is then combined with some sort of local search. The basic algorithm scheme is modified, or extended by integrating further features such as path relinking, forward-backward improvement, self-adapting mechanisms, non-standard crossover techniques, or even other meta-heuristics. Sometimes, various modifications and extensions are applied within the same heuristic. While several of these approaches lead to excellent results, it remains unclear, at least in some papers, if all modifications and extensions really contribute to the performance.

For instance, scatter search was adopted, in a similar way, by Valls et al. [72] combined with their GA algorithm and by Debels et al. [12] combined with their EM algorithm. The forward-backward improvement (also called justification) is used in a surprisingly large number of the best performing algorithms, see for example Alcaraz and Maroto [1, 2], Valls et al. [73, 70, 71] and Debels and Vanhoucke [13, 14], Tormos and Lova [65, 66]. The power of FBI is also demonstrated by Valls et al. [70] who add it to the simplest project scheduling heuristic, i.e. pure random sampling, obtaining much better results than adding a priority rule. In fact, on the set J120, random sampling with FBI (which is still a remarkably simple procedure) obtains better results than several more complex approaches. Moreover, FBI can easily be added to any existing heuristic for the RCPSP because it can be applied to any intermediate schedule. A similar approach was proposed by Ying et al [75], where schedules are randomly extended by using a forward, backward or bidirectional scheme, with equal probability. Although the results reported are poor, they seem to indicate that the inclusion of the hybrid-directional planning scheme in heuristic approaches (GA, SA, and TS) could lead to improved results. In [80] a multiple-pass heuristic, also similar to FBI, is incorporated into PSO-based approach. Merkle et al. [49] hybridize their ACO by using it with a modified latest finish time (LFT) priority rule

Here we review a couple of works that could not be inserted in any of the above categories. The first one, due to Tseng and Chen [68], combines ant colony optimization with a genetic algorithm and also includes a local search strategy. First, ACO searches the solution space and generates activity lists to provide the initial population for GA. Next, GA is executed and the pheromone set in ACO is updated when GA obtains a better solution. When GA terminates, ACO searches again by using a new pheromone set. ACO and GA search alternately and cooperatively in the solution space. The local search procedure is applied whenever ACO or GA obtains a solution. A final search is applied upon the termination of ACO and GA. The local search procedure, for a given activity list, evaluates the forward schedule, then the backward schedule and also derives a standard list based on the backward schedule. To construct an activity list by ACO, the ants use heuristic information as well as pheromone information. The pheromone information is evaluated as proposed by Merkle et al. [49]. The authors compute utilization rates for each resource using partial schedules. When $r - 1$ activities have been selected and scheduled their start times, finish times and the resources allocated to them are known, thus permitting to compute the aforementioned resource utilization rate. Activities are chosen using a weighted sum of heuristic information and pheromone information. Another difference is that, instead of selecting a single activity to assign to position r, they try to find an activity or a combination of two activities from the candidate set to assign to position r or positions r and $r + 1$, respectively, such that the resources can be best utilized. The genetic algorithm used is permutation-based and provides feedback to the ACO by updating the pheromone set when it finds a better solution. The GA initial populations are provided by the ACO algorithm, except for the first time, when half come from a LFT heuristic and half form the ACO algorithm. Experimental results, on the standard sets, have shown the method to be very good and fast, being amongst the 12 best.

Jedrzejowicz and Ratajczak-Ropen in [31] propose a agent-base approach using the concept of A-Teams (asynchronous teams). An A-Team is a collection of "software agents", i.e. optimization algorithms, that cooperate to solve a problem by dynamically evolving a population of solutions. The algorithms create, modify, and remove individuals from the population, improving the solutions quality. Individuals are represented as ordered activity lists and the schedules are then obtained by the serial SGS. The initial population is randomly generated and then evolved through the use of five different heuristics: A local search, that tries to move activities to all other possible places in the solution; A crossing heuristic based on the one point crossover operator; A precedent tree, which enumerates partitions; A TS implementing exchange moves between two activities; and a minimal critical set heuristic. The results reported are very good. However, they cannot be compared, except for $J30$, since the authors report on the average deviation for the best known solution rather than from the CPM lower bound, as commonly done by others.

6. Performance Evaluation and Discussion

The effectiveness of the various algorithms is usually inferred through the comparison of the results obtained. In order to do so benchmark problems are needed. The collection of these problems led to the creation of the PSPLIB. Patterson [55] collected 110 problem instances (generated by a variety of authors) and assembled optimal solutions for these problems for the minimum makespan (using known algorithms). These set became known as the Patterson problems. Latter Alvarez-Valdés and Tamarit [3], recognizing that these problems were generated by several authors and lacked a well defined set of generation parameters, have generated 144 new problem instances. Kolish et al. [43] proposed an instance generator for a general class of PSPs, with which they generated several hundred problems. The authors created and maintain a PSP library[3] with the problems they have generated, Patterson problems, and others. The most commonly used problems are the well known sets $J30$, $J60$, $J90$, and $J120$, see Kolisch and Sprecher [42], however, we only report on $J30$, $J60$, and $J120$, since not many authors report values for set $J90$. Recently, Kolisch and Hartmann [39] have analyzed the computational performance of several methods on these problems. The authors reported their findings in extensive tables containing the best optimality deviations (to the CPM lower bound, except for $J30$ since for these optimal solutions are known) of current literature. These tables are partially reproduced here, since we only report on the best performing methods. In addition, the results of the most recent works are also reported. We report on the 12 best performing methods, including the most recent ones.

In each of the three tables, three results are reported since they correspond to three different values of the maximum allowed number of schedules. The methods are briefly described by including its type (SA, TS, GA, etc. or more than one), the improvements considered (scater-search, FBI, etc.), the representation used (activity list or priorities), and the type of SGS (serial, parallel or both).

From the computational analysis, see Tables 1 to 3 for algorithms computational performance, it can be concluded that the best performing heuristics are population-based metaheuristics. The exceptions found are the parallel adaptive sample SA proposed by Shukla et al. [60], being second twice and third for $J120$; the GA proposed by Mendes et al. [48] that uses a parameterized SGS, which ranks third twice and fourth for $J120$; the sampling method with FBI of Tormos and Lova [66], which is in the twelfth position for $J30$ and $J60$ and the ninth for $J120$; and the self-adapting GA by Hartmann [21], that although not amongst the 12 best for $J30$ and $J60$ sets, it comes up to eighth for the $J120$ set. Since the review by Kolish and Hartmann [39], several new methods, that outperform former benchmark approaches have been developed.

Hartman and Kolisch [22] found that in general the activity list based schedule representation was superior to the random key representation, and until recently most of the research in the field followed these findings. Since Kolisch [38] has demonstrated that the

[3]www.bwl.uni-kiel.de/bwlinstitute/Prod/psplib

Table 1. Average deviation from optimal solution – set $J30$.

Algorithm	SGS	Ref.	Max. no. schedules		
			1000	5000	50000
GA, TS, path relinking	Both	[37]	0.10	0.04	0.00
Fuzzy SA, priority	Serial	[60]	0.19	0.03	0.005
GA, RK	Par. Active	[48]	0.06	0.02	0.01
EM, Scatter Search, RK, FBI	Serial	[12]	0.27	0.11	0.01
EM, random key, FBI	Serial	[14]	-		0.10
DBGA, RK, FBI	Serial	[15]	0.12	0.04	0.02
HGA, activity list, FBI	Serial	[71]	0.27	0.06	0.02
GA, FBI	Serial	[70]	0.34	0.20	0.02
GA, forw.backw., FBI	Both	[1]	0.25	0.06	0.03
ACO, GA, activity list, FBI	Mod. Serial	[68]	0.22	0.09	-
GA, forw.backw.	Serial	[1]	0.33	0.12	-
Sampling, LFT, FBI	Both	[67]	0.25	0.13	0.05

parallel SGS is sometimes unable to produce the optimum schedule solution, which is not the case for the serial SGS, researchers have mainly used parallel SGS. However, Debels and Vanhoucke [12], have found that the drawback of the random key representation is the fact that a single schedule can have many random key representations, which significantly extends the search space. In order to overcome it, they present a standardized random key (SRK) representation. In the most recent literature this trend is actually shifting and currently most of the very best employ priority base representation, in particular random keys, see Tables 1 to 3. Exceptions can be found in some works, e.g. [21, 37, 67, 2], which use both serial and parallel SGs.

From the literature review presented in this chapter it can be seen that a large amount of research has been conducted on the application of the evolutionary algorithm to the RCPSP. Researchers have traveled a long journey since the GA algorithm developed. Several new ideas have been developed such as self-adaptation, iterative forward and backward scheduling techniques and the application of problem specific crossover operators. This development over the last, say 10 years, has produced a steady increase in the accuracy and efficiency of evolutionary algorithms applied to this class of problem. Other types of evolutionary algorithms, such as ACO, PSO, and EM, have also been applied. However, despite the implementations by Merkle et al. [49], Herbots et al. [23], Zhang et al [81, 80], Debels et al. [12, 14] and Jedrzejowicz and Ratajczak-Ropen [31], there have been limited applications of these techniques. Nevertheless, there is no evidence that these methods are any less suited to solution of the RCPSP than are genetic algorithms, thus in our opinion there is still the need to investigate them further.

Table 2. Average deviation from CPM lower bound – set $J60$.

Algorithm	SGS	Ref.	Max. no. schedules		
			1000	5000	50000
GA, TS, path relinking	Both	[37]	11.72	11.04	10.67
Fuzzy SA, priority	Serial	[60]	11.31	10.95	10.68
GA, RK	Par. Active	[48]	11.76	11.08	10.71
EM, Scatter Search, RK, FBI	Serial	[12]	11.73	11.1	10.71
EM, RK, FBI	Serial	[14]	11.56	11.10	10.73
DBGA, RK, FBI	Serial	[15]	11.71	11.17	10.74
HGA, activity list, FBI	Serial	[71]	11.94	11.27	-
GA, FBI	Serial	[70]	12.21	11.27	10.74
GA, forw.backw., FBI	Both	[2]	11.89	11.19	10.84
ACO, GA, activity list, FBI	Mod. Serial	[68]	-	-	11.11
GA, forw.backw.	Serial	[1]	12.21	11.7	11.21
Sampling, LFT, FBI	Both	[67]	12.68	11.89	11.23

7. Conclusion

This paper has summarized an extensive array of research on the various aspects of heuristic procedures for the project scheduling problem. Methodologies for solving project scheduling type problems have definitively moved on from constructive approaches such as simple priority rule-based heuristics and enumeration-based exact procedures. Learning also from the fields of artificial intelligence and computing, more elaborated approaches have been proposed. The recent methods include meta-heuristics, that can escape the problems of lack of feasibility and local optimality, as well as parallel computations.

By looking back to the methodologies proposed, it can be seen that recently proposed heuristics contain more components than earlier ones. Many methods consider both scheduling directions instead of only forward scheduling, both SGS instead of only one, more than one type of local search operator, or even more than one type of meta-heuristic strategy. While recombining merely existing ideas seems to be less creative than developing new ideas, some of the integration efforts have put well known techniques into a new and promising perspective, and the results have often been encouraging. It should also be noticed that several heuristics EM, ACO, PSO, and others (non meta-heuristics) have started to appear in literature. Although, for some, the results reported are not yet sufficiently competitive, we view them as promising, since we believe that when further studied and investigated they may lead to good results.

Hybrid heuristics are based on principles borrowed from various meta-heuristic approaches. Hence, we believe that the incorporation of ideas from GA, EM, ACO, PSO in hybrid frameworks might contribute to the development of better meta-heuristic techniques.

Another research direction that is in need of being explored, as was recently also pointed out by Herroelen [25], is the commercial software issue. It is obvious that much of the research has not yet found its way into practice. However, new and better software is

Table 3. Average deviation from CPM lower bound – set $J120$.

Algorithm	SGS	Ref.	Max. no. schedules		
			1000	5000	50000
DBGA, RK, FBI	Serial	[15]	33.55	32.18	30.69
HGA, activity list, FBI	Serial	[71]	34.07	32.54	31.24
Fuzzy SA, priority	Serial	[60]	34.86	32.54	31.24
GA, RK	Param. Active	[48]	35.87	33.03	31.44
GA, forw.backw., FBI	Both	[2]	36.53	33.91	31.49
EM, Scatter Search, RK, FBI	Serial	[12]	35.22	33.1	31.57
GA, FBI	Serial	[70]	35.39	33.24	31.58
GA, TS, path relinking	Both	[37]	34.74	33.36	32.06
GA, self-adapting	Both	[21]	37.19	35.39	33.21
Sampling, LFT, FBI	Both	[67]	35.01	34.41	33.71
EM, RK, FBI	Serial	[14]	-	-	33.94
ACO, GA, Activity list, FBI	Mod. Serial	[68]	36.39	34.49	-

needed in practice, since many publications have appeared, some very recently, reporting budget and/or completion date over-runs, see e.g. [77].

References

[1] J. Alcaraz and C. Maroto. A robust genetic algorithm for resource allocation in project scheduling. *Annals of Operations Research*, 102:83–109, 2001.

[2] J. Alcaraz, C. Maroto, and R. Ruiz. Improving the performance of genetic algorithms for the RCPS problem. In *Proceedings of the Ninth International Workshop on Project Management and Scheduling*, pages 40–43, Nancy, 2004.

[3] R. Alvarez-Valdés and J.M. Tamarit. Heuristic algorithms for resource-constrained project scheduling: a review and an empirical analysis. In R. Owi Ski and J Glar, editors, *Advances in project scheduling*, pages 113–34. Elsevier, Amsterdam, 1989.

[4] T. Bäck, D. B. Fogel, and Z. Michalewicz, editors. *Handbook of Evolutionary Computation.* IOP Publishing and Oxford University Press, 1997.

[5] F. Ballestýn, V. Valls, and S. Quintanilla. Pre-emption in resource-constrained project scheduling. *European Journal of Operational Research*, 189:1136–1152, 2008.

[6] S.I. Birbil and S.C. Fang. An electromagnetism-like mechanism for global optimization. *Journal of Global Optimization*, 25:263–282, 2003.

[7] J. Blazewicz, J.K Lenstra, and A.H.G. Rinnooy Kan. Scheduling subject to resource constraints: classification and complexity. *Discrete and Applied Mathematics*, 5:11–24, 1983.

[8] F.F. Boctor. Some efficient multi-heuristic procedures for resource-constrained project scheduling. *European Journal of Operational Research*, 49:3–13, 1990.

[9] K. Bouleimen and H. Lecocq. A new efficient simulated annealing algorithm for the resource-constrained project scheduling problem and its multiple mode version. *European Journal of Operational Research*, 149:268–241, 2003.

[10] P. Brucker, A. Drexl, R. Möhring, K. Neumann, and E. Pesch. Resource-constrained project scheduling: Notation, classification, models, and methods. *European Journal of Operational Research*, 112:3–41, 1999.

[11] E.W. Davis and J.H. Patterson. A comparison of heuristic and optimum solutions in resource-constrained project scheduling. *Management Science*, 21:944–955, 1975.

[12] D. Debels, B. De Reyck, R. Leus, and M. Vanhoucke. A hybrid scatter search/electromagnetism meta-heuristic for project scheduling. *European Journal of Operational Research*, 169:638–653, 2006.

[13] D. Debels and M. Vanhoucke. A bi-population based genetic algorithm for the resource-constrained project scheduling problem. Working paper series, Vlerick Leuven Gent Management School, Belgium, 2005.

[14] D. Debels and M. Vanhoucke. *The Electromagnetism Meta-heuristic Applied to the Resource-Constrained Project Scheduling Problem*, volume 3871 of *Lecture Notes in Computer Science*, pages 259–270. Springer-Verlag, Berlin, Heidelberg, 2006.

[15] D. Debels and M. Vanhoucke. A decomposition-based genetic algorithm for the resource-constrained project-scheduling problem. *Operations Research*, 55:457–469, 2007.

[16] E. Demeulemeester and W. Herroelen. *Project Scheduling: A Research Handbook*, volume 49 of *International Series in Operations Research & Management Science*. Kluwer Academic Publishers, Boston, MA, 2002.

[17] M. Dorigo, G. di Caro, and L. Gambardella. Ant algorithms for discrete optimisation. *Artificial Life*, 5:137–172, 1999.

[18] K. Fleszar and K.S. Hindi. Solving the resource-constrained project scheduling problem by a variable neighbourhood search. *European Journal of Operational Research*, 155:402–413, 2004.

[19] F. Glover and M. Laguna. *Tabu Search*. Kluwer Academic Publishers, Boston, MA, 1997.

[20] S. Hartmann. A competitive genetic algorithm for resource-constrained project scheduling. *Naval Research Logistics*, 45:733–750, 1998.

[21] S. Hartmann. A self-adapting genetic algorithm for project scheduling under resource constraints. *Naval Res Logist*, 49:433–448, 2002.

[22] S. Hartmann and R. Kolisch. Experimental evaluation of state-of-the-art heuristics for the resource-constrained project scheduling problem. *European Journal of Operational Research*, 127:394–407, 2000.

[23] J. Herbots, W. Herroelen, and R. Leus. Experimental investigation of the applicability of ant colony optimisation algorithms for project scheduling. Research report 0459, Department of Applied Economics, FETEW, K. U., Leuven, Belgium, 2004.

[24] S. Herroelen, P.V. Dommelen, and E. Demeulemeester. Project network models with discounted cash flows: A guided tour through recent developments. *European Journal of Operational Research*, 100:97–121, 1997.

[25] W. Herroelen. Project scheduling – theory and practice. *Production and Operations Management*, 14:413–432, 2005.

[26] W. Herroelen, E. Demeulemeester, and B. De Reyck. Resource-constrained project scheduling – a survey of recent developments. *Computers & Operations Research*, 25:279–302, 1998.

[27] K.S. Hindi, H. Yang, and K. Fleszar. An evolutionary algorithm for resource-constrained project scheduling. *IEEE Transactions on Evolutionary Computation*, 6:512–518, 2002.

[28] J. H. Holland. *Adaptation in natural and artificial systems.* University of Michigan Press, 1975.

[29] O. Icmeli, S.S. Erenguc, and C.J. Zappe. Project scheduling problems: A survey. *International Journal of Operations & Production Management*, 13:80–91, 1993.

[30] B. Jarboui, N Damak, P. Siarry, and A. Rebai. A combinatorial particle swarm optimization for solving multi-mode resource-constrained project scheduling problems. *Applied Mathematics and Computation*, 195:299–308, 2008.

[31] P. Jedrzejowicz and E. Ratajczak-Ropen. *Agent-based Approach to Solving the resource constrained project scheduling problem*, pages 480–487. Lecture Notes in Computer Science. Springer-Verlag, Berlin, Heidelberg, 2007.

[32] J. Jozefowska, M. Mika, R. Rozycki, G Waligóra, and J. Weglarz. Solving the discrete-continuous project scheduling problem via its discretization. *Mathematical Methods of Operations Research*, 52:489–499, 2000.

[33] J.E. Kelley. The critical path method: Resource planning and scheduling. In J.F. Muth and G.L. Thompson, editors, *Industrial Scheduling*, pages 47–365. Prentice Hall, New Jersey, 1963.

[34] J. Kennedy and R.C. Eberhart. Particle swarm optimization. In *Proceedings of the Conference on Neural Networks*, volume IV, pages 1942–1948, IEEE, Piscataway, NJ, USA, 1995.

[35] S. Kirkpatrick, C. D. Gelatt, and M. P. Vecchi. Optimization by simulated annealing. *Science*, 220:671–680, 1983.

[36] R. Klein. Bidirectional planning: improving priority rule-based heuristics for scheduling resource-constrained projects. *European Journal of Operational Research*, 127:619–638, 2000.

[37] Y. Kochetov and A. Stolyar. Evolutionary local search with variable neighborhood for the resource constrained project scheduling problem. In *Proceedings of the 3rd International Workshop of Computer Science and Information Technologies*, Russia, 2003.

[38] R. Kolisch. Serial and parallel resource-constrained project scheduling methods revisited: theory and computation. *European Journal of Operational Research*, 90:320–333, 1996.

[39] R. Kolisch and S. Hartmann. Experimental investigation of heuristics for resource-constrained project scheduling: An update. *European Journal of Operational Research*, 174:23–37, 2006.

[40] R. Kolisch and R. Padman. An integrated survey of deterministic project scheduling. *Omega*, 29:249–272, 2001.

[41] R. Kolisch, C. Schwindt, and A. Sprecher. Benchmark instances for project scheduling problems. In J. Weβglarz, editor, *Project Scheduling – Recent Models, Algorithms and Applications*, pages 197–212. Kluwer Academic Publishers, Boston, NY, 1999.

[42] R. Kolisch and A. Sprecher. PSLIB - a project scheduling problem library. *European Journal of Operational Research*, 96:205–216, 1996.

[43] R. Kolisch, A. Sprecher, and A. Drexl. Characterization and generation of a general class of resource-constrained project scheduling problems. *Management Science*, 41:1693–1703, 1995.

[44] O. Lambrechts, E. Demeulemeester, and W. Herroelen. A tabu search procedure for developing robust predictive project schedules. *International Journal of Production Economics*, 111:493–508, 2008.

[45] M.A. Lee and H. Takagi. Dynamic control of genetic algorithm using fuzzy logic techniques. In *Proceedings of the 5th International Conference on Genetic Algorithms (ICGA'93)*, pages 76–83, Urbana-Champaign, Illinois, 1993.

[46] B.S.J. Lin and U. Wun. The dominance technique for the resourceconstrained project scheduling problem. *Management and System*, 2:191–203, 1995.

[47] G. Waligóra M. Mika and J. Weglarz. Tabu search for multi-mode resource-constrained project scheduling with schedule-dependent setup times. *European Journal of Operational Research*, 187:1238–1250, 2008.

[48] J.J.M. Mendes, J.F. Goncalves, and M.G.C. Resende. A random key based genetic algorithm for the resource constrained project scheduling problem. *Computers & Operations Research*, 36:92–109, 2009.

[49] D. Merkle, M. Middendorf, and H. Schmeck. Ant colony optimization for resource-constrained project scheduling. *IEEE Transactions on Evolutionary Computation*, 6:333–346, 2002.

[50] M. Mika, G. Waligóra, and J. Weeglarz. Simulated annealing and tabu search for multi-mode resource-constrained project scheduling with positive discounted cash flows and different payment models. *European Journal of Operational Research*, 164:639–668, 2005.

[51] N. Mladenovic and P. Hansen. Variable neighborhood search. *Computers & Operations Research*, 24:1097–1100, 1997.

[52] L. Ö̸zdamar and G. Ulusoy. A survey on the resource constrained project scheduling problem. *IIE Transactions*, 27:574–586, 1995.

[53] M. Palpant, C. Artigues, and P. Michelon. LSSPER: Solving the resource-constrained project scheduling problem with large neighbourhood search. *Annals of Operations Research*, 131:237–257, 2004.

[54] N.-H. Pan, P.-W. Hsaio, and K.-Y. Chen. A study of project scheduling optimization using tabu search algorithm. *Engineering Applications of Artificial Intelligence*, 21:1101–1112, 2008.

[55] J.H. Patterson. A comparison of exact approaches for solving the multiple constrained resource, project scheduling problem. *Management Science*, 30:854–867, 1984.

[56] I. Pesek, A. Schaerf, and J. Žerovnik1. *Hybrid Local Search Techniques for the Resource-Constrained Project Scheduling Problem*, volume 4771 of *Lecture Notes in Computer Science*, pages 57–68. Springer-Verlag, Berlin, Heidelberg, 2007.

[57] C. R. Reeves, editor. *Modern Heuristic Techniques for Combinatorial Problems*. McGraw-Hill, 1995.

[58] S.E. Sampson and E.N. Weiss. Local search techniques for the generalized resource constrained project scheduling problem. *Naval Research Logistics*, 40:365–75, 1993.

[59] M. Schäffter. Scheduling with respect to forbidden sets. *Discrete and Applied Mathematics*, 72:141–154, 1997.

[60] S.K. Shukla, Y.J. Son, and M.K. Tiwari. Fuzzy-based adaptive sample-sort simulated annealing for resource-constrained project scheduling. *International Journal of Advanced Manufacturing Technology*, 36:982–995, 2008.

[61] A. Sprecher. Scheduling resource-constrained projects competitively at modest memory requirements. *Management Science*, 46:710–723, 2000.

[62] E.D. Taillard, L.M. Gambardella, M. Gendreau, and J.-Y Potvin. Adaptive memory programming: A unified view of metaheuristics. *European Journal of Operational Research*, 134:1–16, 2001.

[63] Q. Tang and L. Tang. Heuristic particle swarm optimization for resource-constrained project scheduling problem in chemical industries. In *Proceedings of the Control and Decision Conference*, pages 1475–1480, Yantai, Shandong, China, 2008. IEEE.

[64] D.R. Thompson and G.L. Bilbro. Sample-sort simulated annealing. *IEEE Transactions on Systems, Man, and Cybernetics, Part B*, 35:625–632, 2005.

[65] P. Tormos and A. Lova. A competitive heuristic solution technique for resource-constrained project scheduling. *Annals of Operations Research*, 102:65–81, 2001.

[66] P. Tormos and A. Lova. An efficient multi-pass heuristic for project scheduling with constrained resources. *International Journal of Production Research*, 41:1071–1086, 2003.

[67] P. Tormos and A. Lova. Integrating heuristics for resource constrained project scheduling: One step forward. Technical report, Department of Statistics and Operations Research, Universidad Politécnica de Valencia, Spain, 2003.

[68] L. Tseng and S.-C. Chen. A hybrid metaheuristic for the resource-constrained project scheduling problem. *European Journal of Operational Research*, 175:707–721, 2006.

[69] V. Valls, F. Ballestin, and M.S. Quintanilla. A population-based approach to the resource-constrained project scheduling problem. *Annals of Operations Research*, 131:305–324, 2004.

[70] V. Valls, F. Ballestin, and M.S. Quintanilla. Justification and RCPSP: a technique that pays. *European Journal of Operational Research*, 165:375–386, 2005.

[71] V. Valls, F. Ballestin, and M.S. Quintanilla. A hybrid genetic algorithm for the resource-constrained project scheduling problem. *European Journal of Operational Research*, 185:495–508, 2008.

[72] V. Valls, M.S. Quintanilla, and F. Ballestin. An evolutionary approach to the resource constrained project scheduling problem. In *Proceedings of the 4th Metaheuristics International Conference*, Porto, Portugal, 2001.

[73] V. Valls, M.S. Quintanilla, and F. Ballestin. Resource-constrained project scheduling: A critical reordering heuristic. *European Journal of Operational Research*, 149:282–301, 2003.

[74] G. Waligóra. Discrete-continuous project scheduling with discounted cash flows-a tabu search approach. *Computers & Operations Research*, 35:2141–2153, 2008.

[75] K. Ying, S. Lin, and Z. Lee. Hybrid-directional planning: Improving improvement heuristics for scheduling resource-constrained projects. *International Journal of Advanced Manufacturing Technology*, 2008. DOI 10.1007/s00170-008-1486-5.

[76] S. Yong-Yi. Ant colony algorithm for scheduling resource constrained projects with discounted cash flows. In *Proceedings of the Fifth International Conference on Machine Learning and Cybernetics*, pages 176 – 180. IEEE, 2006.

[77] E. Yourdon. *Death March.* Prentice Hall, New Jersey, 2nd edition edition, 2003.

[78] Y.S. Yun and M. Gen. Advanced scheduling problem using constraint programming techniques in SCM environment. *Computers and Industrial Engeneering*, 43:213–229, 2002.

[79] Q.C. Zeng and Z.Z. Yang. Two-phase tabu search algorithm of unloading operation scheduling project in container wharf. *Journal of Traffic and Transportation Engineering*, J 7:109–112, 2007.

[80] H. Zhang, H. Li, and C.M. Tam. Particle swarm optimization for resource-constrained project scheduling. *International Journal of Project Management*, 24:83–92, 2006.

[81] H. Zhang, X. Li, H. Li, and F. Huang. Particle swarm optimization-based schemes for resource-constrained project scheduling. *Automation in Construction*, 14:393–404, 2005.

[82] X. Zhang, X. Yuan, and Y. Yuan. Improved hybrid simulated annealing algorithm for navigation scheduling for the two dams of the three gorges project. *Computers and Mathematics with Applications*, 56:151–159, 2008.

INDEX

E

H

I

J

K

L

N

O

P

Q

R

S

T

U

V

W

X

Y

Z